International Series in Operations Research & Management Science

Volume 127

Series Editor

Frederick S. Hillier
Stanford University, CA, USA

INT. SERIES IN OPERATIONS RESEARCH & MANAGEMENT SCIENCE
Series Editor: Frederick S. Hillier, Stanford University
Special Editorial Consultant: Camille C. Price, Stephen F. Austin State University
Titles with an asterisk (*) were recommended by Dr. Price

~A list of the early publications in the series is found at the end of the book~

Wieslaw Kubiak

Proportional Optimization
and Fairness

 Springer

Wieslaw Kubiak
Memorial University
Faculty of Administration
John's NL
Canada A1B 3X5
wkubiak@mun.ca

ISBN: 978-0-387-87718-1   e-ISBN: 978-0-387-87719-8
DOI: 10.1007/978-0-387-87719-8

Library of Congress Control Number: 2008934787

Printed on acid-free paper

springer.com

To My Inka i Michał

Preface

If the beginning provides countless possibilities, then why not to start with few questions? Why are cars of different colors spread along an assembly line rather then batched together in a single long sequence of the same color? How to make equal priority jobs progress at the rates proportional to their lengths so that a job twice the length of another one gets a shared resource allocated twice the time of the other job up to any point in time? Or a client who pays three times more for its computations than another client gets its computations to progress three times faster than the other client's by getting more processor and bandwidth allocations? How to make sure that the Internet gateway bandwidth is shared fairly so that the community sharing the network is not reduced to few getting all and most nothing? All these questions deal with proportional representation either according to the demand for particular car color, or according to the job length or its right to resources, or according to the reciprocal of the packet size to name just few. They are fundamental even more so today when we are surrounded by systems enabled by technology to work in a just-in-time mode since this mode very principle requires a steady, smooth, and evenly spread progress of tasks in time. The progress is proportional to the demand for the tasks's outcomes.

As a thinker and futurist Alvin Toffler [1] in his *Financial Times* interview points out "Global positioning satellites are key to synchronising precision time and data streams for everything from mobile phone calls to ATM withdrawals. They allow just-in-time productivity because of precise tracking."

What is somewhat surprising is that all these questions that seem so far apart have similar underlying framework, which is simply speaking to build a finite or infinite often cyclic sequence; we shall refer to it as a just-in-time sequence, on a finite n letter alphabet where each letter is spread "as evenly as possible" and occurs with a given rate or a given number of times. The problem of finding such a sequence is not only a mathematical one since there is no mathematical definition of "as evenly as possible" that would satisfactorily capture the challenge behind this phrase. The problem can find many mathematical formulations, but none will probably satisfy all. Thus, one way of approaching the problem is to use the well-known apportionment theory and especially its house monotone methods to build the desired just-in-time sequence.

The apportionment problem has its roots in the proportional representation system designed for the House of Representatives of the United States where each state receives seats in the House proportionally to its population. The theory has been in the making for more than 200 years now and its exciting story as well as main results can be found in an excellent book by Balinski and Young [2], see also more recent book by Young [3], and Balinski's popular introduction in [4]. The title of Balinski and Young's book speaks for itself: "Fair Representation: Meeting the Ideal of One Man, One Vote." Its main underlying message is that the ideal is not one but many and that we can only hope to agree on one by stating some "obvious" axioms that it must meet and then find a method that would deliver a solution meeting these axioms, or to prove that one does not exist. This process may, however, not save us from falling into various anomalies that do not contradict the axioms yet may be at odds with the commonly accepted sense of fair representation.

This book argues that the apportionment methods, in particular the John Quincy Adams's and the Thomas Jefferson's, have been widely, yet unknowingly, rediscovered and used in resource allocation and sequencing computer, manufacturing, and other real-life technical systems. Sometimes without a clear understanding of what solutions they lead to in terms of their properties. The properties which have been well researched and known from the apportionment literature but missing in the technical one, either computer science or operations research. This lack of proper context may have resulted, as we argue in some parts of this book, in overlooking other apportionment methods, in particular the Daniel Webster's method, that may offer a number of additional attractive properties, like being better balanced than either the Adams's or the Jefferson's.

The axiomatic approach favored by the apportionment theory for the proportional representation systems is preferred over an optimization approach championed by operations research scientists since the problem with the latter approach is in the words of Balinski and Young from [2] as follows: "The moral of this tale is that one cannot choose objective functions with impunity, despite current practices in applied mathematics. The choice of an objective is, by and large an *ad hoc* affair... Of much deeper significance than the formulas that are used are the *properties* they enjoy."

We think, however, that in order to adequately address the proportional representation problems listed at the beginning of this preface and others we need to study them not only through the apportionment theory but through optimization as well. After all the questions of quantifying excess inventory and shortage in just-in-time manufacturing, the throughput error in stride scheduling, or the relative and absolute bounds in fair queueing are clearly important. By doing so, we also realize that the optimization reveals a new role of the well-known apportionment methods, the Webster's method in particular. The optimization moreover reveals connections with the well-known and still open mathematical conjectures as the Fraenkel's Conjecture, see Tijdeman [5] for a brief account and Chap. 6, finally it relates to the multimodular functions minimization, introduced by Hajek [6] and later developed by Altman et al. [7], which aims at evenly spreading the demand and workload in computer and supply chains.

The question of which objective function to choose we settle by choosing either total deviation or maximum deviation objective functions. Our solution method is general enough to include a large class of point deviation functions. The choice of objective functions follows sometime the choice made by Monden who, in his seminal book [8], described the Goal Chasing Method of Toyota by using the square point deviation function which apparently follows the minimization of square error in the least squares method of Carl Friedrich Gauss. The attractive feature of this optimization is that it can be done efficiently, though certain intriguing computational complexity issues remain open, and produce solutions which have many though not all, by the Impossibility Theorem of Balinski and Young [2], desirable properties identified by the theory and practice of apportionment.

The book intends to chart a solid common ground for discussing and solving problems ranging from sequencing mixed-model just-in-time assembly lines, through just-in-time batch production, balancing workloads in event graphs to bandwidth allocation in the Internet gateways and resource allocation in operating systems. From problems in mathematics of social sciences through operations research and computer science problems, it argues that the apportionment theory and the optimization based on deviation functions provide natural benchmarks in this process. However, the process has just started and this book is to provide just a small stepping stone on the way to this common ground. Needless to say it will be a great pleasure for the author if the book's topic finds its followers.

The book includes mostly very recent results – some of them published recently, some of them new and yet unpublished. It includes ten main chapters. Chapter 2 briefly reviews main results of the apportionment theory used in the remainder of the book. It emphasizes the axiomatic approach to the apportionment problem and to the construction of the just-in-time sequences. The approach relies on the divisor methods, in particular parametric methods advocated by Balinski and Young [2], and their desirable properties embedded in the resulting just-in-time sequences. Chapter 3 considers the problems of deviation minimization, the total and the maximum deviation, as tools for obtaining just-in-time sequences. It formulates these problems as nonlinear integer optimization and presents efficient algorithms for their solution. The algorithms are based on the concept of ideal positions, closely related to the Webster's apportionment method. They transform the deviation minimization problems to either the assignment or the bottleneck assignment problem, respectively, and then solve the latter. The algorithms run in time which is polynomial in the length of the outcome just-in-time output sequence. Chapter 4 proves that there exist cyclic solutions that minimize the total deviation for symmetric point deviation functions, the same is shown for the maximum deviation. It also proves that limiting optimization to the sequences with the bottleneck deviation not exceeding 1 renders some functions of point deviation equivalent. The oneness property claims that limiting search for optimal just-in-time sequences to those with bottleneck not exceeding 1 will be optimal in general. However, the chapter shows that all optimal just-in-time sequences for some instances may have the bottleneck deviation higher than 1 – thus showing that the oneness does not hold generally. Chapter 5 gives a more efficient algorithm for the maximum *absolute* deviation (referred to as bottleneck) deviation. The absolute value function of deviation results in optimal bottleneck being always less than 1, and allows to develop strong upper and lower bounds on the optimal bottleneck. These bounds and other properties of the bottleneck optimal just-in-time sequences are used in the application to the Liu–Layland problem, stride scheduling, fair queueing, and others in the subsequent chapters. Chapter 5 also shows that the optimal bottleneck just-in-time sequences for $n = 2$ are in fact Webster's sequences of apportionment and the most regular words at the same time; thus, they optimize the throughput of any two cyclic process sharing a common resource. This new observation underlines again the advantages of the Webster's sequences for other than apportionment problems. Chapter 6 further exploits the properties of just-in-time sequences with small bottleneck deviations, which are understood as those less than $\frac{1}{2}$. The question is what are the instances that admit this small bottleneck deviation? The answer given in the chapter is that there is only one, called the power-of-two instance that results in this small bottleneck deviation for $n \geq 3$. The chapter also shows the connection between the small bottleneck deviation problem and the famous Fraenkel's Conjecture, which states that the only distinct rates for which it is possible to build a balanced word on three or more letters come essentially from the power-of-two instances. Finally, the chapter presents the small bottleneck problem in the broader context of regular sequences and multimodular functions they minimize. The applications of multimodular functions to workload balancing in event graphs (for instance the queues and supply chains) are also discussed in the chapter. Chapter 7 addresses the response time variability minimization problem, where the average response time for a client is a reciprocal of its desirable rate. Thus, being as close as possible to the average response time aims at achieving the "as evenly as possible" goal. The response time variability is one of the main objectives in stride scheduling as well. The chapter shows that the problem is NP-hard, proposes exact and heuristic solutions, and reports computational experiments with the latter. Chapter 8 proves that the optimal bottleneck sequences make tasks progress at the rates close enough to the tasks' processing time to request interval ratios so that they solve the Liu–Layland problem – likely the best known scheduling problem in the hard real-time systems. It also gives necessary conditions for the apportionment divisor methods to solve the Liu–Layland problem, and proves that the quota-divisor methods solve the Liu–Layland problem as well. Finally, the chapter presents solutions to some special cases of the pinwheel scheduling problem given by the bottleneck optimal just-in-time sequences. Chapter 9 focuses on the problem of constructing just-in-time sequences for supply chains so that the temporal capacity constraints imposed by suppliers are respected. The constraints are modeled by giving the limiting, supply-dependent proportions $p:q$ that stipulate that at most p out of any q models delivered by the supply chain must be supplied by a particular supplier. Though the problem of finding such a sequence is NP-hard in the strong sense the chapter discusses a number of approaches: synchronized delivery and periodic synchronized delivery for better balancing workloads in supply chains. Finally, the chapter points out a potential for using tools developed by the combinatorics on words to design the just-in-time sequences having desirable properties, and discusses the class of balanced words in this role in more detail. Chapter 10 looks into the problem of fairness in fair queueing and stride scheduling. It shows that both use the Jefferson's and Adams's method of apportionment, and both are peer-to-peer fair. However, the chapter also argues that the Webster's method could prove a better yet untested choice for fair queueing and stride scheduling. The chapter gives also a closer look at the measures and criteria typically used in the fair queueing and stride scheduling and analyzes them using the apportionment theory and just-in-time optimization tools developed in Chaps. 2, 5, and 7. Finally, Chap. 11 extends the models developed in Chaps. 2, 3, and 9 to manufacturing environments with variable processing and set-up times. This is a departure from the usual assumption of negligible variability resulting in an simplification, often criticized, of unit times and synchronized lines assumed in the applications of just-in-time sequences. The chapter's approach is based on batching to smooth out the variability of processing and set-up times, and then on sequencing the batches to minimize the total deviation or alternatively to gain the advantages of the Webster's method. The approach is applied to a real-life problem arising in an automotive pressure hose manufacturer. The computational experiments with both algorithms are also presented in the chapter.

Special thanks go to my friends and colleagues, listed here in a random order, for their encouragement and support: Prof. Dominique de Werra (École Politechnique Fédérale de Lausanne), Profs. Jan Węglarz and Jacek Błażewicz (Poznań University of Technology), Prof. Albert Corominas (Universitat Politècnica de Catalunya), Prof. Jacques Cariler (Université de Technologie de Compiègne), Prof. Erwin Pesch (University of Siegen), Prof. Moshe Dror (University of Arizona), Prof. Gerd Finke (Université Joseph Fourier), and Prof. Marek Kubale (Gdańsk University of Technology). I am indebted in particular to Dr. Cynthia Philips (Sandia National Laboratories) and Dr. Bruno Gaujal (INRIA-Grenoble) for pointing me to a number of important references.

Finally, I wish to acknowledge the research support of the Natural Sciences and Engineering Research Council of Canada without which many of my research projects on just-in-time would simply not happen.

St. John's, Canada   *Wieslaw Kubiak*

Contents

List of Figures

List of Tables

Chapter 1
Preliminaries

This chapter briefly reviews the basic terminology and notation used in the book. We begin with some notation and terminology borrowed from the formal language theory, see for instance Hopcroft and Ullman [9].

An *alphabet* $\mathcal{A} = \{a_1, \ldots, a_n\}$ is a finite non-empty set of symbols. A *word* (or *sequence*) S (we also use small s to denote sequence) over $\mathcal{A}$ is any finite sequence of symbols from $\mathcal{A}$. The length of S, that is the number of symbol occurrences in S, is denoted by $|S|$. The empty word, denoted by Λ, is the unique word over $\mathcal{A}$ of length 0. The word

$$S = s_1 s_2 \cdots s_m$$

where $s_i \in \mathcal{A}$ for $i = 1, \ldots, m$ will also be denoted as

$$S = s_1 \rightarrow s_2 \rightarrow \cdots \rightarrow s_m.$$

The index i will be called the position of the letter s_i in the word S. If word $S = S_1 S_2$ is the *concatenation* of words S_1 and S_2, then S_1 is called a *prefix* of S, and S_2 is called a *suffix* of S. For $k = 1, \ldots, |S|$, the prefix made up of the first k symbols of a non-empty word S is referred to as the k-prefix. For a word S and a non-negative integer m, the concatenation m times of S will be denoted as follows

$$\underbrace{SS\ldots S}_{m-\text{times}} = (S)^m = S^m.$$

The infinite repetition of word S will be denoted by

$$S^\infty,$$

and then S is called a cycle of S^∞. For

$$S = s_1 s_2 \cdots s_m$$

its *mirror* reflection S^R is

$$S^R = s_m \cdots s_2 s_1.$$

W. Kubiak, *Proportional Optimization and Fairness,* International Series in Operations Research & Management Science 127, DOI 10.1007/978-0-387-87719-8_1, © Springer Science+Business Media LLC 2009

A sequence S is a *palindrome* if there is a sequence W such that

$$S = WW^R.$$

We denote by $\mathcal{A}^*$ the set of all finite words over $\mathcal{A}$. The number of occurrences of a given symbol $a \in \mathcal{A}$ in a word $S \in \mathcal{A}^*$ is denoted by $|S|_a$. Any occurrence of a given symbol a in sequence S will also be referred to as a *copy* of the symbol.

The *Parikh* vector associated with a word $S \in \mathcal{A}^*$ with respect to the alphabet $\mathcal{A} = \{a_1, \ldots, a_n\}$ is

$$(|S|_{a_1}, \ldots, |S|_{a_n}).$$

A *factor* (subsequence) of length (size) $b \geq 0$ of $S = s_1 s_2 \cdots s_m$ is a word x such that $x = s_i \ldots s_{i+b-1}$.

We denote by $\mathbb{R}$, $\mathbb{Z}$, and $\mathbb{N}$ sets of real, integer, and natural numbers, respectively. Let $\{a_1, \ldots, a_n\}^{\mathbb{Z}}$ be the set of infinite sequences on the alphabet $\mathcal{A} = \{a_1, \ldots, a_n\}$. For the letter a_i and an infinite sequence $S \in \{a_1, \ldots, a_n\}^{\mathbb{Z}}$ let $I(S, a_i) \in \{0, 1\}^{\mathbb{Z}}$ be the *indicator* in S of the letter a_i, that is $I(S, a_i)_j = 1$ if and only if $s_j = a_i$.

Let $d_1, \ldots, d_n$ be $n \geq 1$ positive integers called demands (or model demands) and let the alphabet $\mathcal{A} = \{1, \ldots, n\}$. This particular alphabet will be most often used in the book. Any letter $i \in \mathcal{A}$ will also be referred to as a model i, or a client i, or a state i, or a queue i depending on the context of our discussion. The vector

$$\mathbf{d} = (d_1, \ldots, d_n)$$

will be referred to as the vector of demands. We use the bold notation $\mathbf{d}, \mathbf{p}, \mathbf{a}$, etc. for vectors in the book. We define the total demand

$$D = d_1 + \cdots + d_n,$$

and the rates

$$r_i = \frac{d_i}{D}$$

for letters $i = 1, \ldots, n$. The vector of demands is called a standard instance if $0 < d_1 \leq d_2 \leq \cdots \leq d_n$, $n \geq 2$, and the greatest common divisor of $d_1, d_2, \ldots, d_n, D$ is 1, that is $\gcd(d_1, \ldots, d_n, D) = 1$.

Consider the set JIT of words on $\mathcal{A}$ all having their Parikh vectors with respect to $\mathcal{A}$ equal

$$\mathbf{d} = (d_1, \ldots, d_n).$$

Any, $S \in JIT$ will be referred to as a just-in-time sequence or a just-in-time word for demand vector $\mathbf{d}$, or simply just-in-time sequence. For $S \in JIT$, let $x_{i,k}$ be the number of letter $i \in \mathcal{A}$ occurrences in the $k-$ prefix, $k = 1, \ldots, D$ of w. We also write x_{ik} instead of $x_{i,k}$.

The *floor* function $\lfloor x \rfloor$ of x is the greatest integer less than or equal to x. The *ceiling* function $\lceil x \rceil$ is the least integer greater that or equal to x, see Graham et al. [10] for more on these functions. The *nearest integer* function $[x]$ or $[x]_{\frac{1}{2}}$ is the integer closest to x when the fractional part of x is equal to $\frac{1}{2}$ we round downward unless otherwise specified. The $|x|$ denotes the absolute value of any real x. Though, we use the same notation $|.|$ for both the absolute value and the sequence length the context will clearly indicate which of the two applies.

The least common multiple of integers $n_1, \ldots, n_m$, $m \geq 1$, will be denoted by $\mathrm{lcm}(n_1, \ldots, n_m)$. The greatest common divisor of non-zero integers $n_1, \ldots, n_m$, $m \geq 1$, will be denoted by $\gcd(n_1, \ldots, n_m)$. The notation $d_i \nmid D$ for positive integers d_i and D means that d_i does not divide D.

The infimum or greatest lower bound of a subset R of real numbers is denoted by $\inf(R)$ and is defined to be the biggest real number that is smaller than or equal to every number in R. If no such number exists (because R is not bounded below), then we define $\inf(S) = -\infty$.

Further terminology and notation will be introduced through the book.

Chapter 2
The Theory of Apportionment and Just-In-Time Sequences

2.1 Introduction

The apportionment problem and theory have their roots in the proportional representation system intended for the House of Representatives of the United States where each state receives seats in the House proportionally to its population. This chapter reviews these results of the apportionment theory that are most relevant to the topic of just-in-time optimization. It follows the excellent expositions of the basics of the theory presented in the books by Balinski and Young [2], and Young [3]. However, the chapter also includes new results obtained since these publications – especially in the context of the theory's new applications presented in this book.

The apportionment theory has been developed to address the problem of fair representation or "meeting the ideal of one man, one vote" as Balinski and Young put it in the title of their book. This ideal is clearly a fundamental one yet, as one feels, unattainable, and thus the apportionment problem is not just a problem in mathematics.

This book looks at this ideal in a broader than just political context in order to recognize the ideal's universality. For instance, the clients or virtual clients paying for the executions of their jobs in today's distributed computational economies, see for instance Waldspurger et al. [11], expect a fair implementation of these virtual economies – the clients demand a fair representation in terms of resource allocations to their jobs so that the ideal of one currency unit spent equals any other spent in the same distributed economy is met. Thus, a client who pays twice as much for its job execution as another one would like to see its job progressing at twice the rate of the other client's similar job at any time. Another example is a protection mechanism against antisocial behavior of individual hosts on the Internet and in other networks, see Nagle [12]. There, the apportionment methods can be used to establish an accepted norm for a good behavior and can lead to the whole network increased stability. There are volume differences too, the apportionment methods used traditionally in proportional election or representation system are usually called to work every 4–5 years, whereas the same methods would be called millions of times

W. Kubiak, *Proportional Optimization and Fairness,* International Series in Operations Research & Management Science 127, DOI 10.1007/978-0-387-87719-8_2, © Springer Science+Business Media LLC 2009

every minute on the Internet proving their application huge volume. This volume requires such apportionment methods that are computationally extremely efficient and relay on just few data in making online decisions as to who will receive the resources next. Fortunately, most apportionment methods, the divisor methods and in particular parametric methods for instance, satisfy all these conditions. Thus, the apportionment theory is where we feel any discussion of proportional representation should start.

Section 2.2 defines the apportionment problem. Section 2.3 introduces the basic axioms of the apportionment theory. These include the basic exact, anonymous and homogenous apportionments as well as population monotone apportionments introduced to avoid undesirable anomalies. Section 2.4 presents the divisor methods of apportionment. These are the only apportionment methods that deliver population monotone apportionments. Section 2.5 discusses incompatibility of being population monotone and staying within a quota properties of apportionment. Section 2.6 focuses on these features of divisor methods that encourage coalitions and schisms. Section 2.7 shows how to construct the just-in-time sequences using the house monotone apportionment methods. Section 2.8 discusses the desirable properties of just-in-time sequences inherited from the parametric apportionment methods. The properties include periodicity and various symmetries. Finally, Sect. 2.9 discusses the consistency with a standard two-state solution which is unique for the Webster's method of apportionment.

2.2 The Apportionment Problem

The instance of the apportionment problem is defined by the integer house size $h \geq 0$ and a positive real vector of state populations:

$$\mathbf{p} = (p_1, p_2, p_3, \ldots, p_s) > 0. \tag{2.1}$$

An apportionment of h seats among s states is an integer vector

$$\mathbf{a} = (a_1, a_2, a_3, \ldots, a_s) \geq 0 \tag{2.2}$$

such that $\sum_{i=1}^{s} a_i = h$.

Let the total population be

$$P = \sum_{k=1}^{s} p_k.$$

A "fair" share of state i seats is its *quota*

$$q_i = \frac{p_i h}{P}.$$

However, the quota vector

$$\mathbf{q} = (q_1, q_2, q_3, \ldots, q_s)$$

may be fractional and thus *not* an apportionment. We sometimes use the notation $\mathbf{a}^h$ and $\mathbf{q}^h$ instead of $\mathbf{a}$ and $\mathbf{q}$, respectively to emphasize that the latter two vectors correspond to the house of size h. We refer to

$$\lfloor q_i \rfloor = \left\lfloor \frac{p_i h}{P} \right\rfloor \qquad (2.3)$$

as the *lower quota* of state i and to

$$\lceil q_i \rceil = \left\lceil \frac{p_i h}{P} \right\rceil \qquad (2.4)$$

as the *upper quota* of the state.

The solution to the apportionment problem is found by an *apportionment method* $\mathbf{M}$. The method maps the vector $\mathbf{p}$ and the house size h into a set $\mathbf{M}(\mathbf{p}, h)$ of apportionments $\mathbf{a}$ that satisfy the condition (2.2).

2.3 Which Apportionment?

The definition of the apportionment problem given in (2.2) may result into trivial though unacceptable, for instance socially, solutions which need to be ruled out from further consideration. This is done by imposing axioms that define what is socially acceptable as properties of an apportionment. However, we believe that these properties should hold for other applications of the apportionment theory as well. We begin with the basic properties.

2.3.1 The Basics: Exact, Anonymous and Homogeneous Apportionments

We call the method $\mathbf{M}$ *exact* if

$$(q_1, \ldots, q_s) \in \mathbf{M}(\mathbf{p}, h) \text{ whenever quota } q_i = \frac{p_i h}{P} \text{ is an integer for all } i,$$

and this solution is unique. A method is *anonymous* if for any permutation π of the states $1, \ldots, s$ we have

$$(a_1, a_2, a_3, \ldots, a_s) \in \mathbf{M}((p_1, p_2, p_3, \ldots, p_s), h)$$

if and only if

$$(a_{\pi(1)}, a_{\pi(2)}, a_{\pi(3)}, \ldots, a_{\pi(s)}) \in \mathbf{M}((p_{\pi(1)}, p_{\pi(2)}, p_{\pi(3)}, \ldots, p_{\pi(s)}), h)$$

for all population vectors $\mathbf{p}$ and house sizes h. That is permuting the state populations results in apportionments that are permuted the same way.

An apportionment method $\mathbf{M}$ is *homogeneous* if for any $\mathbf{p}$ and h one requires $\mathbf{M}(\mathbf{p}, h) = \mathbf{M}(\lambda \mathbf{p}, h)$ for any positive rational number λ.

We continue the list of axioms with the not-so-obvious ones. These came to the attention of politicians and researchers as a result of infamous paradoxes or anomalies that lead to abandoning some earlier used apportionment methods. The new axioms were then formulated to protect against these paradoxes. We begin with the most famous one, the Alabama paradox, and its remedy, namely the house monotone methods.

2.3.2 House Monotone Apportionments

Any apportionment method $\mathbf{M}(\mathbf{p}, h)$ that gives an apportionment vector $\mathbf{a}$ for the house size h and the population vector $\mathbf{p}$, and an apportionment vector $\mathbf{a}' \geq \mathbf{a}$ for the house of size $h' = h + 1$ and the same population vector $\mathbf{p}$ is said to be *house monotone*. Precisely, a method $\mathbf{M}$ is house monotone if for every $\mathbf{p}$ and h if $\mathbf{a} \in \mathbf{M}(\mathbf{p}, h)$, then there is $\mathbf{a}' \in \mathbf{M}(\mathbf{p}, h + 1)$ such that $\mathbf{a}' \geq \mathbf{a}$. This books relies on apportionment methods for iteratively building sequences which requires the house size h to grow. Thus, only house monotone methods are relevant for our discussion since they allow to extend a sequence without any change to it, that is to what has already been built. All divisor methods defined later in Sect. 2.4.1 are house monotone. There are, however, historically important apportionment methods that are *not* house monotone. The Alexander *Hamilton's* method, known also as the largest reminder method is an example of an apportionment method that is not house monotone. The Hamilton's method lead to the infamous paradox of Alabama in 1882, when a larger size of the House gave fewer seats to the state of Alabama. The method's failure to be house monotone can be illustrated by the following example with the population vector $\mathbf{p} = (6, 6, 1)$. The house size $h = 5$ results in quotas $\frac{6 \times 5}{13} = 2\frac{4}{13}$ for the first two states and quota $\frac{5}{13}$ for the third state, thus according to the Hamilton's method the apportionment is $(2, 2, 1)$ since each state gets its whole number of seats, that is $2, 2$ and 0 respectively, first, and then since the total number of seats apportioned is one less than the house size, the difference goes to the state with the largest reminder, that is to the third state. Now, let us increase the size of the house to $h = 6$. Then, the quotas are $\frac{6 \times 6}{13} = \frac{36}{13} = 2\frac{10}{13}$ for the first two states and $\frac{6}{13}$ for the third. Then, the Hamilton's method results in the apportionment $(3, 3, 0)$. Thus, the third state losses its only seat in a larger house – an example of the Alabama paradox.

An even stronger axiom is the following *population monotone* axiom.

2.3.3 Population Monotone Apportionments

The Alabama paradox reveals *anomalies* that some apportionment methods exhibit whenever the size of the house h grows. However, it is not the only anomaly encountered in the theory of apportionment. Other anomalies may show in case of rapid changes in populations. One such anomaly is the *population paradox*, which happens whenever an apportionment method is not able to ensure that if the state *is* population increases and the state *js* decreases, then state i gets no fewer seats and state j gets no more seats with the new populations than they do with the original ones and the unchanged house size h. To avoid this population paradox as well as other paradoxes the population monotone apportionment methods have been introduced. Formally, the method is population monotone if for any two vectors of populations $\mathbf{p}, \mathbf{p}' > 0$, house sizes h and h', and vectors of apportionments $\mathbf{a} \in \mathbf{M}(\mathbf{p}, h)$, $\mathbf{a}' \in \mathbf{M}(\mathbf{p}', h')$ the following implication holds

$$\frac{p'_\ell}{p'_k} \geq \frac{p_i}{p_j} \Rightarrow \left\{ \begin{array}{c} \neg(a'_\ell < a_i \wedge a'_k > a_j) \\ or \\ \frac{p'_\ell}{p'_k} = \frac{p_i}{p_j} \text{ and } a'_\ell, a'_k \text{ can be substituted for } a_i, a_j \text{ in } \mathbf{a} \end{array} \right\}. \quad (2.5)$$

The population monotone apportionment methods also avoid the *new states paradox* that may arise whenever the sates join the union. The new states paradox consists in the following. Suppose a state $s+1$ joins in the union of s states. Then, this paradox happens whenever there are two states i and j of the old union such that one of them losses seats and the other gains seats. The population monotone apportionment methods avoid also the *seceding states paradox*. Suppose a state k secedes from the union of s states. Then, the paradox happens if there are two states i and j other than k such that one of them losses seats and the other gains seats.

The next class of methods ensures that an apportionment that is satisfactory for all states remains so for any subset of states considered alone. For instance two competing jobs may monitor their own progress and compare it with each other irrespectively of other jobs being present in the system and competing for the same resources, see the peer-to-peer fairness in Chap. 10. The uniform apportionment method ensures their satisfaction irrespective of other jobs.

2.3.4 Uniform Apportionments

An apportionment method is said to be *uniform* if it ensures that an apportionment $\mathbf{a} = (a_1, a_2, \ldots, a_s)$ of h seats of the house among states with populations $\mathbf{p} = (p_1, p_2, \ldots, p_s)$ will stay the same whenever it is restricted to any subset S of these states and the house size $\sum_{i \in S} a_i = h'$. In other words, if for every t, $2 \leq t \leq s$, $(a_1, \ldots, a_s) \in \mathbf{M}((p_1, \ldots, p_s), h)$ implies $(a_1, \ldots, a_t) \in \mathbf{M}((p_1, \ldots, p_t), \sum_{i=1}^{t} a_i)$ and if also $(b_1, \ldots, b_t) \in \mathbf{M}((p_1, \ldots, p_t), \sum_{i=1}^{t} a_i)$, then $(b_1, \ldots, b_t, a_{t+1}, \ldots, a_s) \in$

$\mathbf{M}((p_1,\ldots,p_s),h)$. Each uniform method can be obtained by using a *rank-index* function. The rank-index functions will be defined in Sect. 2.4.3.

Finally, the quota satisfaction property. This property however is at odds with population monotonicity, see the Impossibility Theorem 2.4.

2.3.5 Apportionments Satisfying Quota

The apportionment of a_i seats to state i satisfies the *lower quota* for population vector $\mathbf{p}$ and house size h if and only if

$$\lfloor q_i \rfloor \leq a_i, \tag{2.6}$$

and it satisfies the *upper quota* if and only if

$$a_i \leq \lceil q_i \rceil. \tag{2.7}$$

The a_i *satisfies the quota* if it simultaneously satisfies the lower and the upper quota, that is

$$\lfloor q_i \rfloor \leq a_i \leq \lceil q_i \rceil. \tag{2.8}$$

The apportionment vector $\mathbf{a}$ satisfies the quota if and only if it satisfies simultaneously the lower and the upper quota for all states. The method $\mathbf{M}(\mathbf{p}, h)$ satisfies the quota if each apportionment vector $\mathbf{a} \in \mathbf{M}(\mathbf{p}, h)$ satisfies the quota for any $\mathbf{p}$ and h.

Though it may appear natural to request that an apportionment method stays within the quota this request has been virtually rejected by the apportionment theory since it can not be simultaneously met with the requirement of population monotonicity. The latter is considered more important for the apportionment. However, staying within the quota gains importance in just-in-time applications where it may be argued as being more important than population monotonicity due to its ability to closely track a target value.

We now turn to the apportionment methods themselves.

2.4 Apportionment Methods

2.4.1 Divisor Methods

One can argue that the most successful approach to apportionment is based on a deceptively simple and unquestionably natural idea that calls for calculating the ideal district size x, or a divisor, that is the number of voters a single seat in the house ideally represents, and then checking how many such ideal districts fit into each state population. This latter number called quotient is then rounded to give the state apportionment. The nature of the game thus consists in doing the rounding – the way the rounding is done defines a divisor method. However, irrespective of the rounding the following result underlines the importance of divisor methods.

Theorem 2.1. *An apportionment method is population monotone if and only if it is a divisor method.*

The proof of this theorem can be found in Balinski and Young [2]. Formally, a divisor method finds proportional share of h seats by finding an ideal district size, or divisor x, then computing the quotients q_i^x of each state

$$q_i^x = \frac{p_i}{x}, \tag{2.9}$$

and finally rounding them according to some rounding rule. The sum of all rounded quotients must equal h. There are many ways to round the quotient. Any rounding procedure can be described by specifying a dividing point $d(a)$ in each interval of quotients $[a, a+1]$ for each non-negative integer a.

Define a *divisor function d* to be any monotone, real-valued function defined for all integers $a \geq 0$ such that $a \leq d(a) \leq a+1$ and such that there exists no pair of integers $b \geq 0$ and $c \geq 1$ with $d(b) = b+1$ and $d(c) = c$. This last condition implies that d is strictly increasing, and $k+1 > d(a+k) - d(a) > k-1$ for any $a \geq 0$ and $k \geq 1$. Any divisor function d can be used to define the $d-rounding$ of a real number $y \geq 0$ as follows

$$[y]_d = \begin{cases} 0 & \text{if } 0 \leq y \leq d(0), \\ a & \text{if } d(a-1) \leq y \leq d(a). \end{cases}$$

The $[y]_d$ is unique for any $y \neq d(a)$ for all integer $a \geq 0$. However, the threshold $y = d(a)$ can be either rounded down to a or up to $a+1$.

The *divisor method* based on d is thus defined as follows:

$$\mathbf{M}(\mathbf{p}, h) = \left\{ \mathbf{a} : a_i = \left[\frac{p_i}{z} \right]_d \text{ and } \sum_{i=1}^{s} a_i = h \text{ for some } z > 0 \right\}. \tag{2.10}$$

From (2.10) we have $d(a_i - 1) \leq \frac{p_i}{z} \leq d(a_i)$ for any $a_i > 0$ and $\frac{p_i}{z} \leq d(a_i)$ for $a_i = 0$, for some $z > 0$. Therefore,

$$\min_{a_i > 0} \frac{p_i}{d(a_i - 1)} \geq z \geq \max_{a_i \geq 0} \frac{p_i}{d(a_i)}.$$

Observe that there usually is a closed interval of z values that satisfy these inequalities, however, they all give the same apportionment, see Young [3]. We assume here that $\frac{p_i}{0}$ is defined and arbitrary large. The divisor method based on d can then be defined alternatively by a min–max relationship as follows.

$$\mathbf{M}(\mathbf{p}, h) = \left\{ \mathbf{a} : \min_{a_i > 0} \frac{p_i}{d(a_i - 1)} \geq \max_{a_j \geq 0} \frac{p_j}{d(a_j)} \text{ and } \sum_{i=1}^{s} a_i = h \right\} \tag{2.11}$$

where $p_i > p_j$ implies $\frac{p_i}{0} > \frac{p_j}{0}$.

Table 2.1 The best known divisor methods

Method's name	Adams	Dean	Hill	Webster	Jefferson
$d(a)$	a	$\frac{a(a+1)}{a+1/2}$	$\sqrt{a(a+1)}$	$a+1/2$	$a+1$

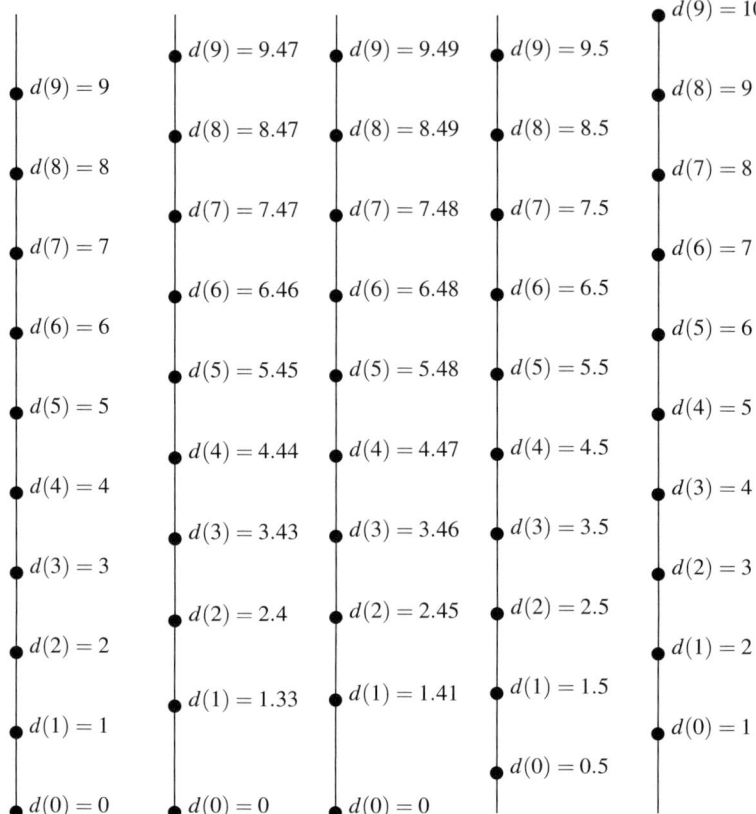

Fig. 2.1 The dividing points for the five divisor methods in Table 2.1. The order of the methods, for the *left* to *right*, follows the order in the table

Table 2.1 shows the $d(a)$ function of the best known divisor methods.

It follows from the table that the Dean's method uses the *harmonic* mean of a and $a+1$ as the divisor function, the Hill's method uses the *geometric* mean of a and $a+1$, and the Webster's uses simply the *arithmetic* mean of a and $a+1$. Figure 2.1 shows the dividing points for the divisor methods from Table 2.1.

Any divisor method is population monotone, and thus house monotone, see Theorem 2.1. Therefore, by the Impossibility Theorem 2.4 no divisor method stays within the quota.

2.4.2 Parametric Methods

Parametric method ϕ^δ is a divisor method with $d(a) = a + \delta$, where $0 \leq \delta \leq 1$. The parametric methods are of interest for two reasons. First, they are clearly computationally very efficient, their divisor functions are linear. Second, they generate cyclic just-in-time sequences whereas the divisor methods which are not parametric do not always generate cyclic just-in-time sequences. This will be shown in Theorem 2.12.

Clearly, the Adams's, Webster's (known also as the Sainte-Lagüe's method) and Jefferson's (known also as the d'Hondt's method) methods are parametric, while the Dean's and Hill's are not.

2.4.3 Rank-Index Methods

The idea behind the rank-index methods is to define a *standard of comparison* or *the rank-index function,* see Young [3], and then use it to obtain an *equitable* allocation relative to that standard. An allocation is equitable relative to a rank-index function if no seat transfer is justified. The seat transfer is justified if there are i and j so that j can give up one seat to i and by doing so reduce the inequity between the two states, that is if the following inequality is satisfied

$$\left| r(p_i, a_i) - r(p_j, a_j) \right| > \left| r(p_i, a_i + 1) - r(p_j, a_j - 1) \right|.$$

The rank-index function is any real-valued function of rational p and integer $a \geq 0$ that is decreasing in a, that is $r(p, a-1) > r(p,a)$ for any p and $a \geq 1$. The rank-index methods begin apportionment of h seats with defining a rank-index function $r(p,a)$ on all possible pairs (p,a) where p is a state population and a is any integer between 0 and h. The pairs (p,a) are then ordered in descending order of their rank-index values. The h seats are then apportioned according with the first h pairs on the ordered list. Thus, the h seats are apportioned to the most deserving, according to the rank-index, states.

An equitable allocation can also be found by the following simple algorithm given in Balinski and Young [2]. Let F be the family of functions f defined as follows.

1. For $h = 0$ let $f(p, 0) = 0$.
2. If $f(p,h) = \mathbf{a}$, then $f(p, h+1)$ is found by giving $a_i + 1$ seats to some state i such that $r(p_i, a_i) \geq r(p_j, a_j)$ for all j, and a_j seats to each $j \neq i$.

The *rank-index* method based on $r(p,a)$ is defined as follows:

$$\mathbf{M}(\mathbf{p}, h) = \{\mathbf{a} \ : \ \mathbf{a} = f(p,h) \ for \ some \ f \in F\}. \tag{2.12}$$

Any function of the form $r(p,a) = \frac{p}{d(a)}$ with a divisor function $d(a)$ is a rank-index function since $\frac{p}{d(a-1)} > \frac{p}{d(a)}$ by definition of divisor functions. Therefore, we can equivalently define the divisor method as follows, see Balinski and Young [2].

Theorem 2.2. *The divisor method based on a divisor function d can be defined as follows*

1. $\mathbf{M}(\mathbf{p}, 0) = 0$,
2. *If* $\mathbf{a} \in \mathbf{M}(\mathbf{p}, h)$ *and* k *satisfies* $\frac{p_k}{d(a_k)} = \max_i \frac{p_i}{d(a_i)}$, *then* $\mathbf{b} \in \mathbf{M}(p, h+1)$ *with* $b_k = a_k + 1$ *for* $i = k$ *and* $b_i = a_i$ *for* $i \neq k$.

Clearly this definition gives us a simple way of constructing the just-in-time sequence as long as the divisor function d itself can be efficiently computed. The above discussion and definitions prove that any rank-index method is uniform. Balinski and Young [2] furthermore prove that any *uniform* method is a rank-index method as long as it is balanced. A method is balanced if $\mathbf{a} \in \mathbf{M}(\mathbf{p} = (p_1, \ldots, p_i = p, \ldots, p = p_j, \ldots), h)$ implies $|a_i - a_j| \leq 1$. That is the balanced method guarantees that the apportionments of states with equal populations differ by at most one seat. We have the following theorem.

Theorem 2.3. *A method is balanced and uniform if and only if it is a rank method.*

2.5 What is Impossible?

Balinski and Young [2] show that any population monotone method is house monotone but not the other way around. Their famous Impossibility Theorem shows that the failure to stay within the quota is the price that any population monotone apportionment method must pay for its desirable quality of being population monotone.

Theorem 2.4 (The Impossibility Theorem). *It is impossible for an apportionment method to be population monotone and stay within quota at the same time.*

Proof. Consider the vector of populations $\mathbf{p} = \left(5 + e, \frac{2}{3}, \frac{2}{3}, \frac{2}{3} - e\right)$, where $\frac{2}{3} > e > 0$ is a rational number.

For $P = h = 7$, the p_i is *is* exact quota for all i. Let $\mathbf{a} \in \mathbf{M}(\mathbf{p}, h)$. Now consider the populations $\mathbf{p}' = \left(4 - e, 2 - \frac{e}{2}, \frac{1}{2} + \frac{e}{2}, \frac{1}{2} + e\right)$. Again $P' = h = 7$, so p_i' is *is* exact quota. Let $\mathbf{a}' \in \mathbf{M}(\mathbf{p}', h)$. Choose any e that meets the condition $e < \frac{1}{61}$, we then have

$$\frac{p_1'}{p_4'} > \frac{p_1}{p_4}.$$

This implies either $a_1' \geq a_1$ or $a_4' \leq a_4$ by population monotonicity, see (2.5). Now assume $a_1 \geq 5$ and $a_4' > 1$. Then $a_1' \geq a_1$ implies $a_1' \geq 5$ and thus $\mathbf{a}'$ fails upper quota. On the other hand, $a_4 \geq a_4'$ implies $a_4 > 1$ and thus $\mathbf{a}$ fails upper quota. Next, assume $a_1 \geq 5$ and $a_4' = 0$. Then, by population monotonicity $a_3' = a_4' = 0$. Thus $a_1' + a_2' = 7$. Therefore $\mathbf{a}'$ fails upper quota. If $a_1 \geq 5$ and $a_4' = 1$, then it remains to consider $a_1' < a_1$ and $a_4' \leq a_4$. Then, $a_1 + a_4 \geq 6$ and thus either $a_2 = 0$ or $a_3 = 0$. This, however leads to a contradiction since $\mathbf{M}$ being population monotone apportions the two seats to more populous states a_2 and a_3 rather than to a_4, the least populous out of the four states. Finally, if $a_1 < 5$ then $\mathbf{a}$ fails lower quota. This ends the proof since the cases considered exhaust all possibilities for $e < \frac{1}{61}$. $\square$

However, we have the following two results.

Theorem 2.5. *Each divisor method stays within the quota for s=2.*

Proof. See Exercise 2.23. $\square$

Theorem 2.6. *The Webster's method stays within the quota for s=3.*

Proof. See Exercise 2.24. $\square$

The following two technical lemmas will be used in Sect. 8.4 of Chap. 8. The lemmas show that if a divisor method fails to satisfy quota, then it does so for some fractional quota.

Lemma 2.7. *If a divisor method $\mathbf{M}$ does not stay above lower quota, then there are the vector of populations $\mathbf{p} = (p_1, p_2, \ldots, p_s)$, the house size h and state k for which the lower quota is not satisfied and the quota $q_k = \frac{p_k h}{P}$ is fractional.*

Proof. If $\mathbf{M}$ does not stay above lower quota, then there exist a vector of populations $\mathbf{p} = (p_1, p_2, \ldots, p_s)$, $s \geq 3$, the house size h and the apportionment $\mathbf{a} = (a_1, \ldots, a_s)$

$$\mathbf{a} \in \mathbf{M}(\mathbf{p}, h) = \{\mathbf{a} : \sum_i \left[\frac{p_i}{x} \right]_d = h \text{ for some divisor } x > 0\} \qquad (2.13)$$

for which the lower quota is not satisfied for some state k, i.e.,

$$a_k < \lfloor q_k \rfloor = \left\lfloor \frac{p_k h}{P} \right\rfloor. \qquad (2.14)$$

In (2.13), the d is the divisor function for $\mathbf{M}$. Let us assume that q_k is integral, otherwise the lemma holds. The proof proceeds by cases.
If there is a state $j \neq k$ such that either

$$d(a_j - 1) < \frac{p_j}{x} < d(a_j)$$

or

$$\frac{p_j}{x} = d(a_j) \text{ and } \left[\frac{p_j}{x} \right]_d = a_j,$$

that is $\frac{p_j}{x} = d(a_j)$ is d-rounded down, then define the populations

$$p_i' = \begin{cases} p_i(1 + \frac{\varepsilon}{x}) & \text{if } i \neq j, \\ p_j & \text{if } i = j, \end{cases}$$

for all i and $\varepsilon > 0$. For the divisor $x + \varepsilon$, we have

$$\frac{p_i'}{x + \varepsilon} = \frac{p_i}{x} \text{ for } i \neq j.$$

Moreover, for sufficiently small $\varepsilon > 0$ we get

$$\left[\frac{p_j}{x}\right]_d = \left[\frac{p_j}{x + \varepsilon}\right]_d.$$

Thus, $\mathbf{a} \in \mathbf{M}(\mathbf{p}', h)$ and

$$q_k' = \frac{p_k(1 + \frac{\varepsilon}{x})h}{(1 + \frac{\varepsilon}{x})\sum_{i \neq j} p_i + p_j} > q_k.$$

Therefore, a_k does not stay above lower quota and furthermore q_k' is fractional for sufficiently small $\varepsilon > 0$. This proves the lemma.

It remains to show what happens if for all $j \neq k$

$$\frac{p_j}{x} = d(a_j - 1) \text{ and } \left[\frac{p_j}{x}\right]_d = a_j, \tag{2.15}$$

that is all $\frac{p_j}{x}$ are $d-$rounded up to a_j. In this case, if

$$d(a_k - 1) \leq \frac{p_k}{x} < d(a_k), \tag{2.16}$$

then define the populations

$$p_i' = \begin{cases} p_i & \text{if } i \neq k \\ p_k + \lambda & \text{if } i = k. \end{cases}$$

Then, the divisor x results in the apportionment $\mathbf{a} \in M(\mathbf{p}', h)$ for sufficiently small $\lambda > 0$. Since q_k' is fractional for sufficiently small $\lambda > 0$ and $q_k' > q_k$, then again the lemma holds.

Otherwise, that is if (2.16) does not hold, we have

$$\frac{p_k}{x} = d(a_k) \text{ and the } \frac{p_k}{x} \text{ is } d-\text{rounded down to } a_k, \text{i.e., } \left[\frac{p_k}{x}\right]_d = a_k. \tag{2.17}$$

Since $\mathbf{M}$ is homogeneous, then $\mathbf{M}(\mathbf{p}, h) = \mathbf{M}((d(a_1 - 1), \ldots, d(a_k), \ldots, d(a_s - 1)), h)$, where $\mathbf{p} = x(d(a_1 - 1), \ldots, d(a_k), \ldots, d(a_s - 1))$ by (2.15) and (2.17). Thus, it suffices to consider the vector $(d(a_1 - 1), \ldots, d(a_k), \ldots, d(a_s - 1))$ of populations and the house size h. By (2.15) and (2.17) it follows that

$$h = \sum_{i=1}^{s} a_i.$$

Now, for $a_k \geq 1$ consider the vector of $s(m + 1)$ populations

$$\mathbf{p}' = (d(a_1-1),\ldots,d(a_k),\ldots,d(a_s-1),\underbrace{d(a_1-1),\ldots,d(a_k-1),\ldots,d(a_s-1)}_{m\text{-times}}),$$

for some $m \geq 1$, and the size of the house $H = h(1+m)$. We have

$$\mathbf{a}' = (a_1,\ldots,a_k,\ldots,a_s,\underbrace{a_1,\ldots,a_k,\ldots,a_s}_{m\text{-times}}) \in \mathbf{M}(\mathbf{p}',H).$$

To show this just take the divisor $x = 1$ and d-round accordingly. However,

$$\frac{d(a_k)h}{\sum_{i=1,i\neq k}^s d(a_i-1)+d(a_k)} \tag{2.18}$$

$$\leq \frac{d(a_k)h(1+m)}{\sum_{i=1,i\neq k}^s d(a_i-1)+d(a_k)+m\sum_{i=1}^s d(a_i-1)}. \tag{2.19}$$

obviously holds since $d(a_k-1) < d(a_k)$ by definition. Furthermore, the quota on the right hand side of (2.18) can be made fractional by the appropriate choice of m. Therefore, k fails its lower quota which is fractional. Thus, the lemma holds for $\mathbf{p}'$ and H.

Finally, for $a_k = 0$ we have $d(0) > 0$ by (2.17). Then, consider the vector of $s(m+1)$ populations

$$\mathbf{p}'' = (d(a_1-1),\ldots,d(a_k),\ldots,d(a_s-1),\underbrace{d(a_1-1),\ldots,\varepsilon,\ldots,d(a_s-1)}_{m\text{-times}}),$$

for some $m \geq 1$, $d(0) > \varepsilon > 0$, and the size of the house $H = h(1+m)$. We have

$$\mathbf{a}'' = (a_1,\ldots,a_k=0,\ldots,a_s,\underbrace{a_1,\ldots,a_k=0,\ldots,a_s}_{m\text{-times}}) \in \mathbf{M}(\mathbf{p}'',H).$$

To show this just take the divisor $x = 1$ and d−round accordingly. However,

$$\frac{d(0)h}{\sum_{i=1,i\neq k}^s d(a_i-1)+d(0)} < \frac{d(0)h(1+m)}{d(0)+m\varepsilon+(1+m)\sum_{i=1,i\neq k}^s d(a_i-1)}. \tag{2.20}$$

obviously holds since $\varepsilon < d(0)$ by definition. Furthermore, the quota on the right hand side of (2.20) can be made fractional by the appropriate choice of m or ε. Therefore, k fails its lower quota which is fractional. Thus, the lemma holds for $\mathbf{p}'$ and H. $\square$

Similarly we can prove the following lemma, the proof will be omitted.

Lemma 2.8. *If a divisor method* $\mathbf{M}$ *does not stay below upper quota, then there are the vector of populations* $\mathbf{p} = (p_1,p_2,\ldots,p_s)$, *the house size h and state k for which the upper quota is not satisfied and the quota $q_k = \frac{p_k h}{P}$ is fractional.*

2.6 Coalitions and Schisms

In the apportionment practice and theory the coalitions and schisms in proportional representation system have always been common tools for gaining and maintaining power, and the apportionment methods were investigated for their ability to encourage, or discourage, coalitions or schisms.

Similar advantages of being either larger or smaller can be found in other contexts as well. For instance, a number of clients sharing resources, for example computer, network or manufacturing resources, may consider merging their individual jobs to create one single composite job being the sum of these individual jobs. The composite job then enters the competition for resources as a single entity belonging to a single "corporate" client. The advantage of such coalition could be that the composite job finishes earlier and, in the worst case, never later than the latest of the individual jobs were they to compete for resources individually. We show in this section that to realize this advantage of being larger the resource allocation must be done according to the Jefferson's method. No other divisor method is capable of ensuring it. This advantage of being larger has been observed in stride scheduling and networks, see Waldspurger and Weihl [13]. The issue of splitting the gains resulting from the advantage of being larger between the individual clients though very important and challenging is beyond the scope of this book.

The opposite situation occurs when a client considers breaking its job into a number of smaller jobs so that doing them all separately and independently results into the completion of the job itself. The smaller jobs are then let to compete for resources between themselves and all other jobs. The advantage of such schism would be that the latest of the smaller jobs finishes earlier and, in the worst case, never later than the whole job were it done as a single job. This section shows that to realize this advantage of being smaller the resource allocation must be done according to the Adams's method. No other divisor method is capable of ensuring it.

We begin with definitions.

Let $C \subseteq \{1, \ldots, s\}$ be a subset of states we refer here to as a coalition. For any vector $\mathbf{p} = (p_1, p_2, p_3, \ldots, p_s)$, let $\mathbf{p}_C$ be the $s - |C| + 1$ dimensional vector obtained from $\mathbf{p}$ by replacing all coordinates in C by a single coordinate $p_C = \sum_{i \in C} p_i$, all the remaining coordinates do not change. We say that the apportionment method $\mathbf{M}$ encourages coalitions if and only for any $\mathbf{a} \in \mathbf{M}(\mathbf{p}, h)$ there always exists $\mathbf{b}_C \in \mathbf{M}(\mathbf{p}_C, h)$ such that $b_C \geq a_C$. We say that the apportionment method $\mathbf{M}$ encourages schisms if and only for any $\mathbf{a} \in \mathbf{M}(\mathbf{p}, h)$ there always exists $\mathbf{b}_C \in \mathbf{M}(\mathbf{p}_C, h)$ such that $b_C \leq a_C$.

Theorem 2.9. *Jefferson's method is the unique method that is population monotone and encourages coalitions.*

Proof. Let us denote the Jefferson's method by $\mathbf{J}$. We have $d(a) = a + 1$ for the Jefferson's method, see Table 2.1. If $\mathbf{a} \in \mathbf{J}(\mathbf{p}, h)$, then for some $x > 0$

$$a_i = d(a_i - 1) \leq \frac{p_i}{x} \leq d(a_i) = a_i + 1 \text{ for all } i.$$

We assume $d(-1) = 0$ by definition. Thus, for any subset $C \subseteq \{1, \ldots, s\}$ we have

$$a_C = \sum_{i \in C} a_i = d\left(\sum_{i \in C} a_i - 1\right) \leq \frac{\sum_{i \in C} p_i}{x} = \frac{p_C}{x} \text{ and } a_i \leq \frac{p_i}{x} \text{ for all } i \notin C.$$

Therefore, for the divisor x

$$x \leq \frac{p_C}{a_C}$$

and consequently there is a divisor y

$$x \leq y \leq \frac{p_C}{a_C}$$

such that

$$d(b_C - 1) \leq \frac{p_C}{y} \leq d(b_C) \text{ and } b_i \leq \frac{p_i}{y} \text{ for all } i \notin C$$

and

$$b_C \geq a_C \text{ and } b_C + \sum_{i \notin C} b_i = h.$$

This proves that the Jefferson's method encourages coalitions. Now, let $\mathbf{M}$ be a population monotone methods. By Theorem 2.1, $\mathbf{M}$ is a divisor method based on some divisor function d. Consider an instance with $s = n + 1$ and

$$\mathbf{p} = (p, \ldots, p) \text{ and } h = n(a+1) + a$$

for some p and a. Then, by (2.11)

$$(a, a+1, \ldots, a+1) \in \mathbf{M}(\mathbf{p}, h).$$

Consider a coalition $C = \{2, \ldots, n+1\}$. If $\mathbf{M}$ encourages coalitions, then

$$(\alpha, \beta) \in \mathbf{M}((p, np), h) \text{ with } \alpha \leq a \text{ and } \beta \geq n(a+1). \tag{2.21}$$

However, by (2.11)

$$(a, na) \in \mathbf{M}((p, np), h - n).$$

Since $\mathbf{M}$, being population monotone, is house monotone then

$$\alpha \geq a.$$

Consequently, by (2.21)

$$(a, na + n) \in \mathbf{M}((p, np), h).$$

Thus

$$d(a-1) \leq \frac{p}{x} \leq d(a) \text{ and } d(an + n - 1) \leq \frac{np}{x} \leq d(an + n)$$

and consequently

$$\frac{1}{d(a)} \leq \frac{n}{d(an + n - 1)},$$

or

$$a+1-\frac{1}{n} \leq \frac{d(an+n-1)}{n} \leq d(a) \leq a+1$$

for any $a \geq 0$ and all $n \geq 1$. Therefore, $d(a) = a+1$ and thus $\mathbf{M}$ is the Jefferson's method. $\square$

Similarly, it can be shown that:

Theorem 2.10. *Adams's method is the unique method that is population monotone and encourages schisms.*

All other divisor methods may result in two states (parties) either gaining a seat or losing a seat sometime. If a method results in a coalition of two states gaining a seat as likely as loosing it, then it is called coalition-neutral method. We have the following theorem.

Theorem 2.11. *Webster's method is the unique method that is population monotone and coalition-neutral.*

We refer the reader to Balinski and Young [2] for details of the proof, however, we would like to give intuition behind this result here. Assume vector of populations $\mathbf{p} = (p_1, p_2, p_3, \ldots, p_s)$ and a fixed divisor x. Let i and j form a coalition. Assume that ε_i and ε_j are the reminders of the quotients

$$\frac{p_i}{x} \text{ and } \frac{p_j}{x}.$$

The Webster's method gives one less seat to the coalition of i and j than to i and j separately provided that

$$\varepsilon_i > 0.5 \text{ and } \varepsilon_j > 0.5 \text{ and } \varepsilon_i + \varepsilon_j < 1.5 \tag{2.22}$$

for the reminders that are independently and uniformly distributed between 0 and 1 the probability of this happening is $\frac{1}{8}$. On the other hand, the coalition gains a seat provided that

$$\varepsilon_i < 0.5 \text{ and } \varepsilon_j < 0.5 \text{ and } \varepsilon_i + \varepsilon_j > 0.5 \tag{2.23}$$

thus the probability of this happening is $\frac{1}{8}$ as well. Therefore, the Webster's method is coalition-neutral. Observe that the probabilities do not change if we replace the strong $<$ inequalities by the weak ones $\leq$ in both (2.22) and (2.23). Finally, observe that the probability that the coalition of i and j gets the same as its two partners is $1 - (\frac{1}{8} + \frac{1}{8}) = \frac{3}{4}$.

2.7 From Apportionments to Just-In-Time Sequences

In the transformation between the just-in-time sequencing and apportionment problems, state i corresponds to model i and the demand d_i for model i corresponds to population p_i of state i. The cumulative number of units $x_{i,k}$ of i completed by k

corresponds to the number a_i of seats apportioned to state i in a house of size k. The following is the summary of the correspondences between the two problems:

$$\text{number of states } s \longleftrightarrow \text{ number of models } n$$
$$\text{state } i \longleftrightarrow \text{ model } i$$
$$\text{population } p_i \text{ of state } i \longleftrightarrow \text{ demand } d_i \text{ for model } i$$
$$\text{vector of populations } \mathbf{p} \longleftrightarrow \text{vector of demands } \mathbf{d}$$
$$\text{size of house } h \longleftrightarrow \text{ position in sequence } k$$
$$\text{for a house of size } h, a_i \longleftrightarrow x_{i,k}$$
$$\text{total population } P = \textstyle\sum_{i=1}^{s} p_i \longleftrightarrow \text{ total demand } D = \textstyle\sum_{i=1}^{n} d_i.$$

Any exact and house monotone apportionment method $\mathbf{M}$ can be used to construct a just-in-time sequence as follow. Let

$$\mathbf{M}(\mathbf{d},1), \mathbf{M}(\mathbf{d},2), \ldots, \mathbf{M}(\mathbf{d},D)$$

be the apportionments for the house sizes $1, 2, \ldots, D$ respectively and the population vector $\mathbf{d}$. The just-in-time sequence s is built as follows. Let vectors $\mathbf{a}^1, \mathbf{a}^2, \ldots, \mathbf{a}^D$ be selected from $\mathbf{M}(\mathbf{d},1), \mathbf{M}(\mathbf{d},2), \ldots, \mathbf{M}(\mathbf{d},D)$ respectively so that $\mathbf{a}^h \leq \mathbf{a}^{h+1}$ for any $h = 1, 2, \ldots, D-1$. Such selection of vectors exists since $\mathbf{M}$ is house monotone. Moreover, there is exactly one i such that $a_i^h + 1 = a_i^{h+1}$ since $h = \sum a_i^h$ and $h+1 = \sum a_i^{h+1}$. The i will be placed in the position $h+1$ of the sequence s. Moreover, the only state that gets a seat in $\mathbf{a}^1$ will be placed in position 1 of s. Finally, the quotas $\frac{Dd_i}{D} = d_i$ for the house size D are all integer numbers, thus there is exactly one vector in $\mathbf{M}(\mathbf{d},D)$ and it is $(d_1, d_2, \ldots, d_n)$. The latter holds since the method is exact. Consequently, the sequence is just-in-time since i occurs exactly d_i times in it. We denote by $S_{\mathbf{M}}(\mathbf{d})$ the set of all just-in-time sequences obtained by method $\mathbf{M}$ for vector $\mathbf{d}$.

2.8 Which Just-In-Time Apportionments?

We now focus on these properties of apportionment methods that make them especially attractive for building just-in-time sequence with desired properties. We begin with the cyclic sequences.

2.8.1 Cycles

Let $g = \gcd(d_1, d_2, \ldots, d_n)$. The method $\mathbf{M}$ is cyclic

$$\text{if } s \in S_{\mathbf{M}}(\mathbf{d}/g) \text{ implies } s^g \in S_{\mathbf{M}}(\mathbf{d}).$$

A powerful advantage of the parametric methods of apportionment is that they are cyclic which is not the case for other divisor methods that are not parametric. The key observation is that the distance between any two consecutive dividing points of a parametric divisor method is constant, equal 1, which is not the case for divisor methods not being parametric. We have the following theorem.

Theorem 2.12. *A divisor method is cyclic if and only if it is parametric.*

Proof. First, we observe that

$$\frac{d_i}{a_i + \delta} \geq \frac{d_j}{a_j + \delta} \text{ if and only if } \frac{d_i}{k\frac{d_i}{g} + a_i + \delta} \geq \frac{d_j}{k\frac{d_j}{g} + a_j + \delta}$$

for $k = 0, 1, \ldots, g-1$. Thus any parametric method ϕ^δ is cyclic. Second, no divisor method which is not parametric is cyclic. The proof is by contradiction. Assume that **M** is a divisor method based on a divisor function d and **M** is not parametric. Then, there exists positive integer k^* such that $d(k) = k + \delta$ for $0 \leq k < k^*$ and $d(k^*) = k^* + \delta^*$ for some $0 \leq \delta \neq \delta^* \leq 1$. Assume $\delta < \delta^*$. Let d_1 and d_2 be any two positive integers that satisfy $0 \leq \delta < \frac{d_2}{d_1} < \delta^* \leq 1$. Consider a vector $\mathbf{d'} = (\alpha d_1 - 1, 1)$ where α is a positive integer. Then $\mathbf{M}(\mathbf{d'}, h = \alpha d_1) = \mathbf{d'}$ since **M**, being a divisor method, is exact. Moreover, $\mathbf{M}(\mathbf{d'}, h = \alpha d_2) = (\alpha d_2 - 1, 1)$ for sufficiently large α, that is state 2 always gets a single seat in the house of size αd_2 provided α is sufficiently large. To show this it suffices to observe that by Theorem 2.2 state 2 gets at least one seat in such house since the following inequality

$$\frac{1}{d(0)} = \frac{1}{\delta} > \frac{d_1 - \frac{1}{\alpha}}{d_2 - \frac{1}{\alpha}} = \frac{\alpha d_1 - 1}{\alpha d_2 - 1} \geq \frac{\alpha d_1 - 1}{d(\alpha d_2 - 1)}, \tag{2.24}$$

holds for sufficiently large α. Furthermore, state 2 does not get more than one seat since $d_2 < d_1$ and **M**, being a divisor method, is house monotone.

Now, consider the vector $(k^* + 1)\mathbf{d'} = (p_1 = (k^* + 1)(\alpha d_1 - 1), p_2 = k^* + 1)$. If **M** is cyclic, then $(a_1 = k^*(\alpha d_1 - 1) + \alpha d_2 - 1, a_2 = k^* + 1) \in \mathbf{M}((k^* + 1)\mathbf{d'}, h = k^* \alpha d_1 + \alpha d_2)$. However, since

$$\frac{p_1}{d(a_1)} = \frac{(k^* + 1)(\alpha d_1 - 1)}{d(k^*(\alpha d_1 - 1) + \alpha d_2 - 1)} \geq \frac{(k^* + 1)(\alpha d_1 - 1)}{k^*(\alpha d_1 - 1) + \alpha d_2}$$

$$= \frac{k^* + 1}{k^* + \frac{\alpha d_2}{\alpha d_1 - 1}} > \frac{k^* + 1}{k^* + \delta^*} = \frac{k^* + 1}{d(k^*)} = \frac{p_2}{d(a_2 - 1)}$$

the min–max condition in (2.11) is violated. Thus, we get a contradiction. This proves the theorem for $\delta < \delta^*$. The proof for $\delta > \delta^*$ proceeds in a similar fashion and will be omitted. $\square$

We stay with the parametric divisor methods to investigate the influence of the parameter δ on the just-in-time sequences being built.

2.8.2 Advancing and Delaying

A parametric divisor apportionment method used to built just-in-time sequence could advance some jobs at the cost of retarding others in comparison to some other parametric method. Since all methods we discuss in this book are anonymous the only factor that affect this behavior is the size of the job and the parameter δ used by the method. Table 2.2 shows all possible just-in-time sequences for the vector of populations (or demands) $\mathbf{d} = (7, 6, 4, 2, 1)$ obtained by the parametric methods ϕ^δ for $0 \le \delta \le 1$. The longest job, that is job 1, completes there at $t = 20$ for $\delta = 0$ and at $t = 16$, at the earliest, for $\delta = 1$. At the same time the shortest job, that is job 5, completes there at $t = 5$ for $\delta = 0$ and at $t = 20$, at the latest, for $\delta = 1$.

The longest job starts advancing its positions in the sequence starting from its second position $t = 6$ for $\delta = 0$ and never stops advancing. The second longest job starts advancing its positions in the sequence starting from its second position $t = 7$ for $\delta = 0$ and all its positions gets advanced. The third longest never gets advanced or retarded. The losers are the two shortest jobs. The second shortest, that is job 4, starts retarding from its first position $t = 4$ for $\delta = 0$ and never stops retarding. The shortest starts retarding from its first position $t = 5$ for $\delta = 0$ and never stops retarding. Thus as δ grows the three longest jobs benefit by advancing their progress whereas the two shortest lose by retarding their progress. Though the second longest job completes only slightly earlier, at $t = 19$ for $\delta = 0$ and at $t = 17$, at the earliest,

Table 2.2 The just-in-time sequences for different δ

$\delta \in$	ϕ^δ
$\{0\}$	$1 \to 2 \to 3 \to 4 \to 5 \to 1 \to 2 \to 3 \to 1 \to 2 \to 1 \to 4 \leftrightarrow 3 \leftrightarrow 2 \to 1 \to 2 \to 1 \to 3 \to 2 \to 1$
$(0, \frac{1}{6})$	$1 \to 2 \to 3 \to 4 \to 5 \to 1 \to 2 \to 3 \to 1 \to 2 \to 1 \to \underline{2} \to \underline{3} \to \underline{4} \to 1 \to 2 \to 1 \to 3 \to 2 \to 1$
$\{\frac{1}{6}\}$	$1 \to 2 \to 3 \to 4 \to 5 \leftrightarrow 1 \to 2 \to 3 \to 1 \to 2 \to 1 \to 2 \to 3 \to 4 \to 1 \to 2 \to 1 \to 3 \to 2 \to 1$
$(\frac{1}{6}, \frac{1}{5})$	$1 \to 2 \to 3 \to 4 \to \underline{1} \to \underline{5} \to 2 \to 3 \to 1 \to 2 \to 1 \to 2 \to 3 \to 4 \to 1 \to 2 \to 1 \to 3 \to 2 \to 1$
$\{\frac{1}{5}\}$	$1 \to 2 \to 3 \to 4 \to 1 \to 5 \leftrightarrow 2 \to 3 \to 1 \to 2 \to 1 \to 2 \to 3 \to 4 \leftrightarrow 1 \to 2 \to 1 \to 3 \to 2 \to 1$
$(\frac{1}{5}, \frac{1}{3})$	$1 \to 2 \to 3 \to 4 \to 1 \to \underline{2} \to \underline{5} \to 3 \to 1 \to 2 \to 1 \to 2 \to 3 \to \underline{1} \to \underline{4} \to 2 \to 1 \to 3 \to 2 \to 1$
$\{\frac{1}{3}\}$	$1 \to 2 \to 3 \to 4 \to 1 \to 2 \to \underline{5} \leftrightarrow 3 \leftrightarrow \underline{1} \to 2 \to 1 \to 2 \to 3 \to 1 \to 4 \to 2 \to 1 \to 3 \to 2 \to 1$
$(\frac{1}{3}, \frac{2}{5})$	$1 \to 2 \to 3 \to 4 \to 1 \to 2 \to \underline{1} \to 3 \to \underline{5} \to 2 \to 1 \to 2 \to 3 \to 1 \to 4 \to 2 \to 1 \to 3 \to 2 \to 1$
$\{\frac{2}{5}\}$	$1 \to 2 \to 3 \to 4 \leftrightarrow 1 \to 2 \to 1 \to 3 \to 5 \leftrightarrow 2 \to 1 \to 2 \to 3 \to 1 \to 4 \to 2 \to 1 \to 3 \to 2 \to 1$
$(\frac{2}{5}, \frac{1}{2})$	$1 \to 2 \to 3 \to \underline{1} \to 4 \to 2 \to 1 \to 3 \to \underline{2} \to \underline{5} \to 1 \to 2 \to 3 \to 1 \to 4 \to 2 \to 1 \to 3 \to 2 \to 1$
$\{\frac{1}{2}\}$	$1 \to 2 \to 3 \to 1 \to 4 \leftrightarrow 2 \to 1 \to 3 \to 2 \to 1 \leftrightarrow 5 \to 2 \to 3 \to 1 \to 2 \leftrightarrow 4 \to 1 \to 3 \to 2 \to 1$
$(\frac{1}{2}, \frac{3}{5})$	$1 \to 2 \to 3 \to 1 \to \underline{2} \to \underline{4} \to 1 \to 3 \to 2 \to \underline{1} \to \underline{5} \to 2 \to 3 \to 1 \to \underline{2} \to \underline{4} \to 1 \to 3 \to 2 \to 1$
$\{\frac{3}{5}\}$	$1 \to 2 \to 3 \to 1 \to 2 \to 4 \to 1 \to 3 \to 2 \to 1 \to 5 \leftrightarrow 2 \to 3 \to 1 \to 2 \to 4 \leftrightarrow 1 \to 3 \to 2 \to 1$
$(\frac{3}{5}, \frac{2}{3})$	$1 \to 2 \to 3 \to 1 \to 2 \to 4 \to 1 \to 3 \to 2 \to 1 \to \underline{2} \to \underline{5} \to 3 \to 1 \to 2 \to \underline{1} \to \underline{4} \to 3 \to 2 \to 1$
$\{\frac{2}{3}\}$	$1 \to 2 \to 3 \to 1 \to 2 \to 4 \to 1 \to 3 \to 2 \to 1 \to 2 \to \underline{5} \leftrightarrow 3 \leftrightarrow \underline{1} \to 2 \to 1 \to 4 \to 3 \to 2 \to 1$
$(\frac{2}{3}, \frac{4}{5})$	$1 \to 2 \to 3 \to 1 \to 2 \to 4 \to 1 \to 3 \to 2 \to 1 \to 2 \to \underline{1} \to 3 \to \underline{5} \to 2 \to 1 \to 4 \to 3 \to 2 \to 1$
$\{\frac{4}{5}\}$	$1 \to 2 \to 3 \to 1 \to 2 \to 4 \leftrightarrow 1 \to 3 \to 2 \to 1 \to 2 \to 1 \to 3 \to 5 \leftrightarrow 2 \to 1 \to 4 \to 3 \to 2 \to 1$
$(\frac{4}{5}, \frac{5}{6})$	$1 \to 2 \to 3 \to 1 \to 2 \to \underline{1} \to 4 \to 3 \to 2 \to 1 \to 2 \to 1 \to 3 \to \underline{2} \to \underline{5} \to 1 \to 4 \to 3 \to 2 \to 1$
$\{\frac{5}{6}\}$	$1 \to 2 \to 3 \to 1 \to 2 \to 1 \to 4 \to 3 \to 2 \to 1 \to 2 \to 1 \to 3 \to 2 \to 5 \leftrightarrow 1 \to 4 \to 3 \to 2 \to 1$
$(\frac{5}{6}, 1)$	$1 \to 2 \to 3 \to 1 \to 2 \to 1 \to 4 \to 3 \to 2 \to 1 \to 2 \to 1 \to 3 \to 2 \to \underline{1} \to \underline{5} \to 4 \to 3 \to 2 \to 1$
$\{1\}$	$1 \to 2 \to 3 \to 1 \to 2 \to 1 \to \underline{2} \leftrightarrow \underline{3} \leftrightarrow \underline{4} \to 1 \to 2 \to 1 \to 3 \to 2 \to 1 \to \underline{1} \leftrightarrow \underline{2} \leftrightarrow \underline{3} \leftrightarrow \underline{4} \leftrightarrow \underline{5}$

for $\delta = 1$ and the third longest job completes at $t = 18$ for $\delta = 0$ and at $t = 18$, at the earliest, for $\delta = 1$.

We now investigate this phenomenon formally beginning with the following definition, see Balinski and Rachev [14], and Balinski and Ramirez [15].

Method $\mathbf{M}$ gives up to method $\mathbf{M}^*$ if

$$\mathbf{a} \in \mathbf{M}(\mathbf{p},h), \ \mathbf{a}^* \in \mathbf{M}^*(\mathbf{p},h), \ \text{and} \ p_i > p_j \ \text{implies} \ a_i \leq a_i^* \ \text{or} \ a_j \geq a_j^*.$$

That is for any two states (jobs) of different sizes, the bigger gets no worse treatment relative to the smaller by $\mathbf{M}^*$ than it does by $\mathbf{M}$.

Theorem 2.13. *A parametric method ϕ^α gives-up to another parametric method ϕ^β if and only if $\alpha < \beta$.*

Proof. Let $\alpha < \beta$. We show that ϕ^α gives-up to ϕ^β. For any divisor $x > 0$ we have

$$\left[\frac{p_i}{x}\right]_\alpha \geq \left[\frac{p_i}{x}\right]_\beta \quad \text{for all } i.$$

Thus, if $\mathbf{a} \neq \mathbf{a}^*$, then the largest feasible divisor for $\phi^\beta(\mathbf{p},h)$ is smaller that the smallest feasible divisor for $\phi^\alpha(\mathbf{p},h)$. Hence, if the former is x, then the latter is $x - \varepsilon, \varepsilon > 0$, and we have

$$d(a_i^*) = a_i^* + \beta \geq \frac{p_i}{x-\varepsilon} > \frac{p_i}{x} \geq a_i - 1 + \alpha = d(a_i - 1). \tag{2.25}$$

Moreover, if

$$a_j < a_j^*,$$

then

$$\frac{p_j}{x} \leq d(a_j) = a_j + \alpha < a_j^* - 1 + \beta = d(a_j^* - 1) \leq \frac{p_j}{x-\varepsilon}. \tag{2.26}$$

Therefore, if

$$a_i > a_i^* \ \text{and} \ a_j < a_j^*,$$

then it follows from (2.26) that

$$\frac{\varepsilon p_j}{x(x-\varepsilon)} \geq \beta - \alpha,$$

and at the same time it follows from (2.25)

$$\frac{\varepsilon p_i}{x(x-\varepsilon)} \leq \beta - \alpha.$$

But these two inequalities lead to a contradiction as long as $p_i > p_j$. Therefore, we must have

$$a_i \leq a_i^* \ \text{or} \ a_j \geq a_j^*,$$

and thus ϕ^α gives-up to ϕ^β.

Now, assume $\mathbf{a} \in \phi^{\alpha}(\mathbf{p}, h)$, $\mathbf{a}^* \in \phi^{\beta}(\mathbf{p}, h)$, and $p_i > p_j$ implies $a_i \leq a_i^*$ or $a_j \geq a_j^*$, that is ϕ^{α} gives-up to ϕ^{β}. We need to show that then $\alpha < \beta$. By contradiction, assume $\alpha \geq \beta$. Let $\alpha > \beta$. Define $\mathbf{p} = (p_i = a_i + \alpha, p_j = a_j + \alpha)$, where $a_i = ma_j$ for some integer $m > 1$, and the house size

$$h = a_i + a_j.$$

We have

$$\frac{p_i}{a_i - 1 + \alpha} \geq \frac{p_j}{a_j + \alpha},$$

thus by (2.11) $(a_i, a_j) \in \phi^{\alpha}(\mathbf{p}, h)$, but also

$$\frac{p_i}{a_i - 1 + \beta} < \frac{p_j}{a_j + \beta},$$

for $m > \frac{\alpha - \beta + 2}{\alpha - \beta}$ and $a_j \geq 1$. Then, however either

$$a_i^* < a_i \text{ or } a_j^* > a_j,$$

for $(a_i^*, a_j^*) \in \phi^{\beta}(\mathbf{p}, h)$. Thus, if $a_i^* < a_i$, then $a_j^* > a_j$ since $a_i^* + a_j^* = a_i + a_j = h$. Also, if $a_j^* > a_j$, then $a_i^* < a_i$. Therefore, we get $a_i^* < a_i$ and $a_j^* > a_j$, which leads to a contradiction, since ϕ^{α} gives-up to ϕ^{β}.

If $1 > \alpha = \beta > 0$, then consider an instance with

$$\mathbf{p} = (1 + \alpha, 2 + \alpha, m - m\alpha, \underbrace{1 - \alpha, \alpha, \ldots, \alpha}_{m-1\text{-times}})$$

where $m > \max\{1, \frac{1-\alpha}{\alpha}\}$. We have $P = h = 4 + m$. Then, both vectors $(1, 3, a_3, \ldots, a_{m+4})$ and $(2, 2, a_3, \ldots, a_{m+4})$ belong to $\phi^{\alpha}(\mathbf{p}, h)$ as long as $\frac{m}{m+1} \geq \alpha$. To show this, we observe, that if $m > \max\{1, \frac{1-\alpha}{\alpha}\}$ and $\frac{m}{m+1} \geq \alpha$, then we can always have $1 \leq a_3 \leq m - 1$. Thus, the remaining $m - 1 \geq h - (4 + a_3) \geq 1$ can always be apportioned according to ϕ^{α}. Finally, the first two states get their seats by appropriately rounding their quotients $1 + \alpha$ and $2 + \alpha$ according to the divisor $d(a) = a + \alpha$. Therefore, we again get a contradiction since ϕ^{α} gives-up to ϕ^{α}. The case $\alpha = 1$ or 0 is left for Exercise 2.21. □

The last property stipulates that the just-in-time sequence should be roughly the same as its mirror reflection, that is they should be quasi-palindromes.

2.8.3 Symmetry and Quasi-Palindromes

Theorem 2.13 suggests that as far as parametric methods are concerned this can only happen for the parameter equal $\frac{1}{2}$, that is for the Webster's method. The following section proves this observation.

Theorem 2.14. *The method of Webster produces a sequence SPS^R, where P is a permutation of the states with odd populations, and S^R is a mirror reflection of S. Moreover, if it produces a sequence SPS^R and Q is a different permutation of the states with odd populations, then it produces SQS^R as well.*

Proof. Consider D numbers $\frac{d_i}{k-\frac{1}{2}}$ for $k = 1, \ldots, d_i$ and $i = 1, \ldots, n$ ordered in descending order. Build a sequence according to this order as follows: if $\frac{d_i}{k-\frac{1}{2}}$ is in position t of the order then place an i in position t of the sequence. The sequence so built can in fact be obtained by the Webster's parametric method $\phi^{\frac{1}{2}}$. To show this let us consider position t in the order. Let y_{jt}, $1 \le t \le D, 1 \le j \le n$ be the smallest k such that $\frac{d_j}{k-\frac{1}{2}}$ is in one of the positions $t, \ldots, D$ in the order. The i selected for position t of the sequence satisfies the following inequalities

$$\frac{d_i}{y_{it} - \frac{1}{2}} \ge \frac{d_j}{y_{jt} - \frac{1}{2}} \quad \text{for all } j. \tag{2.27}$$

However, $y_{jt} = x_{jt} + 1$, where x_{jt} is the number of allocations obtained by j up to t exclusively. Therefore, the i selected for position t of the sequence satisfies the inequalities

$$\frac{d_i}{x_{it} + \frac{1}{2}} \ge \frac{d_j}{x_{jt} + \frac{1}{2}} \quad \text{for all } j,$$

which is exactly the test the parametric method $\phi^{\frac{1}{2}}$ does to allocate a seat for position t of the sequence, see Theorem 2.2. We now show that the sequence is of the form SPS^R, where P is a permutation of all states with odd populations and S^R is a mirror reflection of S provided that the ties in the order are broken appropriately, for instance according to the order $d_1 \ge \cdots \ge d_n$. Suppose that the number $\frac{d_i}{k-\frac{1}{2}}$, $k \le \lfloor \frac{d_i}{2} \rfloor$, is in position $t \le \lfloor \frac{D}{2} \rfloor$ of the order. We claim that, then the number $\frac{d_i}{d_i+1-k-\frac{1}{2}}$ is in position $D+1-t$. This holds since

$$\frac{d_i}{k-\frac{1}{2}} \le \frac{d_j}{l-\frac{1}{2}} \quad \text{if and only if} \quad \frac{d_i}{d_i+1-k-\frac{1}{2}} \ge \frac{d_j}{d_j+1-l-\frac{1}{2}},$$

for $l \le \lfloor \frac{d_j}{2} \rfloor$. Finally, we notice that for any state i with an odd population $d_i = 2b_i + 1$ we have

$$\frac{d_i}{b_i+1-\frac{1}{2}} = \frac{d_i}{b_i+\frac{1}{2}} = \frac{2d_i}{d_i} = 2.$$

Thus they all fall in the middle of the order since for any state i with an even population $d_i = 2b_i$ we have

$$\frac{d_i}{b_i-\frac{1}{2}} > 2 \quad \text{and} \quad \frac{d_i}{b_i+\frac{1}{2}} < 2.$$

Table 2.3 The Webster's just-in-time sequences

δ	ϕ^{δ}
$[0,\frac{1}{2})$	$1 \longrightarrow 2 \longrightarrow 1 \longrightarrow 1$
$\frac{1}{2}$	$1 \longrightarrow 2 \longleftrightarrow 1 \longrightarrow 1$
$(\frac{1}{2},1)$	$1 \longrightarrow 1 \longrightarrow 2 \longrightarrow 1$
1	$1 \longrightarrow 1 \longrightarrow 1 \longleftrightarrow 2$

This proves the theorem since any permutation of the states with odd populations can be obtained in the middle of the sequence.   □

The Webster's method is the only parametric method for which the property from theorem holds. The following example with $\mathbf{d} = (3,1)$ given in Table 2.3 is due to Balinski and Shahidi [16].

In this example

$$\{1 \longrightarrow 2 \longrightarrow 1 \longrightarrow 1, 1 \longrightarrow 1 \longrightarrow 2 \longrightarrow 1\} \subseteq S_{\phi^{\delta}}(\mathbf{d})$$

only for $\delta = \frac{1}{2}$.

We have proven that the Webster's method is the only divisor method that produces cyclic quasi-palindromes. We study other desirable properties of this method in the next section.

2.9 The Consistency of Webster's Method

The Webster's parametric method will return in our discussions in this book on many occasions. Therefore, we further underline its importance in this section. We begin by investigating the Webster's apportionments for two states and begin with defining the standard two-state solution, Young [3]. We denote the Webster's method by $\mathbf{W}$. Let $\mathbf{a} = (a_1, a_2) \in \mathbf{W}(\mathbf{p} = ((p_1, p_2), h = a_1 + a_2)$. By the min–max condition of (2.11), for positive a_1 and a_2 we have

$$p_1 a_2 \geq p_2 a_1 - \frac{1}{2}(p_1 + p_2)$$

and

$$p_2 a_1 \geq p_1 a_2 - \frac{1}{2}(p_1 + p_2)$$

which implies

$$a_1 - \frac{1}{2} \leq d(a_1 - 1) \leq \frac{p_1 h}{p_1 + p_2} \leq d(a_1) = a_1 + \frac{1}{2}$$

and

$$a_2 - \frac{1}{2} = d(a_2 - 1) \leq \frac{p_2 h}{p_1 + p_2} \leq d(a_2) = a_2 + \frac{1}{2}.$$

Therefore,

$$a_1 = \left\lfloor \frac{p_1 h}{p_1 + p_2} \right\rfloor_{\frac{1}{2}} \text{ and } a_2 = \left\lfloor \frac{p_1 h}{p_1 + p_2} \right\rfloor_{\frac{1}{2}},$$

which proves that the Webster's solution for any two-state $\mathbf{p} = (p_1, p_2)$ and the house size h simply calculates the quota

$$q_1 = \frac{p_1 h}{p_1 + p_2} \text{ and } q_2 = \frac{p_2 h}{p_1 + p_2}$$

and then rounds them to the nearest integer numbers. If the fractional parts of q_1 and q_2 are both equal $\frac{1}{2}$, then both $(q_1 + \frac{1}{2}, q_2 - \frac{1}{2})$ and $(q_1 - \frac{1}{2}, q_2 + \frac{1}{2})$ are Webster's solutions. Finally, if either $a_1 = 0$ or $a_2 = 0$, then the proof is similar. The Webster's solutions for two-states will be refereed to as the *standard two-state solutions*.

Let $\mathbf{M}$ be an apportionment method. For any population vector $\mathbf{p}$, house size h and apportionment $\mathbf{a} = (a_1, \dots, a_s) \in \mathbf{M}(\mathbf{p}, h)$ define

$$h_{i,j} = a_i + a_j$$

for any two different states i and j. For the $h_{i,j}$ define the set

$$H_{i,j} = \{(a_i, a_j) : h_{i,j} = a_i + a_j, \mathbf{a} = (a_1, \dots, a_s) \in \mathbf{M}(\mathbf{p}, h)\}$$

The method is refereed to as pairwise consistent with the standard two-state solution if and only if $H_{i,j} = \mathbf{W}((p_i, p_j), h_{i,j})$ for all $i \neq j$. We have the following theorem.

Theorem 2.15. *Webster's method is the unique apportionment method that is pairwise consistent with the standard two-state solution.*

Proof. If $\mathbf{a} = (a_1, \dots, a_s) \in \mathbf{W}(\mathbf{p} = ((p_1, \dots, p_s), h)$, then we can repeat the reasoning presented at the beginning of this section for any couple of states $i \neq j$ with the vector of populations (p_i, p_j), apportionment (a_i, a_j) and house size $h_{i,j} = a_i + a_j$ to prove that

$$a_i = \left\lfloor \frac{(a_i + a_j) p_i}{p_i + p_j} \right\rfloor_{\frac{1}{2}} \text{ and } a_j = \left\lfloor \frac{(a_i + a_j) p_j}{p_i + p_j} \right\rfloor_{\frac{1}{2}}$$

which implies that the Webster's method is pairwise consistent with the standard two-state solution.

We now prove that the Webster's method is the only method pairwise consistent with the standard two-state solution. The proof is by contradiction. Let $\mathbf{M}$ be pairwise consistent with the standard two-state solution but not the Webster's method. Let $\mathbf{a} \in \mathbf{M}(\mathbf{p}, h)$ and $\mathbf{a} \notin \mathbf{W}(\mathbf{p}, h)$. Let $\mathbf{a}' \in \mathbf{W}(\mathbf{p}, h)$ be chosen so it differs from $\mathbf{a}$ in a minimal number of coordinates. There are states i and j such that $a_i' < a_i$ and

$a'_j > a_j$. Without loss of generality $a'_i + a'_j \geq a_i + a_j$. Since both $\mathbf{M}$ and $\mathbf{W}$ are pairwise consistent with the standard two-state solution, then we get a contradiction by the following inequality

$$a'_i = \left[q'_i = \frac{p_i\left(a'_i + a'_j\right)}{p_i + p_j} \right]_{\frac{1}{2}} \geq \left[q_i = \frac{p_i(a_i + a_j)}{p_i + p_j} \right]_{\frac{1}{2}} = a_i, \qquad (2.28)$$

unless $a'_i + a'_j = a_i + a_j$ and the fractional parts of q_i and q'_i are both equal $\frac{1}{2}$. Then,

$$a'_i = a_i - 1 \text{ and } a_j = a'_j - 1$$

and the apportionment $\mathbf{a}''$ obtained from $\mathbf{a}'$ by substituting a'_i and a'_j by a_i and a_j, respectively also belongs to $\mathbf{W}(\mathbf{p},h)$. This leads to a contradiction since $\mathbf{a}''$ differs from $\mathbf{a}$ in fewer components than $\mathbf{a}'$. This proves that $\mathbf{a} \in \mathbf{W}(\mathbf{p},h)$ consequently $\mathbf{M}(\mathbf{p},h) \subseteq \mathbf{W}(\mathbf{p},h)$. Now, let $\mathbf{a} \in \mathbf{W}(\mathbf{p},h)$ and $\mathbf{a} \notin \mathbf{M}(\mathbf{p},h)$. Let $\mathbf{a}' \in \mathbf{M}(\mathbf{p},h)$ be chosen so it differs from $\mathbf{a}$ in a minimal number of coordinates. There are states i and j such that $a'_i < a_i$ and $a'_j > a_j$. Without loss of generality $a'_i + a'_j \geq a_i + a_j$. By assumption, we get a contradiction since

$$a'_i = \left[q'_i = \frac{p_i\left(a'_i + a'_j\right)}{p_i + p_j} \right]_{\frac{1}{2}} \geq \left[q_i = \frac{p_i(a_i + a_j)}{p_i + p_j} \right]_{\frac{1}{2}} = a_i,$$

unless $a'_i + a'_j = a_i + a_j$ and the fractional parts of q_i and q'_i are both equal $\frac{1}{2}$. Then,

$$a'_i = a_i - 1 \text{ and } a_j = a'_j - 1$$

and the apportionment $\mathbf{a}''$ obtained from $\mathbf{a}'$ by substituting a'_i and a'_j by a_i and a_j, respectively also belongs to $\mathbf{M}(\mathbf{p},h)$. This leads to a contradiction again since again $\mathbf{a}''$ differs from $\mathbf{a}$ in fewer components than $\mathbf{a}'$. This proves that $\mathbf{a} \in \mathbf{M}(\mathbf{p},h)$ consequently $\mathbf{W}(\mathbf{p},h) \subseteq \mathbf{M}(\mathbf{p},h)$. Therefore, $\mathbf{W}(\mathbf{p},h) = \mathbf{M}(\mathbf{p},h)$ which ends the proof. $\qquad \square$

We now turn our attention to the just-in-time sequences built by the Webster's method and extend the pairwise consistency with the standard two-state solution to the pairwise consistency with the standard two-model sequence. Any sequence s in $S_{\mathbf{W}}((d_1, d_2))$ with $\gcd(d_i, d_j) = 1$ is called the standard two-model sequence. For example $\mathbf{d} = (3, 5)$ yields two standard two-model sequences

$$2 \to 1 \to 2 \to 1 \to 2 \to 2 \to 1 \to 2 \text{ and } 2 \to 1 \to 2 \to 2 \to 1 \to 2 \to 1 \to 2.$$

A sequence $S \in S_{\mathbf{W}}(\mathbf{d})$ is consistent with the standard two-model sequence if for any two models $i \neq j$ its projection on models i and j is the sequence

$$s^{\gcd(d_i, d_j)},$$

where $s \in S_\mathbf{W}((d_i,d_j)/\gcd(d_i,d_j))$ is the standard two-model sequence. The projection of S on models i and j is obtained by deleting from S all models except i and j.

Consider an instance $\mathbf{d} = (d_1,\ldots,d_n)$. Let $\mathbf{W}(\mathbf{d},1), \mathbf{W}(\mathbf{d},2), \ldots, \mathbf{W}(\mathbf{d},D)$ be the apportionments for house sizes $1,2,\ldots,D$ respectively and the population vector $\mathbf{d}$ obtained by the Webster's method. Consider

$$\mathbf{W}(\mathbf{d}) = \bigcup_{h=1}^{D} \mathbf{W}(\mathbf{d},h).$$

For any two $1 \le i \ne j \le n$, let $\mathbf{W}_{i,j}(\mathbf{d})$ be the set of projections of vectors in $\mathbf{W}(\mathbf{d})$ on the coordinates i and j and let $\mathbf{W}_{i,j}(\mathbf{d},h)$ be the set of projections of vectors in $\mathbf{W}(\mathbf{d})$ on the coordinates i and j that sum up to h, $h = 1,\ldots,d_i + d_j$. By Theorem 2.15 $\mathbf{W}_{i,j}(\mathbf{d},h) = \mathbf{W}((d_i,d_j),h)$. By Theorem 2.12

$$s \in S_\mathbf{W}((d_i,d_j)/\gcd(d_i,d_j)) \text{ implies } s^{\gcd(d_i,d_j)} \in S_\mathbf{W}((d_i,d_j)).$$

Thus, there is a sequence in

$$S_\mathbf{W}(\mathbf{d})$$

that is pairwise consistent with the standard two-model sequence.

Example 2.16. For instance, consider $\mathbf{d} = (3,5,15)$ the Webster's sequence is

$$32313323331233323313233.$$

It results in the projections

$$21212212,\ 331333331333331333,\ \text{and}\ 32333233323332333233$$

for models 1 and 2, 1 and 3, and 2 and 3 respectively. The first is consistent with the standard two-model sequence 21212212 for $(3,5)$, the second is consistent with the standard two-model sequence 331333 for $(1,5)$, and the third is consistent with the standard two-model sequence 3233 for $(1,3)$. Thus, this Webster's sequence is consistent with the standard two-model sequence.

We shall return to the standard two-model sequences in Sect. 10.6 in the context of peer-to-peer fair solutions.

2.10 Exercises

Exercise 2.17. Find the smallest apportionment problem instance with Alabama paradox.

Exercise 2.18. Show a population monotone apportionment method that stays within quota for $s \le 3$.

Exercise 2.19. Prove Lemma 2.8.

Exercise 2.20. Prove the case $\delta > \delta^*$ in Theorem 2.12.

Exercise 2.21. Prove that $(p-1,q,1)$, $(p,q-1,1)$, and $(p,q,0)$ all belong to $\phi^1((p,q,1),h = p+q)$. Similarly, $(p-1,q,1)$, $(p,q-1,1)$, and $(p-1,q-1,2)$ all belong to $\phi^0((p,q,1),h = p+q)$.

Exercise 2.22. The Hamilton's method minimizes any L_p norm of the difference $\mathbf{a}^h - \mathbf{q}^h$. Why then it does not minimize $\sum_{h=1}^{P} L_p(\mathbf{a}^h - \mathbf{q}^h)$?.

Exercise 2.23. Prove Theorem 2.5.

Exercise 2.24. Prove Theorem 2.6.

2.11 Comments and References

The presentation of the apportionment theory in this chapter is based on an excellent book by Balinski and Young [2] and a chapter in Young [3]. Both focus on the ideal of representative democracy – one man, one vote and on achieving it by the fair representation according to the theory of apportionment. The latter addresses the fundamental question of how to divide the seats of a legislature fairly according to the populations of states. The theory is based on axioms and apportionment methods presented in this chapter. The axiomatic method of apportionment theory relies on some *socially* desirable characteristics that one requires an apportionment to posses. These characteristics include, for instance house and population monotonicity but also many others. They have been shown crucial for the solutions of the apportionment problem. The famous *Impossibility Theorem* of Balinski and Young puts a clear limitation on which characteristics do not contradict one another by showing that having solutions that satisfy quota and that are at the same time population monotone is generally impossible. Other good introductions to the apportionment theory are presented in Leyvraz [17], and in Balinski [4]. Grilli di Cortona et al. [18], and Ibaraki and Katoh [19] discuss integer optimization approaches to the apportionment problem.

   The analysis of the divisor methods in the context of coalitions and the schisms belongs to Balinski and Young [2]. Bautista et al. [20] observe a link between the just-in-time sequencing and the apportionment problem. The proof of cyclic theorem for the parametric divisor methods is based on Balinski and Ramirez [15] and the advancing and retarding by the parametric methods on Balinski and Ramirez [15], Balinski and Rachev [14], and Balinski and Shahidi [16]. Table 2.2 is from Balinski and Shahidi [16]. The instance $\mathbf{p} = (6,6,1)$ is shown to suffer from the Alabama paradox by Miltenburg [21]. The discussion of the consistency of the Webster's method with the standard two state solution is based on Young [3].

Chapter 3
Minimization of Just-In-Time Sequence Deviation

3.1 Introduction

Chapter 2 presented an axiomatic approach to constructing just-in-time sequences. The axioms used there followed the ones well established by the apportionment practice and theory, however, some new axioms like periodicity, palindrome symmetry and the consistency with a standard two-model just-in-time sequences were added and shown to be satisfied by the Webster's method of apportionment. This axiomatic approach by definition constructs a just-in-time sequence which gives as fair a representation to each state (or model) in any house size (or sequence prefix) as the axioms permit. More precisely, a fair representation may be understood as the representation free of *known* paradoxes. Thus, the focus of this axiomatic approach is solely on individual states (models) and consequently no population paradox can be tolerated by a fair representation. Therefore, the divisor methods are the only choice. However, there are situations where the focus moves from the individual states (models) towards their *collection*. This may require the existence of an entity on a level higher than the states (models) which may not be that unusual and may even be natural. A just-in-time manufacturer often relies on the assumption that demand for its product models is uniformly distributed over a planning horizon, and hence, its goal of meeting the model demand with minimum inventories and shortages (or simply just-in-time) calls for distributing the production of each model i as uniformly as possible over the planning horizon as well. Ideally, the manufacturer would produce a model with a constant demand rate, equal r_i for model i, thereby making the cumulative production of the model to progress along a straight cumulative demand line with its slope equal to the demand rate for the model. The total deviation of the actual cumulative production from the ideal is then proportional to the total inventory and shortage costs that may be charged at different rates for different models. The bottleneck deviation does the same for the maximum charge. Therefore, the manufacturer would strive to smooth the production rate variations by using functions of *deviations* of the actual cumulative model production levels from their ideal levels over the time horizon. The total deviation, defined in

W. Kubiak, *Proportional Optimization and Fairness,* International Series in Operations Research & Management Science 127, DOI 10.1007/978-0-387-87719-8_3,
© Springer Science+Business Media LLC 2009

Sect. 3.2, and the maximum deviations, defined in Sect. 3.7, have been widely used as the primary objectives in the just-in-time production smoothing literature, see Monden [22]. The total deviation minimization is also known as the product rate variation problem, Kubiak [23]. This production smoothing has a clearer goal, often better achieved by optimization, than the more elusive goal of one man one vote ideal of the apportionment theory of proportional representation.

Though optimization has been almost abandoned in the apportionment theory long time ago, see Balinski and Young [2], we just argued that whenever the problem emphasis moves from the individual model to the whole collection of models, then the optimization methods have much clearer objective and thus might provide a better approach in solving the smoothing problem, for instance, than the axiomatic approach. Consequently, this chapter presents an efficient approach to minimizing the total deviation. This approach constructs an equivalent assignment problem which then can be solved by the well known assignment algorithms. The approach is presented in detail in Sect. 3.3. Sections 3.4 and 3.5 show its correctness. Finally, solution is given in Sect. 3.6. A similar approach is then presented for the maximum deviation minimization in Sects. 3.8 and 3.9.

Finally, having argued that the axioms and the apportionment methods designed to satisfy them may not ensure an optimal solution, if a certain objective function of deviations is to be optimized, we after all realize that the apportionment methods can provide solutions with a lot of potential for improving operations management as we shall continue arguing in Chap. 11 where we return to the production smoothing problem in a real-life setting. These improvements though more elusive and thus difficult to measure in terms of cost savings may however in a long run lead to a stronger and sustainable competitive position.

3.2 Minimization of Total Deviation

Following the motivating example of production smoothing given in the introduction to this chapter and the notation introduced in the preliminary chapter, we consider a set of n, different models $1, 2, \ldots, n$ to produce, with demands $d_1, \ldots, d_n$, units respectively. For a sequence $S = s_1 \cdots s_D$ of length $D = \sum_{i=1}^{n} d_i$ over the alphabet $1, 2, \ldots, n$ where i occurs exactly d_i times, let x_{ik} be the number, sometimes referred to as the cumulative production , of times the i occurs in the $k-$prefix $s_1 \cdots s_k$ of S, $k = 1, 2, \ldots, D$, $i = 1, \ldots, n$. The total deviation minimization problem is formulated as follows.

$$\min_{S} \left\{ F(S) = \sum_{i=1}^{n} \sum_{k=1}^{D} F_i(x_{ik} - kr_i) \right\} \tag{3.1}$$

subject to

$$\sum_{i=1}^{n} x_{ik} = k \quad k = 1, \ldots, D \tag{3.2}$$

$$0 \le x_{ik+1} - x_{ik} \le 1 \quad i = 1, \ldots, n; k = 1, \ldots, D-1 \tag{3.3}$$

$$x_{iD} = d_i \quad i = 1, \ldots, n. \tag{3.4}$$

where $F_i(\cdot)$ is a non-negative convex function satisfying

$$F_i(0) = 0 \text{ and } F_i(x) > 0 \text{ for } x \neq 0, \ i = 1, \ldots, n. \tag{3.5}$$

3.3 The Transformation to the Assignment Problem

The algorithm for minimizing (3.1) subject to constraints (3.2–3.5) transforms the
original problem into an equivalent assignment problem. The key to this transformation is the calculation of the assignment costs to ensure the equivalence. The
transformation and the assignment cost calculations relay on the concept of level
curves, ideal positions and partitions.

The $d_i + 1$ *level curves* f_j^i for i are defined as follows

$$f_j^i(k) = F_i(j - kr_i), \ \ j = 0, 1, \ldots, d_i, \ \ k = 0, \ldots, D. \tag{3.6}$$

Note that $f_0^i(0) = f_{d_i}^i(D) = 0$. For simplicity in exposition we conduct our discussion
assuming the absolute value of deviation for the time being, $F_i = |\cdot|$, $i = 1, \ldots, n$,
that is

$$f_j^i(k) = |j - kr_i|, \ \ j = 0, 1, \ldots, d_i, \ \ k = 0, \ldots, D, \tag{3.7}$$

however, this assumption will be later relaxed to general F_i, $i = 1, \ldots, n$, defined in
(3.5). The level curves for $F_i = |\cdot|$, $i = 1, \ldots, n$ are shown in Fig. 3.1.

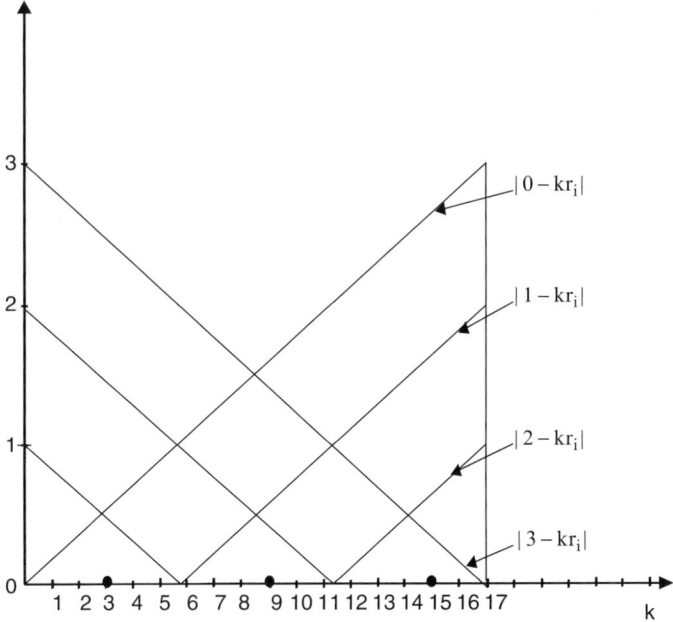

Fig. 3.1 The level curves for $d_i = 3$, $D = 17$, and $F_i = |\cdot|$ with the ideal positions $Z_1^i = 3$, $Z_2^i = 9$,
and $Z_3^i = 15$

For $\mathbf{d} = (d_1, \ldots, d_n)$ let $\pi = (\pi_1, \ldots, \pi_n)$ be a partition of $\{1, \ldots, D\}$ such that $\cup_{i=1}^{n} \pi_i = \{1, \ldots, D\}$, $\pi_i \cap \pi_j = \emptyset$, for $i \neq j$, and $|\pi_i| = d_i$, for $i = 1, \ldots, n$. We refer to such partition π as a $\mathbf{d}-partition$ of $\{1, \ldots, D\}$. Each just-in-time sequence S for $\mathbf{d}^a = (d_1, \ldots, d_n)$ uniquely defines a $\mathbf{d}-$partition $\pi(S) = (\pi_1(S), \ldots, \pi_n(S))$ of $\{1, \ldots, D\}$ by taking $\pi_i(S) = \{k : s_k = i\}$, for $i = 1, \ldots, n$. Conversely, each $\mathbf{d}-$partition uniquely defines a sequence $S(\pi)$ by taking $s(\pi)_k = i$ if and only if $k \in \pi_i$.

Let $\pi = (\pi_1, \ldots, \pi_n)$ be a $(d_1, \ldots, d_n)$-partition of $\{1, \ldots, D\}$. Suppose that $\pi_i = \{Y_1^i, \ldots, Y_{d_i}^i\}$ and that this set is ordered in ascending order, that is $Y_1^i < \cdots < Y_{d_i}^i$. Then, the cost in (3.1) is charged according to f_0^i at each $k = 1, \ldots, Y_1^i - 1$, then according to f_1^i at each $k = Y_1^i, \ldots, Y_2^i - 1$, and so on until $k = Y_{d_i}^i$, where the cost starts being charged according to $f_{d_i}^i$ and continues doing so until $k = D$. The total cost of $\pi_i = \{Y_1^i, \ldots, Y_{d_i}^i\}$, or in other words the total cost for the copies of i being sequenced in positions $Y_1^i, \ldots, Y_{d_i}^i$ of the sequence $S(\pi)$, is thus equal

$$J_2^i(\pi_i) = \sum_{k=0}^{Y_1^i - 1} f_0^i(k) + \sum_{k=Y_1^i}^{Y_2^i - 1} f_1^i(k) + \cdots + \sum_{k=Y_{d_i}^i}^{D} f_{d_i}^i(k)$$

$$= \sum_{j=0}^{d_i} \sum_{k=Y_j^i}^{Y_{j+1}^i - 1} f_j^i(k), \tag{3.8}$$

where by definition $Y_0^i = 0$ and $Y_{d_i+1}^i - 1 = D$.

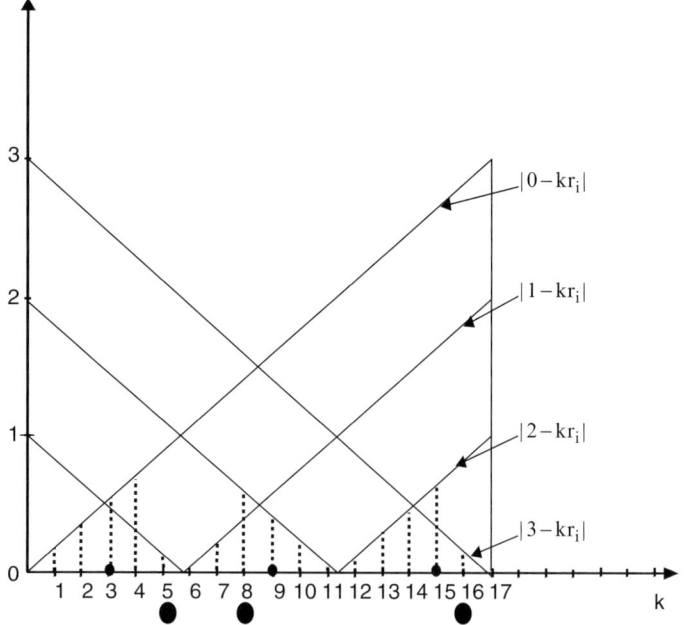

Fig. 3.2 The total deviation cost calculation for i with $d_i = 3$, sequenced in positions 5, 8, and 16

Now, suppose that π_i is not given but rather we would like to find out a $\pi_i = \{Y_1^i, \ldots, Y_{d_i}^i\}$ that minimizes (3.8) in isolation from all other $j \neq i$. This can be can be achieved by solving the following integer program $(P2i)$:

$$\min\left\{J_2^i(\pi_i)\right\} \tag{3.9}$$

subject to

$$\begin{aligned}
Y_{j+1}^i &\geq Y_j^i + 1, \quad j = 1, \ldots, d_i - 1, \\
1 &\leq Y_j^i \leq D, \quad j = 1, \ldots, d_i, \\
Y_j^i &\text{ are non-negative integers.}
\end{aligned} \tag{3.10}$$

It turns out that obtaining an optimal solution for $(P2i)$ is quite straightforward. The solution is given by the following theorem.

Theorem 3.1. *An optimal solution $\pi_i^* = (Z_1^i, \ldots, Z_{d_i}^i)$ to $(P2i)$ is given by*

$$Z_j^i = \left\lceil \frac{2j-1}{2r_i} \right\rceil, \quad j = 1, \ldots, d_i, \tag{3.11}$$

for which the value of objective function J_2^i in (3.9) equals

$$J_2^i(\pi_i^*) = \sum_{k=1}^{D} \inf_j f_j^i(k).$$

Proof. Observe that $\frac{2j-1}{2r_i}$ is the unique crossing point of the level curves f_{j-1}^i and f_j^i for $j = 2, \ldots, d_i$. Thus, $Z_j^i = \left\lceil \frac{2j-1}{2r_i} \right\rceil$ is the smallest integer k for which $f_j^i(k) \leq f_{j'}^i(k)$ for all $j' \neq j$, see Fig. 3.1, where $Z_1^i = 3, Z_2^i = 9$ and $Z_3^i = 15$. Consequently, the solution (3.11) attains the lower bound $\sum_{k=1}^{D} \inf_j f_j^i(k)$ on J_2^i which proves the theorem. $\square$

The values

$$\frac{2j-1}{2r_i}, \tag{3.12}$$

for $i = 1, \ldots, n$ and $j = 1, \ldots, d_i$, will play a central role in our further discussions thus we reserve special names for them in this book. They will be called *ideal vertices*. The ideal vertices in fact define the Webster's method of apportionment, see Chap. 2, in a sense that their ascending order results in the same just-in-time sequences as those obtained by the Webster's method. We leave details to Chap. 5. We shall refer to Z_j^i in (3.11) as the *ideal position* for copy j of i. Clearly, if $\pi_i^* \cap \pi_{i'}^* = \emptyset$, for $i \neq i'$, then all Z_j^i are different and placing copy j of i in position Z_j^i results in a feasible sequence minimizing (3.1). This will unfortunately not be the case in general, and we must somehow come to terms with the resolution of conflicts between the copies of i and i' for the positions in $\pi_i^* \cap \pi_{i'}^*$ whenever $\pi_i^* \cap \pi_{i'}^* \neq \emptyset$. Therefore, our goal is to solve the following linear combination, refereed to as program $(P2)$, of programs $(P2i)$ for $i = 1, \ldots, n$:

$$\min\left\{J_2(\pi) = \sum_{i=1}^{n} J_2^i(\pi_i)\right\} \tag{3.13}$$

subject to

$$Y_{j+1}^i \geq Y_j^i + 1, \quad i = 1,\ldots,n, \quad j = 1,\ldots,d_i - 1, \tag{3.14}$$

$$1 \leq Y_j^i \leq D, \quad i = 1,\ldots,n, \quad j = 1,\ldots,d_i, \tag{3.15}$$

$$Y_j^i \neq Y_{j'}^{i'} \text{ for } (i,j) \neq (i',j') \tag{3.16}$$

$$Y_j^i \text{ are non-negative integers.} \tag{3.17}$$

The constraint (3.16) is to ensure that $\pi_i \cap \pi_{i'} = \emptyset$, for $i \neq i'$, that is that exactly one copy occupies each of D positions.

The idea for the resolution of the conflicts is as follows. Let us set C_{jk}^i to be the additional cost incurred by copy j of i whenever the copy is assigned to position k rather than to its ideal position Z_j^i. What is rather surprising is that this simple heuristic idea turns out also to provide solutions which are both feasible and optimal for $(P2)$ as we shall prove in Sect. 3.5. More precisely, let us define

$$C_{jk}^i = \sum_{l=\min(k,Z_j^i)}^{\max(k,Z_j^i)-1} \Psi_{jl}^i = \begin{cases} \sum_{l=k}^{Z_j^i-1} \Psi_{jl}^i, & \text{if } k < Z_j^i, \\ 0, & \text{if } k = Z_j^i, \\ \sum_{l=Z_j^i}^{k-1} \Psi_{jl}^i, & \text{if } k > Z_j^i, \end{cases} \tag{3.18}$$

where

$$\Psi_{jl}^i = \left|f_j^i(l) - f_{j-1}^i(l)\right| = \begin{cases} f_j^i(l) - f_{j-1}^i(l) & \text{if } l < Z_j^i, \\ f_{j-1}^i(l) - f_j^i(l) & \text{if } l \geq Z_j^i, \end{cases} \tag{3.19}$$

and by definition $\sum_{l=k}^{k'} a_l = 0$ whenever $k' < k$.

One interpretation of the assignment costs in (3.19) can be given in terms of just-in-time manufacturing as follows. The cost $f_j^i(l)$ represents the *inventory* or *shortage* cost in period l if exactly j copies of i have been produced by period l in a just-in-time production sequence. The cost Ψ_{jl}^i represents the excess cost, either excess inventory ($l < Z_j^i$) or excess shortage ($l \geq Z_j^i$) costs of having j copies of i produced by period l *instead of* having $j-1$ copies of i by l. Consequently, if copy j of i is produced too *early,* that is $k < Z_j^i$, then the excess inventory costs Ψ_{jl}^i are incurred in periods $l = k,\ldots,Z_j^i - 1$, adding up to C_{jk}^i. On the other hand if copy j of i is produced too *late,* that is $k > Z_j^i$, then the excess shortage costs Ψ_{jl}^i are incurred in periods $l = Z_j^i,\ldots,k-1$, again adding up to C_{jk}^i. Finally, if copy j of i is produced in its ideal position Z_j^i then no excess costs are incurred and thus $C_{jk}^i = 0$. The cost calculation is illustrated in the following example.

Example 3.2. Consider the instance $\mathbf{d} = (5,3,2)$ and $F_i = |\cdot|$ for $i = 1,\ldots,n$. Thus, $D = 10$, $r_1 = 0.5$, $r_2 = 0.3$, and $r_3 = 0.2$. The ideal positions computed by using (3.11) are as follows: $Z_1^1 = 1$, $Z_2^1 = 3$, $Z_3^1 = 5$, $Z_4^1 = 7$, and $Z_5^1 = 9$ for $i = 1$; $Z_1^2 = 2$, $Z_2^2 = 5$, and $Z_3^2 = 9$ for $i = 2$; $Z_1^3 = 3$, and $Z_2^3 = 8$ for $i = 3$. Thus, $\pi_1^* = \{1,3,5,7,9\}$, $\pi_2^* = \{2,5,9\}$, and $\pi_3^* = \{3,8\}$. Therefore, the conflicts arise three times: for position 3 between copy 2 of 1 and copy 1 of 3; for position 5 between copy 3 of 1 and copy 2 of 2; and for position 9 between copy 5 of 1 and copy 3 of 2. The values of Ψ_{jl}^i for $i = 1,2,3$, $j = 1,\ldots,d_i$ and $l = 1,\ldots,D$ calculated according to (3.19) are shown in Table 3.20. Finally, the costs C_{jk}^i for $i = 1,2,3$, $j = 1,\ldots,d_i$ and $k = 1,\ldots,D$ calculated according to (3.18) are given in Table 3.21. Therefore, the conflicts can easily be solved at no additional cost by moving copy 2 of 1 to position 4, by moving copy 3 of 1 to position 6, and by moving copy 5 of 1 to position 10. These changes result in the partition $\pi_1 = \{1,4,6,7,10\}$, $\pi_2^* = \{2,5,9\}$, and $\pi_3^* = \{3,8\}$. This partition translates into the sequence

$$1 \to 2 \to 3 \to 1 \to 2 \to 1 \to 1 \to 3 \to 2 \to 1$$

with no additional cost. Notice that the conflict for position 5 between copy 3 of 1 and copy 2 of 2 can also be solved at no additional cost by moving copy 2 of 2 to position 6 thus obtaining $\pi_1' = \{1,4,5,7,10\}$, $\pi_2 = \{2,6,9\}$, and $\pi_3^* = \{3,8\}$, and an optimal sequence

$$1 \to 2 \to 3 \to 1 \to 1 \to 2 \to 1 \to 3 \to 2 \to 1.$$

Note that the latter schedule coincides with that in Monden [8, p. 184] for this instance. $\square$

Table 3.20. The costs Ψ_{jl}^i

	$j \backslash l$	1	2	3	4	5	6	7	8	9	10
	1	0	1	1	1	1	1	1	1	1	1
	2	1	1	0	1	1	1	1	1	1	1
$i = 1$	3	1	1	1	1	0	1	1	1	1	1
	4	1	1	1	1	1	1	0	1	1	1
	5	1	1	1	1	1	1	1	1	0	1

	$j \backslash l$	1	2	3	4	5	6	7	8	9	10
	1	0.4	0.2	0.8	1	1	1	1	1	1	1
$i = 2$	2	1	1	1	0.6	0	0.6	1	1	1	1
	3	1	1	1	1	1	1	0.8	0.2	0.4	1

	$j \backslash l$	1	2	3	4	5	6	7	8	9	10
	1	0.6	0.2	0.2	0.6	1	1	1	1	1	1
$i = 3$	2	1	1	1	1	1	0.6	0.2	0.2	0.6	1

$$(3.20)$$

Table 3.21. The costs C^i_{jk}

$i = 1$

$j\backslash k$	1	2	3	4	5	6	7	8	9	10
1	0	0	1	2	3	4	5	6	7	8
2	2	1	0	0	1	2	3	4	5	6
3	4	3	2	1	0	0	1	2	3	4
4	6	5	4	3	2	1	0	0	1	2
5	8	7	6	5	4	3	2	1	0	0

$i = 2$

$j\backslash k$	1	2	3	4	5	6	7	8	9	10
1	0.4	0	0.2	1	2	3	4	5	6	7
2	3.6	2.6	1.6	0.6	0	0	0.6	1.6	2.6	3.6
3	7	6	5	4	3	2	1	0.2	0	0.4

$i = 3$

$j\backslash k$	1	2	3	4	5	6	7	8	9	10
1	0.8	0.2	0	0.2	0.8	1.8	2.8	3.8	4.8	5.8
2	5.8	4.8	3.8	2.8	1.8	0.8	0.2	0	0.2	0.8

$$(3.21)$$

The costs $\{C^i_{jk}\}$ defined in (3.18) and (3.19) define the following assignment problem (P3):

$$\min\left\{J_3(\pi) = \sum_{i=1}^n J^i_3(\pi_i) = \sum_{i=1}^n \left(\sum_{j=1}^{d_i}\sum_{k=1}^{D} C^i_{jk}y^i_{jk}\right)\right\} \qquad (3.22)$$

subject to

$$\sum_{(i,j)\in I} y^i_{jk} = 1, \ k = 1,\ldots,D, \qquad (3.23)$$

$$\sum_{k=1}^{D} y^i_{jk} = 1, \ (i,j) \in I, \qquad (3.24)$$

where $I = \{(i.j) : i = 1,\ldots,n, \ j = 1,\ldots,d_i\}$.

Let $\{y^i_{jk}\}$ be a feasible solution to (P3). We say that $\{y^i_{jk}\}$ preserves order if

$$(y^i_{jk} = 1 \text{ and } y^i_{j+1l} = 1) \Rightarrow k < l \qquad (3.25)$$

for all i, j, k and l. The assignments that preserve orders assign copy j always in front of copy $j+1$ for any i and thus can be turned into feasible $\mathbf{d}$−partitions as follows:

Theorem 3.3. Let $\{y^i_{jk}\}$ be a feasible solution to (P3) that preserves order. Then,

$$Y^i_j = k \text{ if and only if } y^i_{jk} = 1, \ \text{ for } i = 1,\ldots,n; \ k = 1,\ldots,D \qquad (3.26)$$

is a feasible solution for (P2).

Proof. Equation (3.23) implies (3.16). Since $\{y^i_{jk}\}$ preserves order, then $\{Y^i_j\}$ satisfies (3.14). Finally, (3.24) implies (3.15) and (3.17). This ends the proof. $\square$

We now show that the order preserving solution to $(P3)$ always exists.

3.4 The Monge Property of Assignment Costs

The existence of order preserving solutions to the assignment problem $(P3)$ is ensured if only the assignment costs $C^i_{j,k}$ in (3.18) have the local Monge property which is shown in the following lemma.

Lemma 3.4. *If each matrix*

$$M_i = \left[C^i_{jk} \right]_{d_i \times D} \quad i = 1, \ldots, n,$$

is a Monge matrix, then there is an optimal solution $\{y^i_{jk}\}$ to $(P3)$ that preserves order.

Proof. By contradiction. Suppose that no optimal solution $\{y^i_{jk}\}$ to $(P3)$ preserves order. Consider an optimal solution $\{y^i_{jk}\}$ with the smallest number of the quadruples i, j, k and l that violate (3.25). For any violation we have,

$$y^i_{jl} = 1 \text{ and } y^i_{j+1k} = 1 \text{ and } k < l.$$

However,

$$C^i_{jk} + C^i_{j+1l} \leq C^i_{jl} + C^i_{j+1k},$$

since M_i is a Monge matrix . Then, swapping j and $j+1$ does not increase the cost of the assignment J_3 and at the same time it reduces the number of condition (3.25) violations. Thus, we get a contradiction since the solution $\{y^i_{jk}\}$ was chosen with the smallest number of violations and thus the theorem holds. $\square$

It remains to show that the assignment costs $C^i_{j,k}$ in (3.22) indeed have the local Monge property. We shall do this first for the absolute value functions of deviations, that is $F_i = |\cdot|$, $i = 1, \ldots, n$ and than for any functions F_i, $i = 1, \ldots, n$ that meet (3.5). We begin with preliminary observations in Lemmas 3.5 and 3.6.

Lemma 3.5. *The following statements hold true:*

1. $j - kr_i > 0 \Leftrightarrow k < \left\lceil \frac{j}{r_i} \right\rceil$;
2. $\left\lceil \frac{j-1}{r_i} \right\rceil \leq Z^i_j \leq \left\lceil \frac{j}{r_i} \right\rceil$;
3. $p > j \Rightarrow Z^i_p \geq Z^i_j$ and $\left\lceil \frac{p}{r_i} \right\rceil \geq \left\lceil \frac{j}{r_i} \right\rceil$;
4. $p > j \Rightarrow p - kr_i > 0$, for $k \leq Z^i_j$.

Proof. (1) is obvious. (2) follows immediately from (3.11). The fact $r_i \leq 1$ implies (3). To prove (4), we notice that from (3) and (2),

$$\left\lceil \frac{p}{r_i} \right\rceil > \left\lceil \frac{j}{r_i} \right\rceil \geq Z_j^i.$$

Thus from (1)

$$j - r_i Z_j^i > 0$$

and hence

$$j - k r_i > 0 \quad \text{for} \quad Z_j^i \geq k.$$

$\square$

Lemma 3.6. $0 \leq \Psi_{jk}^i \leq 1$ *for all* j, k.

Proof. For any real number a,

$$-1 \leq |a+1| - |a| \leq 1.$$

By definition

$$\Psi_{jk}^i = \left| |j - k r_i| - |j - 1 - k r_i| \right|,$$

thus $0 \leq \Psi_{jk}^i \leq 1$. $\square$

The following two Lemmas 3.7 and 3.8 show that the incremental cost of moving away from the ideal position is non-decreasing.

Lemma 3.7. *For* $p > j$ *and* $k < Z_j^i$, $\Psi_{pk}^i \geq \Psi_{jk}^i$.

Proof. From Lemma 3.5 (1) and (2), $p - k r_i > j - k r_i > 0$ for $k = 1, 2, \ldots, Z_j^i - 1$. Therefore, $\Psi_{pk}^i = 1, k = 1, 2, \ldots, Z_j^i - 1$. Now, the lemma immediately follows from Lemma 3.6. $\square$

Lemma 3.8. *For* $p > j$ *and* $k \geq Z_p^i$, $\Psi_{pk}^i \leq \Psi_{jk}^i$.

Proof. From Lemma 3.5 (3) and (2),

$$k \geq Z_p^i \geq Z_{j+1}^i \geq \left\lceil \frac{j}{r_i} \right\rceil.$$

Thus, from Lemma 3.5 (1), $j - k r_i \leq 0$. Then $\Psi_{jk}^i = 1$ for $k \geq Z_p^i$, and the lemma follows from Lemma 3.6. $\square$

We are now ready to show the local Monge property for M_i.

Theorem 3.9. *The matrices* $M_i = \left[C_{jk}^i \right]_{d_i \times D}$ $i = 1, \ldots n$, *are Monge matrices.*

Proof. It suffices to show that

$$\Delta_{jk} = C_{jk+1}^i + C_{j+1k}^i - C_{jk}^i - C_{j+1k+1}^i \geq 0$$

for any $j = 1, \ldots, d_i$ and $k = 1, \ldots, D - 1$. From (3.18) and (3.19), we have,

$$\Delta_{jk} = \sum_{l=\min(k+1,Z_j^i)}^{\max(k+1,Z_j^i)-1} \Psi_{jl}^i + \sum_{l=\min(k,Z_{j+1}^i)}^{\max(k,Z_{j+1}^i)-1} \Psi_{j+1l}^i$$

$$- \sum_{l=\min(k,Z_j^i)}^{\max(k,Z_j^i)-1} \Psi_{jl}^i - \sum_{l=\min(k+1,Z_{j+1}^i)}^{\max(k+1,Z_{j+1}^i)-1} \Psi_{j+1l}^i$$

$$= \sum_{l=\min(k,Z_{j+1}^i)}^{\min(k+1,Z_{j+1}^i)-1} \Psi_{j+1l}^i - \sum_{l=\max(k,Z_{j+1}^i)}^{\max(k+1,Z_{j+1}^i)-1} \Psi_{j+1l}^i$$

$$+ \sum_{l=\max(k,Z_j^i)}^{\max(k+1,Z_j^i)-1} \Psi_{jl}^i - \sum_{l=\min(k,Z_j^i)}^{\min(k+1,Z_j^i)-1} \Psi_{jl}^i.$$

We need to verify the following four cases:

For $k + 1 \leq Z_j^i < Z_{j+1}^i$, we obtain by Lemma 3.7,

$$\Delta_{jk} = \Psi_{j+1k}^i - \sum_{l=Z_{j+1}^i}^{Z_{j+1}^i-1} \Psi_{j+1l}^i + \sum_{l=Z_j^i}^{Z_j^i-1} \Psi_{jl}^i - \Psi_{jk}^i = \Psi_{j+1k}^i - \Psi_{jk}^i \geq 0.$$

For $Z_j^i \leq k < k + 1 \leq Z_{j+1}^i$, we obtain by (3.19),

$$\Delta_{jk} = \Psi_{j+1k}^i - \sum_{l=Z_{j+1}^i}^{Z_{j+1}^i-1} \Psi_{j+1l}^i + \Psi_{jk}^i - \sum_{l=Z_j^i}^{Z_j^i-1} \Psi_{jl}^i = \Psi_{j+1k}^i + \Psi_{jk}^i \geq 0.$$

For $k = Z_{j+1}^i < k + 1$, we obtain by Lemma 3.8,

$$\Delta_{jk} = \sum_{l=k}^{k-1} \Psi_{j+1l}^i - \Psi_{j+1k}^i + \Psi_{jk}^i - \sum_{l=Z_j^i}^{Z_j^i-1} \Psi_{jl}^i = \Psi_{jk}^i - \Psi_{j+1k}^i \geq 0.$$

For $Z_j^i < Z_{j+1}^i < k$, we obtain by Lemma 3.8,

$$\Delta_{jk} = \sum_{l=Z_{j+1}^i}^{Z_{j+1}^i-1} \Psi_{j+1l}^i - \Psi_{j+1k}^i + \Psi_{jk}^i - \sum_{l=Z_j^i}^{Z_j^i-1} \Psi_{jl}^i = \Psi_{jk}^i - \Psi_{j+1k}^i \geq 0.$$

Therefore, we always have $\Delta_{jk} \geq 0$, and thus the theorem holds.   □

Theorem 3.9 holds generally for any functions F_i, $i = 1, \ldots, n$ that meet (3.5). To prove this it is sufficient to show that Lemmas 3.7 and 3.8 hold for these functions but first we need to show how to calculate the ideal position for the general case. We have the following theorem.

Theorem 3.10. *An optimal solution $\pi_i^* = (Z_1^i, \ldots, Z_{d_i}^i)$ to $(P2i)$ for a convex function $F_i(\cdot)$ satisfying (3.5) is given by*

$$Z_j^i = \lceil k_j^i \rceil, \quad j = 1, \ldots, d_i, \tag{3.27}$$

where k_j^i is the unique crossing point satisfying

$$F_i(j - k_j^i r_i) = F_i(j - 1 - k_j^i r_i).$$

The value of objective function J_2^i in (3.9) for π_i^* equals

$$J_2^i(\pi_i^*) = \sum_{k=1}^{D} \inf_j f_j^i(k).$$

Notice that for any symmetric $F_i(\cdot)$, that is satisfying $F_i(y) = F_i(-y)$ for any y, we have $k_j^i = Z_j^i$, where $Z_j^i = \left\lceil \frac{2j-1}{2r_i} \right\rceil$ as in (3.11). We now show the counterparts of Lemmas 3.7 and 3.8 for convex functions $F_i(\cdot)$ satisfying (3.5).

Lemma 3.11. *For $p > j$ and $k < Z_j^i$, $\Psi_{pk}^i \geq \Psi_{jk}^i$.*

Proof. By Lemma 3.5 (2), $k < Z_j^i$ implies $k < \left\lceil \frac{j}{r} \right\rceil$. Then, by Lemma 3.5 (1), $j - kr_i > 0$. Therefore, $l - kr_i > 0$ for $l = j, j+1, \ldots, p$ as shown in Fig. 3.3a, b. We need to consider two cases:

1. $j - 1 - kr_i \geq 0$, see Fig. 3.3a. Then, the lemma holds since $l - kr_i > 0$ for $l = j - 1, j, j+1, \ldots, p$ and $F_i(\cdot)$ is convex.

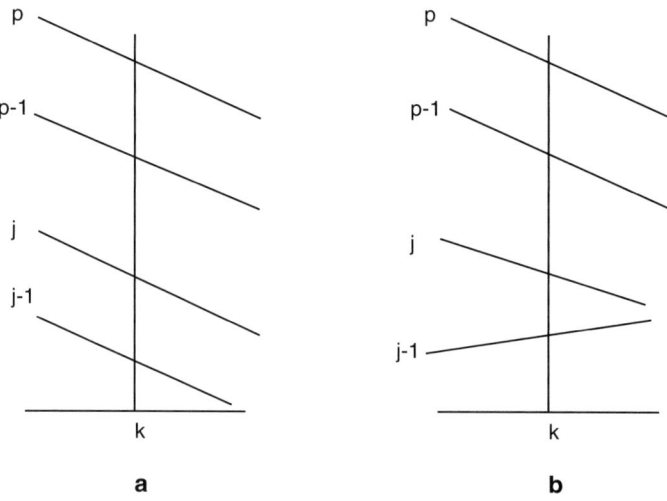

Fig. 3.3 Plots of $|j - kr_i|$ for $k < Z_j^i$: (**a**) $j - 1 - kr_i \geq 0$ and (**b**) $j - 1 - kr_i < 0$

2. $j - 1 - kr_i < 0$, see Fig. 3.3b. Using, (3.6) and (3.19), we obtain

$$\Psi^i_{pk} \geq \cdots \geq \Psi^i_{j+2k} \geq \Psi^i_{j+1k} = F_i(j+1-kr_i) - F_i(j-kr_i)$$
$$\geq F_i(j-kr_i) - F_i(0) = F_i(j-kr_i).$$

Thus, the lemma follows since $\Psi^i_{jk} = F_i(j-kr_i) - F_i(j-1-kr_i) \geq 0$ and $F_i(j - 1 - kr_i) \geq 0$.

This completes the proof. $\square$

Lemma 3.12. *For $p > j$ and $k \geq Z^i_p$, $\Psi^i_{pk} \leq \Psi^i_{jk}$.*

Proof. By Lemma 3.5 (3), $k \geq Z^i_p \geq \left\lceil \frac{p-1}{r_i} \right\rceil$. Then, by Lemma 3.5 (1), $p - 1 - kr_i \leq 0$. Therefore, $l - kr_i \leq 0$ for $l = j, \ldots p - 1$ as shown in Fig. 3.4a, b. We need to consider two cases:

1. $p - kr_i \leq 0$, see Fig. 3.4a. This implies $l - kr_i \leq 0$ for $l = j, \ldots, p-1, p$. Thus, $\Psi^i_{pk} \leq \Psi^i_{jk}$ follows from the convexity of $F_i(\cdot)$.
2. $p - kr_i > 0$, see Fig. 3.4b. Using, (3.6) and (3.19), we obtain

$$\Psi^i_{jk} \geq \cdots \geq \Psi^i_{p-2k} \geq \Psi^i_{p-1k} = F_i(p-1-kr_i) - F_i(p-kr_i)$$
$$\geq F_i(p-1-kr_i) - F_i(0) = F_i(p-1-kr_i).$$

Thus, the lemma follows since $\Psi^i_{pk} = F_i(p-1-kr_i) - F_i(p-kr_i) \geq 0$ and $F_i(p - kr_i) \geq 0$.

This completes the proof. $\square$

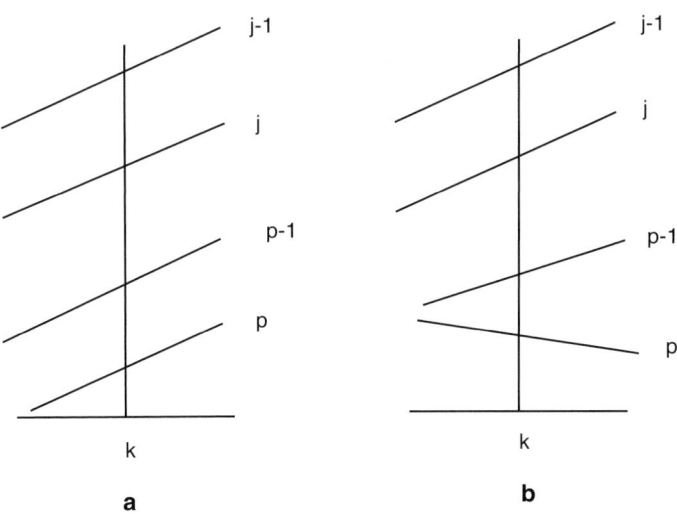

Fig. 3.4 Plots of $|j - kr_i|$ for $k \geq Z^i_p$: (**a**) $p - kr_i \leq 0$ and (**b**) $p - kr_i > 0$

3.5 The Equivalence

We now show that the minimization of (P2) is equivalent to minimization of (P3). We begin with the following lemma.

Lemma 3.13. *For any partition $\pi_i = \{Y_1^i, \ldots, Y_{d_i}^i\}$, we have*

$$J_2^i(\pi_i) = J_3^i(\pi_i) + \sum_{k=1}^{D} \inf_j f_j^i(k). \tag{3.28}$$

Proof. The proof is by induction on

$$\alpha(\pi_i) = \sum_{j=1}^{d_i} \left| Y_j^i - Z_j^i \right|.$$

For π_i with $\alpha = 0$, we have

$$Y_j^i = Z_j^i, \text{ for } j = 1, \ldots, d_i,$$

thus by Theorem 3.1 or generally Theorem 3.10,

$$J_2^i(\pi_i) = \sum_{k=1}^{D} \inf_j f_j^i(k),$$

and by (3.22) and (3.18)

$$J_3^i(\pi_i) = 0.$$

Thus the lemma holds for any π_i with $\alpha = 0$. Suppose that the lemma holds for any π_i with $\alpha(\pi_i) = k \geq 0$. We now prove that then it also holds for any solution π_i with $\alpha(\pi_i) = k+1$. Consider a solution π_i with $\alpha(\pi_i) = k+1 > 0$. Let m be such that

$$\left| Y_m^i - Z_m^i \right| = \max_j \left\{ \left| Y_j^i - Z_j^i \right| \right\} \geq 1. \tag{3.29}$$

Clearly, such m exists since $\alpha(\pi_i) = k+1 \geq 1$. Assume first that

$$Y_m^i - Z_m^i \geq 0. \tag{3.30}$$

If there are more that one m that satisfy both (3.29) and (3.30), then take the smallest of them. We have $Y_m^i \geq Z_m^i + 1$. Define $\pi_i' = \{X_1^i, \ldots, X_{d_i}^i\}$, where

$$X_j^i = \begin{cases} Y_j^i & \text{if } j \neq m, \\ Y_j^i - 1 & \text{if } j = m. \end{cases}$$

Clearly, $\alpha(\pi_i') = k$, thus, by the inductive assumption

$$J_2^i(\pi_i') = J_3^i(\pi_i') + \sum_{k=1}^{D} \inf_j f_j^i(k). \tag{3.31}$$

Moreover,

$$J_3^i(\pi_i') = J_3^i(\pi_i) + C_{m,Y_m^i}^i - C_{m,X_m^i}^i, \tag{3.32}$$

and

$$J_2^i(\pi_i') - J_2^i(\pi_i) = f_{m-1}^i(X_m^i) - f_m^i(X_m^i).$$

For $Y_m^i > X_m^i \geq Z_m^i$, using (3.18) we obtain

$$C_{m,Y_m^i}^i - C_{m,X_m^i}^i = \Psi_{m,X_m^i}^i,$$

and using (3.19) we obtain

$$f_{m-1}^i(X_m^i) - f_m^i(X_m^i) = \Psi_{m,X_m^i}^i.$$

Thus, by (3.31) we have

$$J_2^i(\pi_i) = J_2^i(\pi_i') + \Psi_{m,,X_m^i}^i = J_3^i(\pi_i') + \Psi_{m,,X_m^i}^i + \sum_{k=1}^D \inf_j f_j^i(k)$$

$$= J_3^i(\pi_i) + \sum_{k=1}^D \inf_j f_j^i(k).$$

Thus, it remains to show that π_i' is feasible. We need only to prove that if $m > 1$, then $Y_{m-1}^i < Y_m^i - 1$ in Y. Otherwise, $Y_{m-1}^i + 1 = Y_m^i$ and we have

$$Y_{m-1}^i - Z_{m-1}^i = (Y_{m-1}^i - Z_m^i) + (Z_m^i - Z_{m-1}^i)$$
$$= (Y_m^i - Z_m^i) + (Z_m^i - Z_{m-1}^i) - 1 \geq Y_m^i - Z_m^i.$$

The last inequality follows from Lemma 3.5 (3), see also Exercise 3.19. This leads to a contradiction since $m - 1$ satisfies (3.29) and (3.30), and it is smaller than m. Thus if $m > 1$, then $Y_{m-1}^i < Y_m^i - 1$ in π_i, which implies that π_i' is a feasible solution. Therefore, the lemma holds for $\alpha(\pi_i) = k + 1$.

Assume now that

$$Z_m^i - Y_m^i > 0. \tag{3.33}$$

If there are more that one m that satisfy both (3.29) and (3.33) take the largest of them. We have $Z_m^i \geq Y_m^i + 1$. Define a solution $\pi_i' = \{X_1^i, \ldots, X_{d_i}^i\}$, where

$$X_j^i = \begin{cases} Y_j^i & \text{if } j \neq m, \\ Y_j^i + 1 & \text{if } j = m. \end{cases}$$

Clearly, $\alpha(\pi_i') = k$, thus by the inductive assumption

$$J_2^i(\pi_i') = J_3^i(\pi_i') + \sum_{k=1}^D \inf_j f_j^i(k). \tag{3.34}$$

For $Z_m^i \geq X_m^i > Y_m^i$, using (3.19) and (3.18) we obtain

$$J_3^i(\pi_i) = J_3^i(\pi_i') + C_{m,Y_m^i}^i - C_{m,X_m^i}^i = \Psi_{m,,Y_m^i}^{i\prime}, \tag{3.35}$$

and

$$J_2^i(\pi_i) - J_2^i(\pi_i') = f_m^i(Y_m^i) - f_{m-1}^i(Y_m^i) = \Psi_{m,,Y_m^i}^{i\prime}.$$

Thus, by (3.34) we have

$$J_2^i(\pi_i) = J_2^i(\pi_i') + \Psi_{m,,X_m^i}^i = J_3^i(\pi_i') + \Psi_{m,,X_m^i}^i + \sum_{k=1}^{D} \inf_j f_j^i(k)$$

$$= J_3^i(\pi_i) + \sum_{k=1}^{D} \inf_j f_j^i(k),$$

and the lemma holds for $\alpha(Y) = k+1$. Thus, it remains to show that π_i' is feasible. We need only to prove that if $m < d_i$, then $Y_m^i + 1 < Y_{m+1}^i$ in π_i. Otherwise, $Y_m^i + 1 = Y_{m+1}^i$ and we have

$$Z_{m+1}^i - Y_{m+1}^i = (Z_m^i - Y_{m+1}^i) + (Z_{m+1}^i - Z_m^i)$$

$$= (Z_m^i - Y_m^i) + (Z_{m+1}^i - Z_m^i) - 1 \geq Z_m^i - Y_m^i.$$

The last inequality follows from Lemma 3.5 (3), see also Exercise 3.19. This leads to a contradiction since $m+1$ satisfies (3.29) and (3.33), and it is larger than m. Thus if $m < d_i$, then $Y_m^i + 1 < Y_{m+1}^i$ in π_i, which implies that π_i' is feasible. The lemma holds for $\alpha(\pi_i) = k+1$ and by induction for any $k \geq 0$. □

The equivalence can now readily be established.

Theorem 3.14. *For any* $\mathbf{d} = (d_1, \ldots, d_n) - partition$ π *of* $\{1, \ldots, D\}$,

$$F(S) = J_2(\pi) = J_3(\pi) + \sum_{i=1}^{n} \sum_{k=1}^{D} \inf_j f_j^i(k).$$

Proof. Let $\pi = (\pi_1, \ldots, \pi_n)$. By (3.13), (3.22) and Lemma 3.13, we have

$$J_2(\pi) = \sum_{i=1}^{n} J_2(\pi_i) = \sum_{i=1}^{n} J_3^i(\pi_i) + \sum_{i=1}^{n} \sum_{k=1}^{D} \inf_j f_j^i(k)$$

$$= J_3(\pi) + \sum_{i=1}^{n} \sum_{k=1}^{D} \inf_j f_j^i(k),$$

which proves the theorem. □

3.6 The Solution

By Theorem 3.14 a partition π that minimizes $J_3(\pi)$ minimizes $J_2(\pi)$ at the same time. The following theorem shows how the π can be found.

Theorem 3.15. *Let $\{y^i_{jk}\}$ be an optimal solution for $(P3)$ that preserves order. Then*

$$x_{ik} = \sum_{l=1}^{k} \sum_{j=1}^{d_i} y^i_{jl} \quad \text{for } i = 1,\ldots,n; \ k = 1,\ldots,D, \tag{3.36}$$

is an optimal solution for $(P1)$ and

$$Y^i_j = k \text{ if and only if } y^i_{jk} = 1, \quad \text{for } i = 1,\ldots,n; \ k = 1,\ldots,D$$

is an optimal solution for $(P2)$.

Proof. The solution $\{x_{ik}\}$ satisfies constraints (3.3)–(3.4). By (3.23)

$$0 \le x_{ik+1} - x_{ik} = \sum_{l=1}^{k+1} \sum_{j=1}^{d_i} y^i_{jl} - \sum_{l=1}^{k} \sum_{j=1}^{d_i} y^i_{jl} = \sum_{j=1}^{d_i} y^i_{jk+1} \le \sum_{(i,j)\in I} y^i_{jk+1} = 1,$$

again by (3.23) Z^i_j

$$\sum_{i=1}^{n} x_{ik} = \sum_{i=1}^{n} \sum_{l=1}^{k} \sum_{j=1}^{d_i} y^i_{jl} = \sum_{l=1}^{k} \sum_{(i,j)\in I} y^i_{jl} = k,$$

by (3.24)

$$x_{iD} = \sum_{l=1}^{D} \sum_{j=1}^{d_i} y^i_{jl} = \sum_{j=1}^{d_i} \sum_{l=1}^{D} y^i_{jl} = d_i.$$

The (3.23) implies (3.16). Since $\{x_{ik}\}$ preserves order, then $\{Y^i_j\}$ satisfies (3.14). Finally, (3.24) implies (3.15) and (3.17). By Theorem (3.14), the optimality of $\{y^i_{jk}\}$ implies optimality of $\{Y^i_j\}$. Finally, since $(P1)$ and $(P2)$ are equivalent, the $\{x_{ik}\}$ is optimal for $(P1)$. This ends the proof. $\quad\square$

Let us close this section with a brief discussion of the time complexity of our approach. There are D^2 values Ψ^i_{jk} to calculate, and then there are D^2 values C^i_{jk} to calculate, each takes $O(D)$ steps to calculate. Therefore, the calculation of the assignment costs takes $O(D^3)$ steps. The Hungarian method takes $O(D^3)$ steps to solve the assignment problem $(P3)$ and the possible swapping to make the assignment order preserving can be done in $O(D)$ steps. Therefore, the optimal solution to the original problem can be obtained in $O(D^3)$ steps. The step here is meant as a simple arithmetic operation, addition and multiplication, or the calculation of the value of function $F_i(x)$ at x. Therefore, the algorithm runs in time polynomial in D as long as the value $F_i(x)$ can be calculated in time polynomial in D for each x.

3.7 Minimization of Maximum Deviation

The maximum deviation minimization problem is defined as follows.

$$\min_{S} \left\{ G(S) = \max_{i,k} F_i(x_{ik} - kr_i) \right\} \tag{3.37}$$

subject to

$$\sum_{i=1}^{n} x_{ik} = k \quad k = 1, \ldots, D \tag{3.38}$$

$$0 \leq x_{ik+1} - x_{ik} \leq 1 \quad i = 1, \ldots, n; k = 1, \ldots, D-1 \tag{3.39}$$

$$x_{iD} = d_i \quad i = 1, \ldots, n. \tag{3.40}$$

where $F_i(\cdot)$ is a non-negative convex function satisfying

$$F_i(0) = 0 \text{ and } F_i(x) > 0 \text{ for } x \neq 0, \ i = 1, \ldots, n. \tag{3.41}$$

We reserve the term *bottleneck* deviation problem, or just the bottleneck problem, for the minimization of the maximum deviation with all $F_i(\cdot)$ being the absolute value functions, $|\cdot|$. We now show that our approach to the minimization of total deviation based on the level curves works for the maximum deviation minimization problem as well. The approach transforms the maximum deviation minimization to an equivalent bottleneck assignment problem which we define in the next section.

3.8 The Bottleneck Assignment

The solution to (3.37) subject to constraints (3.38–3.41) can be found by solving the following bottleneck assignment problem

$$\min \left\{ \max_{i,j,k} B^i_{jk} y^i_{jk} \right\} \tag{3.42}$$

subject to

$$\sum_{(i,j) \in I} y^i_{jk} = 1, \ k = 1, \ldots, D, \tag{3.43}$$

$$\sum_{k=1}^{D} y^i_{jk} = 1, \ (i,j) \in I, \tag{3.44}$$

where $I = \{(i,j) : i = 1, \ldots, n, \ j = 1, \ldots, d_i\}$ and the costs

$$B^i_{j,k} = \max\{f^i_{j-1}(k-1), f^i_j(k)\} \tag{3.45}$$

$$= \max\{F_i(j - 1 - (k-1)r_i), F_i(j - kr_i)\}. \tag{3.46}$$

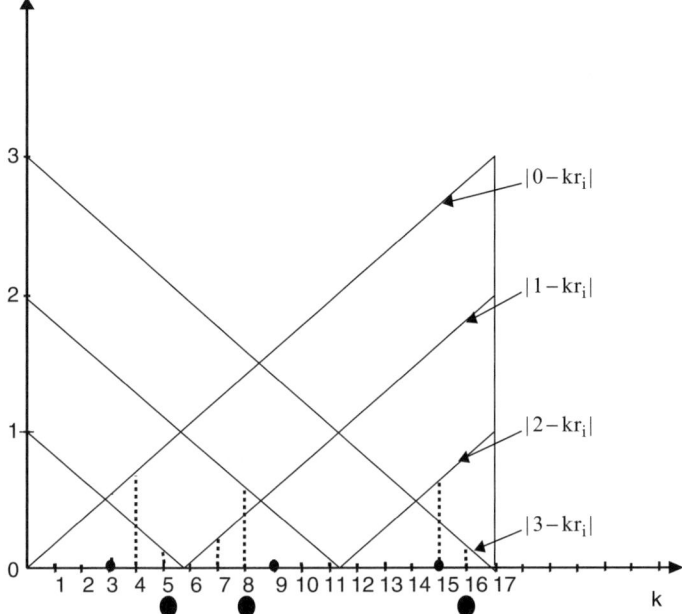

Fig. 3.5 The bottleneck penalty for i with $d_i = 3$, sequenced in positions 5, 8, and 16

Example 10.6 illustrates the calculations of B^i_{jk}. The idea behind calculating the costs B^i_{jk} is follows closely the idea behind the calculation of C^i_{jk} in Sect. 3.3. Suppose that $\pi_i = \{Y^i_1, \ldots, Y^i_{d_i}\}$ and that this set is ordered in ascending order, that is $Y^i_1 < \cdots < Y^i_{d_i}$. Then, the cost in (3.1) is charged according to f^i_0 at each $k = 1, \ldots, Y^i_1 - 1$, then according to f^i_1 at each $k = Y^i_1, \ldots, Y^i_2 - 1$, and so on until $k = Y^i_{d_i}$ where the cost starts being charged according to $f^i_{d_i}$ and continues doing so until $k = D$. The highest cost along f^i_0 is $f^i_0(Y^i_1 - 1)$, the highest cost along f^i_1 is $\max\{f^i_1(Y^i_1), f^i_1(Y^i_2 - 1)\}$, and finally the highest cost along $f^i_{d_i}$ is $f^i_{d_i}(Y^i_{d_i})$. Therefore, by assigning the costs $\max\{f^i_0(Y^i_1 - 1), f^i_1(Y^i_1)\}$, $\max\{f^i_1(Y^i_2 - 1), f^i_2(Y^i_2)\}$ and $\max\{f^i_{d_i-1}(Y^i_{d_i} - 1), f^i_{d_i}(Y^i_{d_i})\}$ to the points Y^i_1, Y^i_2 and $Y^i_{d_i}$ respectively we obtain the required cost of (3.45) or equivalently the cost (3.37). Thus, the solution to this bottleneck assignment problem can be turned in the solution to the original maximum deviation minimization problem in a way similar to the one described earlier in this chapter for the minimization of total deviation provided that there exist order preserving solutions to the bottleneck assignment problem. Their existence will be shown in the next section.

Example 3.16. Consider the instance $\mathbf{d} = (5,3,2)$ and $F_i = |\cdot|$ for $i = 1,\ldots,n$ from Example 3.2. The costs B^i_{jk} for $i = 1,2,3$, $j = 1,\ldots,d_i$ and $k = 1,\ldots,D$ calculated according to (3.45) are given in Table 3.47. $\square$

Table 3.47. The costs B^i_{jk}

$i = 1$

$j\backslash k$	1	2	3	4	5	6	7	8	9	10
1	0.5	0.5	1	1.5	2	2.5	3	3.5	4	4.5
2	1.5	1	0.5	0.5	1	1.5	2	2.5	3	3.5
3	2.5	2	1.5	1	0.5	0.5	1	1.5	2	2.5
4	3.5	3	2.5	2	1.5	1	0.5	0.5	1	1.5
5	4.5	4	3.5	5	2.5	2	1.5	1	0.5	0.5

$i = 2$

$j\backslash k$	1	2	3	4	5	6	7	8	9	10
1	0.7	0.4	0.6	0.9	1.2	1.5	1.8	2.1	2.4	2.7
2	1.7	1.4	1.1	0.8	0.5	0.5	0.8	1.1	1.4	1.7
3	2.7	2.4	2.1	1.8	1.5	1.2	0.9	0.6	0.4	0.7

$i = 3$

$j\backslash k$	1	2	3	4	5	6	7	8	9	10
1	0.8	0.6	0.4	0.6	0.8	1	1.2	1.4	1.6	1.8
2	1.8	1.6	1.4	1.2	1	0.8	0.6	0.4	0.6	0.8

$$(3.47)$$

3.9 The Bottleneck Monge Property

The following theorem proves the local bottleneck Monge property.

Theorem 3.17. *The matrices* $L_i = \left[B^i_{j,k} \right]_{d_i \times D}$ $i = 1, \ldots n$, *are bottleneck Monge matrices.*

Proof. It suffices to show that

$$\max\{B^i_{jk}, B^i_{j+1,l}\} \leq \max\{B^i_{j+1,k}, B^i_{jl}\},$$

by (3.45)

$$\max\{\max\{f^i_{j-1}(k-1), f^i_j(k)\}, \max\{f^i_j(l-1), f^i_{j+1}(l)\}\}$$
$$\leq \max\{\max\{f^i_j(k-1), f^i_{j+1}(k)\}, \max\{f^i_{j-1}(l-1), f^i_j(l)\}\}. \qquad (3.48)$$

This inequality holds since clearly

$$\max\{f^i_j(k), f^i_j(l-1)\} \leq \max\{f^i_j(k-1), f^i_j(l)\}$$

and

$$f^i_{j-1}(k-1) \leq f^i_{j-1}(l-1)$$

for $k - 1 \geq \frac{j-1}{r_i}$ and

$$f^i_{j-1}(k-1) \leq f^i_j(k-1)$$

otherwise,
and

$$f^i_{j+1}(l) \le f^i_{j+1}(k)$$

for $l \le \frac{j+1}{r_i}$ and

$$f^i_{j+1}(l) \le f^i_j(l)$$

otherwise. □

Thus, we have.

Theorem 3.18. *There is an optimal solution $\{y^i_{jk}\}$ to (P4) that preserves order.*

Proof. By contradiction. Suppose that no optimal solution $\{y^i_{jk}\}$ to (P4) preserves order. Consider an optimal solution $\{y^i_{jk}\}$ with the smallest number of the quadruples i, j, k and l that violate (3.25). For any violation we have,

$$y^i_{jl} = 1 \text{ and } y^i_{j+1k} = 1 \text{ and } k < l.$$

However,

$$\max\{B^i_{jk}, B^i_{j+1l}\} \le \max\{B^i_{jl}, B^i_{j+1k}\},$$

since L_i is a Monge matrix. Then, swapping j and $j+1$ does not increase the cost (3.42) and at the same time it reduces the number of condition (3.25) violations. Thus, we get a contradiction since the solution $\{y^i_{jk}\}$ was chosen with the smallest number of violations and thus the theorem holds. □

The optimal solution to $P4$ and thus to the original problem can be obtained in $O(D^3)$ steps, see Burkard et al. [24], and also Burkard et al. [25] for a comprehensive review of assignment algorithms. The step here is again meant as a simple arithmetic operation, addition and multiplication, or the calculation of the value of function $F_i(x)$ at x. Therefore, the algorithm runs in time polynomial in D as long as the value $F_i(x)$ can be calculated in time polynomial in D for each x. Chapter 5 will show a more efficient algorithm for the bottleneck deviation minimization for a special case of the absolute value of deviation, $F_i = |\cdot|$, $i = 1, \ldots, n$.

3.10 Exercises

Exercise 3.19. Let $Z^i_j = \left\lceil k^i_j \right\rceil$, where k^i_j is the unique crossing point satisfying

$$F_i(j - k^i_j r_i) = F_i(j - 1 - k^i_j r_i)$$

for $F_i(\cdot)$ being a non-negative convex function satisfying

$$F_i(0) = 0 \text{ and } F_i(x) > 0 \text{ for } x \ne 0, \ i = 1, \ldots, n.$$

Show that the following statements hold true:

1. $\left\lceil \frac{j-1}{r_i} \right\rceil \leq Z_j^i \leq \left\lceil \frac{j}{r_i} \right\rceil$;
2. $p > j \Rightarrow Z_p^i > Z_j^i$ and $\left\lceil \frac{p}{r_i} \right\rceil > \left\lceil \frac{j}{r_i} \right\rceil$;
3. $p > j \Rightarrow p - kr_i > 0$, for $k \leq Z_j^i$.

Exercise 3.20. Show that the cost matrix $[C_{jk}^i]$ is not a Monge matrix.

Exercise 3.21. Show that the cost matrix $[B_{jk}^i]$ is not a bottleneck Monge matrix.

Exercise 3.22. Show that the optimal solutions to the total deviation problem may not be population monotone. Hint: See Józefowska et al. [26].

Exercise 3.23. Show that for any i there are $O(D)$ costs B_{jk}^i in the cost matrix $[B_{jk}^i]$ of the bottleneck assignment problem (3.42) that are less than 1. Thus there are $O(nD)$ costs B_{jk}^i in the cost matrix $[B_{jk}^i]$ of the bottleneck assignment problem (3.42) that are less than 1.

3.11 Comments and References

The minimization of total deviation problem was formulated in Monden [8] and further refined in Miltenburg [21]. These early formulations assumed the absolute and the square functions of point deviations. The problem was solved by reduction to the assignment problem by Kubiak and Sethi [27, 28] for the general case of unimodal convex functions. The solution to the bottleneck deviation problem by its reduction to the bottleneck assignment problem was suggested by Kubiak [23] and subsequently formulated by Bautista et al. [29]. This solution works for any unimodal function of deviation. Chapter 5 will present a different solution proposed by Steiner and Yeomans [30] for the absolute deviations. For an introduction to Monge matrices and their applications in optimization see Burkard et al. [31]. Inman and Bulfin [32] independently introduced the concept of ideal positions.

Chapter 4
Optimality of Cyclic Sequences and the Oneness

4.1 Introduction

This chapter addresses the question whether there always exist *cyclic* optimal solutions to the total deviation problem and the maximum deviation problem studied in Chap. 3. This question can formally be stated as follows:

Let S be an optimal sequence for $\mathbf{d} = (d_1, \ldots, d_i, \ldots, d_n)$. *Is* S^m, *for any integer* $m \geq 1$, *an optimal sequence for* $m\mathbf{d} = (md_1, \ldots, md_i, \ldots, md_n)$, *where* S^m *is a concatenation of m copies of S?*

An affirmative answer to this question given in this chapter supports the usual for just-in-time manufacturing systems practice of repeating relatively short sequence to build a sequence for a longer time horizon, Monden [22] and Miltenburg [21].

This answer has also obvious consequences for the computational time complexity of any optimization algorithm for just-in-time sequences. This time complexity depends on the magnitude of demands $d_1, \ldots, d_n$ and consequently on the magnitude of total demand D. The only known polynomial time, with respect to D and n, optimization algorithms for just-in-time sequences have time complexity $O(D^3)$, see Chap. 3. The cyclic optimal solutions make it possible to reduce each of these demands by the factor of m, where m is the greatest common divisor of numbers $d_1, \ldots, d_n$, in the computations of optimal just-in-time sequences. The Euclid's algorithm can find the m in $O(n \log D)$ steps, see for instance Graham et al. [10]. Furthermore, the cyclic optimal solutions make a small step forward in tackling theoretically intriguing question of how succinct the encoding of optimal just-in-time sequence can be? The answer to this question also pertains to the computational complexity of the total deviation and the maximum deviation problems since the input of these problems can be made very short by the binary encoding of the demands $d_1, \ldots, d_n$ using $O(\sum_{i=1}^{n} \log(d_i + 1))$ bits. This encoding, however, makes all polynomial time, with respect to D and n, algorithms for the total deviation problem and the maximum deviation problem pseudopolynomial time algorithms, see Grigoriev [33] for a review of the high multiplicity problems. Therefore, the question whether

W. Kubiak, *Proportional Optimization and Fairness,* International Series in Operations Research & Management Science 127, DOI 10.1007/978-0-387-87719-8_4,
© Springer Science+Business Media LLC 2009

there is an algorithm with time complexity bounded by a polynomial function of $\log D$ and n remains open.

The affirmative answer to the main question of this chapter typically relies on two crucial observations. First, for the concatenation ST of sequences S and T for $a\mathbf{d} = (ad_1,\ldots,ad_n)$ and $b\mathbf{d} = (bd_1,\ldots,bd_n)$ respectively, with a and b being positive integers, we have $F(ST) = F(S) + F(T)$ for the total deviation problem, and $G(ST) = \max\{G(T),G(S)\}$ for the bottleneck deviation problem. Second, even if one relaxes the constraints $x_{iD} = d_i$, $i = 1,\ldots,n$, there still exists an optimal sequence S^* such that $x_{iD}^* = d_i$ for all i. The latter would usually require to prove that a simple exchange of two copies of different models in a given sequence does not increase either the total or the maximum deviation of the sequence. However, since the copies exchanged might be of different models this technique may increase these deviations in general whenever the F_is differ. Thus the simple exchange method fails to work in a general case unless some exchanges are forbidden. Consequently, a more sophisticated exchange method will be developed in Sect. 4.3 of this chapter to prove the existence of cyclic optimal solutions. This method limits the exchanges to the copies of models that occupy positions at the same distance from the ends 1 and D of a sequence, we assume for the time being that all demands are even. That is the exchange will only be allowed between positions 1 and D, 2 and $D-1$, 3 and $D-2$, etc., we refer to this exchange as *shuffling*. We show that the shuffling does not increase either the total deviation or the maximum deviation of the sequence, and that the shuffling can be done to ensure that the resulting sequence has the number of copies of each model equally split between its two halves. The existence of this desired distribution of model copies will be guaranteed by the Hall's Theorem for bipartite regular graphs, see for instance Bondy and Murty [34]. Our method will rely on the assignment problems equivalent to the total deviation problem and the bottleneck deviation problems developed in Chap. 3. The crucial symmetries embedded in this assignment problem costs are proven in Sect. 4.2 for the total deviation problem and in Sect. 4.5 for the maximum deviation problem. The optimality of cyclic solutions is shown in Sects. 4.4 and 4.5.

Finally, Sect. 4.6 addresses the question whether optimal solutions to the total deviation problem can always be found among the just-in-time sequences with the bottleneck deviation not exceeding 1, more precisely in the following set of just-in-time sequences

$$\Omega = \{S : |x_{ik} - kr_i| \le 1 \text{ for } i = 1,\ldots,n; k = 1,\ldots,D\},$$

thus the term *oneness*. Interestingly, this set is non-empty for any instance $\mathbf{d}$, which we show in Chap. 5. Moreover, it will be shown in this chapter that the minimization on Ω makes some total deviation problems with different F_is equivalent. This is the case for the absolute value $F_i = |\cdot|$ and the square $F_i = (\cdot)^2$ for all i functions. However, this question itself has a negative answer in general.

4.2 Symmetries of C^i_{jk}s for Symmetric F_is

This section studies various symmetries inherited by the cost coefficients C^i_{jk} from the symmetric F_i for $i = 1, \ldots, n$. The symmetries will be crucial in the proof that there exist optimal solutions that are cyclic. We begin with the following standard feature of convex functions, see Rockafellar [35].

Lemma 4.1. *For a convex function f and four distinct points $\alpha_1, \alpha_2, \alpha_3,$ and α_4 such that*

$$\alpha_2 - \alpha_1 = \alpha_4 - \alpha_3 \text{ and } \alpha_1 < \alpha_2 < \alpha_4, \qquad (4.1)$$

we have the following inequality

$$f(\alpha_1) + f(\alpha_4) \geq f(\alpha_2) + f(\alpha_3).$$

Proof. We observe that $\alpha_1 < \alpha_3 < \alpha_4$. By (4.1), there is λ, $0 < \lambda < 1$, such that

$$\alpha_2 = \lambda \alpha_1 + (1 - \lambda)\alpha_4 \text{ and } \alpha_3 = (1 - \lambda)\alpha_1 + \lambda \alpha_4.$$

Therefore for a convex f, we have

$$f(\alpha_2) \leq \lambda f(\alpha_1) + (1 - \lambda)f(\alpha_4) \text{ and}$$
$$f(\alpha_3) \leq (1 - \lambda)f(\alpha_1) + \lambda f(\alpha_4).$$

By summing up these two inequalities side by side, we obtain the required inequality

$$f(\alpha_2) + f(\alpha_3) \leq f(\alpha_1) + f(\alpha_4).$$

$\square$

The following lemma shows that the cost of being early grows at the same rate as the cost of being late whenever an actual position of a copy moves away from the copy's ideal position. Thus the penalty for deviating from the ideal position can be expected to be symmetric. However, this is not exactly the case. The penalty for finishing $l > 0$ positions after the ideal position is not necessarily equal the penalty for finishing l positions before. Thus, somewhat surprisingly, the symmetry is not ensured by the symmetry of functions F_i for the ideal vertices $\frac{2j-1}{2r_i}$ do not necessarily fall in the middle of two neighboring integers $Z^i_j - 1$ and Z^i_j, where the latter is the ideal position of copy j. Here are the details.

Lemma 4.2. *Let $1 \leq k < Z^i_j < m \leq D$ for some $i = 1, \ldots, n$ and $j = 1, \ldots, d_i$. We have.*

$$\begin{cases} C^i_{jm} \geq C^i_{jk} & \text{if } m + k > 2Z^i_j \\ C^i_{jm} \leq C^i_{jk} & \text{if } m + k < 2Z^i_j \\ C^i_{jm} \leq C^i_{jk} & \text{if } m + k = 2Z^i_j \text{ and } 0 \leq \varepsilon_{ij} < \frac{1}{2} \\ C^i_{jm} = C^i_{jk} & \text{if } m + k = 2Z^i_j \text{ and } \varepsilon_{ij} = \frac{1}{2} \\ C^i_{jm} \geq C^i_{jk} & \text{if } m + k = 2Z^i_j \text{ and } \frac{1}{2} < \varepsilon_{ij} < 1 \end{cases}$$

where

$$0 \le \left\lceil \frac{2j-1}{2r_i} \right\rceil - \frac{2j-1}{2r_i} = \varepsilon_{ij} < 1.$$

Proof. We have, for $m > Z^i_j$ and $k < Z^i_j$

$$C^i_{jm} = \sum_{l=Z^i_j}^{m-1} \Psi^i_{jl} = \sum_{l=Z^i_j}^{m-1} [F_i(j-1-lr_i) - F_i(j-lr_i)]$$

and

$$C^i_{jk} = \sum_{l=k}^{Z^i_j-1} \Psi^i_{jl} = \sum_{l=k}^{Z^i_j-1} [F_i(j-lr_i) - F_i(j-1-lr_i)].$$

First, let $m + k > 2Z^i_j$. Consider $C^i_{jm} - C^i_{jk}$, and couple $\Psi^i_{jZ^i_j+a}$ from C^i_{jm} with $\Psi^i_{jZ^i_j-a}$ from C^i_{jk}, for $a = 1, \ldots, Z^i_j - k$. Since $m > 2Z^i_j - k$ we get

$$\Delta_{mk} = C^i_{jm} - C^i_{jk} \ge \sum_{a=1}^{Z^i_j-k} (\Psi^i_{jZ^i_j+a} - \Psi^i_{jZ^i_j-a}).$$

Thus, it suffices to prove that

$$\Psi^i_{jZ^i_j+a} - \Psi^i_{jZ^i_j-a} = F_i(j-1-Z^i_jr_i - ar_i)$$
$$-F_i(j-Z^i_jr_i - ar_i) - F_i(j-Z^i_jr_i + ar_i)$$
$$+F_i(j-1-Z^i_jr_i + ar_i) \ge 0.$$

for an arbitrary a, $a = 1, \ldots, Z^i_j - k$. By definition of Z^i_j

$$Z^i_jr_i = \left\lceil \frac{2j-1}{2r_i} \right\rceil r_i = j - \frac{1}{2} + \varepsilon_{ij}r_i,$$

for some $0 \le \varepsilon_{ij} < 1$. For a symmetric $F_i(\cdot)$, we get

$$\Psi^i_{jZ^i_j+a} - \Psi^i_{jZ^i_j-a} = F_i(\alpha_4) - F_i(\alpha_1) - F_i(\alpha_3) + F_i(\alpha_2),$$

where

$$\alpha_1 = \frac{1}{2} - \varepsilon_{ij}r_i - ar_i,$$

$$\alpha_2 = \frac{1}{2} + \varepsilon_{ij}r_i - ar_i,$$

$$\alpha_3 = \frac{1}{2} - \varepsilon_{ij}r_i + ar_i,$$

$$\alpha_4 = \frac{1}{2} + \varepsilon_{ij}r_i + ar_i.$$

We have $\alpha_1 \leq \alpha_2 < \alpha_3 \leq \alpha_4$. If $\alpha_1 \geq 0$, then clearly

$$\Psi^i_{jZ^i_j+a} - \Psi^i_{jZ^i_j-a} \geq 0.$$

Assume $\alpha_1 < 0$. Let also $\alpha_2 \leq 0$. Then, by the symmetry of $F_i(\cdot)$, we get

$$\Psi^i_{jZ^i_j+a} - \Psi^i_{jZ^i_j-a} = F_i(\alpha_4) - F_i(\beta_1) - F_i(\alpha_3) + F_i(\beta_2),$$

where

$$\beta_1 = -\alpha_1 = -\frac{1}{2} + \varepsilon_{ij}r_i + ar_i$$

$$\beta_2 = -\alpha_2 = -\frac{1}{2} - \varepsilon_{ij}r_i + ar_i.$$

We have $0 \leq \beta_2 < \beta_1 < \alpha_4$. Since

$$\beta_1 - \beta_2 = 2\varepsilon_{ij}r_i = \alpha_4 - \alpha_3,$$

then by Lemma 4.1 we get

$$F_i(\beta_1) + F_i(\alpha_3) \leq F_i(\alpha_4) + F_i(\beta_2).$$

Hence,

$$\Psi^i_{jZ^i_j+a} - \Psi^i_{jZ^i_j-a} \geq 0.$$

Now, consider $\alpha_2 > 0$ and set

$$\gamma_4 = \alpha_4$$
$$\gamma_3 = \alpha_3$$
$$\gamma_2 = \beta_1$$
$$\gamma_1 = \alpha_2$$

Then, if $\gamma_1 < \gamma_2$, then $0 < \gamma_1 < \gamma_2 < \gamma_4$ and

$$\beta_1 - \alpha_2 = 2ar_i - 1 < \alpha_4 - \alpha_3 = 2\varepsilon_{ij}r_i. \tag{4.2}$$

The inequality (4.2) holds since $0 < \alpha_2$ implies

$$2(a - \varepsilon_{ij})r_i < 1. \tag{4.3}$$

Then, there is $\gamma^* > \alpha_3 > 0$ such that $\alpha_4 - \gamma^* = \beta_1 - \alpha_2$. Hence, by Lemma 4.1 we have

$$F_i(\alpha_1) + F_i(\alpha_3) \leq F_i(\beta_1) + F_i(\gamma^*) \leq F_i(\alpha_4) + F_i(\alpha_2).$$

Thus, again

$$\Psi^i_{jZ^i_j+a} - \Psi^i_{jZ^i_j-a} \geq 0.$$

If $\gamma_1 \geq \gamma_2$, then
$$F_i(\gamma_2) \leq F_i(\gamma_1) \text{ and } F_i(\gamma_3) \leq F_i(\gamma_4),$$

hence,
$$F_i(\alpha_1) + F_i(\alpha_3) \leq F_i(\alpha_4) + F_i(\alpha_2)$$

and $\Psi^i_{jZ^i_j+a} - \Psi^i_{jZ^i_j-a} \geq 0$. Therefore, if $m + k > 2Z^i_j$, then $C^i_{jm} \geq C^i_{jk}$.

Now, let $m + k < 2Z^i_j$. Consider $C^i_{jk} - C^i_{jm}$, and couple $\Psi^i_{jZ^i_j-a-2}$ from C^i_{jk} with $\Psi^i_{jZ^i_j+a}$ from C^i_{jm}, for $a = 0, \ldots, m - Z^i_j - 1$. Since $k < 2Z^i_j - m$ we get

$$\Delta_{km} = C^i_{jk} - C^i_{jm} \geq \sum_{a=0}^{m-Z^i_j-1} (\Psi^i_{jZ^i_j-a-2} - \Psi^i_{jZ^i_j+a}).$$

Thus, it suffices to prove that

$$\Psi^i_{jZ^i_j-a-2} - \Psi^i_{jZ^i_j+a} = F_i(\alpha_3 + 2r_i) - F_i(\alpha_2 - 2r_i) - F_i(\alpha_4) + F_i(\alpha_1) \geq 0$$

for an arbitrary a, $a = 0, \ldots, m - Z^i_j - 1$. However, $\alpha_3 + 2r_i > \alpha_4$ and $\alpha_1 > \alpha_2 - 2r_i$, thus the inequality clearly holds if $\alpha_2 - 2r_i \geq 0$. It remains to consider the case $\alpha_2 - 2r_i < 0$. Then, first assume $\alpha_1 \geq 0$ which implies

$$1 \geq 2(\varepsilon_{ij} + a)r_i. \tag{4.4}$$

Set

$$\gamma_4 = \alpha_3 + 2r_i$$
$$\gamma_3 = \alpha_4$$
$$\gamma_2 = -\alpha_2 + 2r_i$$
$$\gamma_1 = \alpha_1.$$

We have by (4.4)

$$\gamma_4 - \gamma_3 = -2\varepsilon_{ij}r_i + 2r_i \geq -1 + 2ar_i + 2r_i = \gamma_2 - \gamma_1.$$

and by Lemma 4.1

$$F_i(\gamma_4) + F_i(\gamma_1) \geq F_i(\gamma^*) + F_i(\gamma_2) \geq F_i(\gamma_3) + F_i(\gamma_2)$$

for γ^* such that $\gamma_4 - \gamma^* = \gamma_2 - \gamma_1$. Therefore, $\Psi^i_{jZ^i_j-a-2} - \Psi^i_{jZ^i_j-a} \geq 0$. Now, assume $\alpha_1 < 0$. Set

$$\gamma_4 = \alpha_3 + 2r_i$$
$$\gamma_3 = \alpha_4$$
$$\gamma_2 = -\alpha_2 + 2r_i$$
$$\gamma_1 = -\alpha_1.$$

We have

$$\gamma_4 - \gamma_3 = -2\varepsilon_{ij}r_i + 2r_i = \gamma_2 - \gamma_1$$

and by Lemma 4.1

$$F_i(\gamma_4) + F_i(\gamma_1) \geq F_i(\gamma_3) + F_i(\gamma_2).$$

Thus again $\Psi^i_{jZ^i_j-a-2} - \Psi^i_{jZ^i_j-a} \geq 0$. Therefore, if $m+k < 2Z^i_j$, then $C^i_{jm} \leq C^i_{jk}$.

Finally, let $m+k = 2Z^i_j$. Consider $C^i_{jm} - C^i_{jk}$, and couple $\Psi^i_{jZ^i_j+a}$ from C^i_{jm} with $\Psi^i_{jZ^i_j-a-1}$ from C^i_{jk}, for $a = 0,\ldots,(m-1)-Z^i_j$. Since $k = 2Z^i_j - m$ we get

$$\Delta_{mk} = C^i_{jm} - C^i_{jk} = \sum_{a=0}^{(m-1)-Z^i_j} (\Psi^i_{jZ^i_j+a} - \Psi^i_{jZ^i_j-a-1}).$$

For a symmetric $F_i(\cdot)$, we get

$$\Psi^i_{jZ^i_j+a} - \Psi^i_{jZ^i_j-a-1} = F_i(\alpha_4) - F_i(\alpha_1) - F_i(\alpha_3 + r_i) + F_i(\alpha_2 - r_i).$$

For $0 \leq \varepsilon_{ij} < \frac{1}{2}$. If $\alpha_1 \leq 0$, then $\Psi^i_{jZ^i_j+a} - \Psi^i_{jZ^i_j-a-1} \leq 0$. If $\alpha_1 > 0$ and $-\alpha_2 + r_i \leq \alpha_1$, then $\Psi^i_{jZ^i_j+a} - \Psi^i_{jZ^i_j-a-1} \leq 0$. If $\alpha_1 > 0$ and $-\alpha_2 + r_i > \alpha_1$, then set

$$\gamma_4 = \alpha_3 + r_i$$
$$\gamma_3 = \alpha_4$$
$$\gamma_2 = -\alpha_2 + r_i$$
$$\gamma_1 = \alpha_1$$

We have by (4.4)

$$\gamma_4 - \gamma_3 = -2\varepsilon_{ij} + r_i \geq -1 + 2ar_i + r_i = \gamma_2 - \gamma_1,$$

and by Lemma 4.1

$$F_i(\gamma_4) + F_i(\gamma_1) \geq F_i(\gamma^*) + F_i(\gamma_2) \geq F_i(\gamma_3) + F_i(\gamma_2)$$

for γ^* such that $\gamma_4 - \gamma^* = \gamma_2 - \gamma_1$. Therefore, $\Psi^i_{jZ^i_j-a} - \Psi^i_{jZ^i_j-a-1} \leq 0$. Thus, for $0 \leq \varepsilon_{ij} < \frac{1}{2}$, $\Psi^i_{jZ^i_j+a} - \Psi^i_{jZ^i_j-a-1} \leq 0$ for any a. Hence, $\Delta_{mk} \leq 0$.

For $\varepsilon_{ij} = \frac{1}{2}$, clearly $\Delta_{mk} = 0$.

For $\frac{1}{2} < \varepsilon_{ij} < 1$. If $\alpha_2 - r_i \leq 0$, then $\Psi^i_{jZ^i_j+a} - \Psi^i_{jZ^i_j-a-1} \geq 0$. If $\alpha_2 - r_i > 0$ and $\alpha_2 - r_i \geq -\alpha_1$, then again clearly $\Psi^i_{jZ^i_j+a} - \Psi^i_{jZ^i_j-a-1} \geq 0$. If $\alpha_2 - r_i > 0$ and $\alpha_2 - r_i < -\alpha_1$, then set

$$\gamma_4 = \alpha_4$$
$$\gamma_3 = \alpha_3 + r_i$$

$$\gamma_2 = -\alpha_1$$
$$\gamma_1 = \alpha_2 - r_i$$

Since $\alpha_2 - r_i > 0$, we have

$$\gamma_4 - \gamma_3 = 2\varepsilon_{ij} r_i - r_i \geq -1 + 2ar_i + r_i = \gamma_2 - \gamma_1,$$

and by Lemma 4.1

$$F_i(\gamma_4) + F_i(\gamma_1) \geq F_i(\gamma^*) + F_i(\gamma_2) \geq F_i(\gamma_3) + F_i(\gamma_2)$$

for γ^* such that $\gamma_4 - \gamma^* = \gamma_2 - \gamma_1$. Therefore, $\Psi^i_{jZ^i_j - a} - \Psi^i_{jZ^i_j - a - 1} \geq 0$. Thus, for $\frac{1}{2} < \varepsilon_{ij} < 1$, $\Psi^i_{jZ^i_j + a} - \Psi^i_{jZ^i_j - a - 1} \geq 0$ for any a. Hence, $\Delta_{mk} \geq 0$. □

The lemma will be illustrated in the following example based on Example 3.2.

Example 4.3. The reminders ε_{ij} for an instance d $= (5,3,2)$ of Example 3.2 are all equal 0 for $i = 1$ and $j = 1, \ldots, 5$ which means that the penalty for being early is never lower than the penalty for being late at the same distance from an ideal position. This is also confirmed by the costs in Table 3.21, where the costs for being early are actually higher than for being late at equal distances from the ideal positions. Both reminder for $i = 3$ are equal $\frac{1}{2}$ thus the costs are perfectly symmetric, see also Table 3.21. Finally, the reminders for $i = 2$ and $j = 1, 2, 3$ are equal $\frac{1}{3}$, 0 and $\frac{2}{3}$ respectively. This means that the penalty for being early is higher than the penalty for being late at the same distance for the first copy with its ideal position at 2, however, the situation reverses for the third copy with its ideal position at 9. The cost of being late is higher than the cost of being early for this copy.

The following lemma proves an important symmetry embedded in the cost matrices, namely rows j and $d_i + 1 - j$ are essentially the same if the former is followed from 1 to D and the latter in the opposite direction from D to 1, see for instance Table 3.21.

Lemma 4.4. *We have*

$$C^i_{(d_i+1-j)(D+1-k)} = C^i_{jk} \tag{4.5}$$

for any $i = 1, \ldots, n$, $k = 1, \ldots, D$ and $j = 1, \ldots, d_i$.

Proof. By definition Ψ^i_{jk} and the symmetry of $F_i(\cdot)$ we have

$$\begin{aligned}
\Psi^i_{(d_i+1-j)(2D-k)} &= |f^i_{d_i+1-j}(D-k) - f^i_{d_i-j}(D-k)| \tag{4.6} \\
&= |F_i(1 - j + kr_i) - F_i(-j + kr_i)| \\
&= |F_i(j - kr_i) - F_i(j - 1 - kr_i)| \\
&= |f^i_j(k) - f^i_{j-1}(k)| = \Psi^i_{jk}
\end{aligned}$$

Now, let us consider

$$C^i_{(d_i+1-j)(D+1-k)} = \sum_{l=\min\{D+1-k,Z^i_{d_i+1-j}\}}^{\max\{D+1-k,Z^i_{d_i+1-j}\}-1} \Psi^i_{(d_i+1-j)l}.$$

By definitions of Z^i_j, we have

$$Z^i_{d_i+1-j} = \begin{cases} D - Z^i_j & \text{if } \frac{2j-1}{2r_i} \text{ is an integer,} \\ D - Z^i_j + 1 & \text{otherwise.} \end{cases}$$

Suppose that $\frac{2j-1}{2r_i}$ is fractional. Then,

$$C^i_{(d_i+1-j)(D+1-k)} = \sum_{l=D+1-\max\{k,Z^i_j\}}^{D-\min\{k,Z^i_j\}} \Psi^i_{(d_i+1-j)l}. \tag{4.7}$$

By substituting l by $D - l$ on the right hand side of (4.7), we obtain

$$C^i_{(d_i+1-j)(D+1-k)} = \sum_{l=\min\{k,Z^i_j\}}^{\max\{k,Z^i_j\}-1} \Psi^i_{(d_i+1-j)(D-l)}.$$

By (4.6), we have

$$C^i_{(d_i+1-j)(D+1-k)} = \sum_{l=\min\{k,Z^i_j\}}^{\max\{k,Z^i_j\}-1} \Psi^i_{jl} = C^i_{jk},$$

and the lemma holds for a fractional $\frac{2j-1}{2r_i}$. Now, for integral $\frac{2j-1}{2r_i}$,

$$C^i_{(d_i+1-j)(D+1-k)} = \sum_{l=D-\max\{k-1,Z^i_j\}}^{D-\min\{k-1,Z^i_j\}-1} \Psi^i_{(d_i+1-j)l}. \tag{4.8}$$

By substituting l by $D - l$ on the right hand side of (4.8), we obtain

$$C^i_{(d_i+1-j)(D+1-k)} = \sum_{l=\min\{k-1,Z^i_j\}+1}^{\max\{k-1,Z^i_j\}} \Psi^i_{(d_i+1-j)(D-l)}.$$

By (4.6), we have

$$C^i_{(d_i+1-j)(D+1-k)} = \sum_{l=\min\{k,Z^i_j+1\}}^{\max\{k,Z^i_j+1\}-1} \Psi^i_{jl} = \sum_{l=\min\{k,Z^i_j\}}^{\max\{k,Z^i_j\}-1} \Psi^i_{jl} + \Psi^i_{jZ^i_j}$$

$$= C^i_{jk} + \Psi^i_{jZ^i_j}.$$

However, integral $\frac{2j-1}{2r_i}$ and symmetric $F_i(\cdot)$ imply

$$\Psi^i_{jZ^i_j} = |f^i_j(Z^i_j) - f^i_{j-1}(Z^i_j)| = \left|F_i\left(j - Z^i_j r_i\right) - F_i\left(j - 1 - Z^i_j r_i\right)\right|$$

$$= \left|F_i\left(j - (j - \tfrac{1}{2})\right) - F_i\left(j - 1 - (j - \tfrac{1}{2})\right)\right|$$

$$= \left|F_i\left(\tfrac{1}{2}\right) - F_i\left(-\tfrac{1}{2}\right)\right| = 0,$$

and the lemma holds for an integral $\frac{2j-1}{2r_i}$. This ends the proof of the lemma. □

By replacing k by $2D + 1 - k$ in (4.5) of Lemma 4.4 we obtain.

Lemma 4.5. *We have*

$$C^i_{(d_i+1-j)k} = C^i_{j(D+1-k)}$$

for any $i = 1, \ldots, n$, $k = 1, \ldots, D$ *and* $j = 1, \ldots, d_i$.

Finally, the top rows $j = 1, \ldots, \lfloor \frac{d_i}{2} \rfloor$ of the cost matrix M_i for any i that correspond to the first half of copies have higher costs in the second half of the sequence $D - k + 1, k = 1, \ldots, \lfloor \frac{D}{2} \rfloor$ than they do in the corresponding positions of the first half $k = 1, \ldots, \lfloor \frac{D}{2} \rfloor$, again check Table 3.21 for an example.

Lemma 4.6. *We have*

$$C^i_{j(D+1-k)} \geq C^i_{jk}$$

for any $i = 1, \ldots, n$, $k = 1, \ldots, \lceil \frac{D}{2} \rceil$ *and* $j = 1, \ldots, \lfloor \frac{d_i}{2} \rfloor$.

Proof. We have

$$Z^i_{\lfloor \frac{d_i}{2} \rfloor} = \left\lceil \left\lfloor \frac{d_i}{2} \right\rfloor \frac{D}{d_i} - \frac{D}{2d_i} \right\rceil \leq \left\lceil \left\lfloor \frac{d_i}{2} \right\rfloor \frac{D}{d_i} - \frac{1}{2} \right\rceil \leq \left\lceil \frac{D-1}{2} \right\rceil,$$

which implies $D + 1 - k > Z^i_j$ for $k = 1, \ldots, \lceil \frac{D}{2} \rceil$ and $j = 1, \ldots, \lfloor \frac{d_i}{2} \rfloor$. Thus,

$$C^i_{j(D+1-k)} = \sum_{l=\min\{D+1-k,Z^i_j\}}^{\max\{D+1-k,Z^i_j\}-1} \Psi^i_{jl.} = \sum_{l=Z^i_j}^{D-k} \Psi^i_{jl} \geq \sum_{l=Z^i_j}^{k-1} \Psi^i_{jl}$$

$$= \sum_{l=\min\{k,Z^i_j\}}^{\max\{k,Z^i_j\}-1} \Psi^i_{jl} = C^i_{jk}$$

as long as $k \geq Z^i_j$. Thus, it remains to show the lemma for $k < Z^i_j$. We have $m = D+1-k > Z^i_j > k$ and $m+k = D+1 > 2Z^i_j$. The latter inequality holds since

$$2Z^i_{\lfloor \frac{d_i}{2} \rfloor} = 2\left\lceil \left\lfloor \frac{d_i}{2} \right\rfloor \frac{D}{d_i} - \frac{D}{2d_i} \right\rceil \leq 2\left\lceil \left\lfloor \frac{d_i}{2} \right\rfloor \frac{D}{d_i} - \frac{1}{2} \right\rceil \leq 2\left\lceil \frac{D-1}{2} \right\rceil \leq D.$$

Therefore, Lemma 4.2 implies $C^i_{j(D+1-k)} \geq C^i_{jk}$ which completes the proof. $\square$

We now have now shown all symmetries and other essential properties of the costs C^i_{jk} sufficient to prove that there are optimal solutions that are cyclic.

4.3 The Folding, Shuffling, and Unfolding of Sequences

Let us assume that all demands $d_1, \ldots, d_n$ are even in this section. The d_i ideal positions Z^i_j for a model i with a symmetric function F_i are regularly spaced between 1 and D so that at exactly $\frac{d_i}{2}$ of them fall between 1 and $\frac{D}{2}$, since

$$Z^i_{\frac{d_i}{2}} \leq \frac{D}{2},$$

and exactly $\frac{d_i}{2}$ of them fall between $\frac{D}{2}+1$ and D, since

$$Z^i_{\frac{d_i}{2}+1} \geq \frac{D}{2} + 1.$$

Therefore, one can reasonably conjecture that whenever F_i for all i are symmetric, then there are optimal solutions to the total deviation problem (3.1) that are *half-balanced*. A sequence is half-balanced if it has exactly $\frac{d_i}{2}$ copies of each i in its first half, that is in positions 1 to $\frac{D}{2}$, and exactly $\frac{d_i}{2}$ copies of each i in its second half, that is in positions $\frac{D}{2}+1$ to D. This in turn implies that there are optimal solutions that are cyclic. This section proves that this conjecture holds true. Our approach simply turns any sequence into a half-balanced one without cost increasing. Moreover, this approach works under the sole assumption that functions F_i for all i are symmetric. That is, it does not require any more assumptions about these functions, which would be required for instance, if a simple swapping copies of different *i*s was used. The idea is as follows. Consider a feasible sequence

$$S = s_1 \cdots s_{\frac{D}{2}} s_{\frac{D}{2}+1} \cdots s_D$$

for an instance with demands $d_1, \ldots, d_n$. We show how to construct a feasible, half-balanced sequence S' for $d_1, \ldots, d_n$ without cost increasing. The construction goes through three steps: the *folding*, the *shuffling*, and the *unfolding*. The folding replaces S by a sequence of $\frac{D}{2}$ ordered pairs

$$(s_1, s_D) \cdots (s_{\frac{D}{2}}, s_{\frac{D}{2}+1}).$$

The shuffling shuffles copies inside of each pair producing a sequence of ordered pairs

$$(s_1', s_D') \cdots (s_{\frac{D}{2}}', s_{\frac{D}{2}+1}'),$$

such that $\{s_k, s_{D+1-k}\} = \{s_k', s_{D+1-k}'\}$ for $k = 1, \ldots, \frac{D}{2}$. Finally, the unfolding unfolds the outcome of the shuffle into a sequence

$$S' = s_1' \cdots s_{\frac{D}{2}}' s_{\frac{D}{2}+1}' \cdots s_D'.$$

The shuffling uses the Hall's Theorem for bipartite regular graphs to ensure that each i occurs exactly $\frac{d_i}{2}$ times in each of the two halves of S'. We then show that this three-step construction does not increase the cost of the assignment. This proof is based on a crucial observation which is that the construction does not push the copies further from their ideal positions in S' than they were in S, and thus the assignment cost does not increase. We begin with an example.

Example 4.7. Consider the sequence

$$1 \to 2 \to 3 \to 2 \to 1 \to 2 \to 2 \to 3 \to 1 \to 1 \qquad (4.9)$$
$$\to 1 \to 1 \to 2 \to 3 \to 3 \to 2 \to 1 \to 1 \to 1 \to 1$$

for $d = (10, 6, 4) = 2(5, 3, 2)$. The sequence has too many copies of $i = 2, 4$ instead of 3, and too few copies of $i = 1$, 4 instead of 5, in the first half. The the folding operation produces a 2-regular bipartite multigraph in Fig. 4.1.

Then, the shuffling produces the matchings M and M^c shown in Figs. 4.2 and 4.3 respectively.

Finally, the unfolding produces the following sequence for the two matchings:

$$1 \to 1 \to 1 \to 2 \to 2 \to 3 \to 2 \to 3 \to 1 \to 1$$
$$\to 1 \to 2 \to 3 \to 1 \to 1 \to 2 \to 3 \to 2 \to 1 \to 1.$$

Notice that each half of the sequence has exactly 5 copies of $i = 1$, 3 copies of $i = 2$, and 2 copies if $i = 3$ as required.

Let us now define the folding, shuffling and unfolding operations on the corresponding assignments. Let S be an initial sequence for $d_1, \ldots, d_n$ and $\{y_{jl}^i\}$ the assignment, see Chap. 3, that corresponds to S. Define the assignment $\{y_{jl}^i\}$ index set as follows

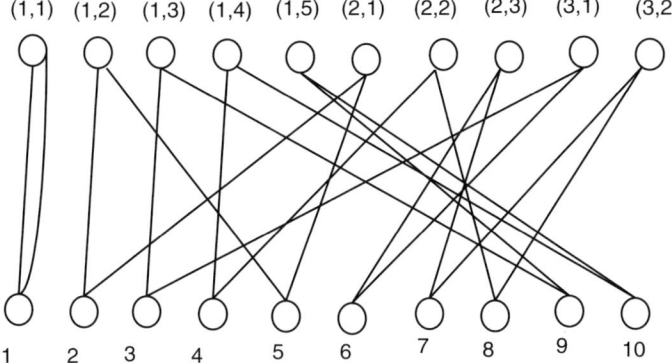

Fig. 4.1 The bipartite multigraph $G = (V_1 \cup V_2, E)$ being the result of the folding operation on the sequence (4.9)

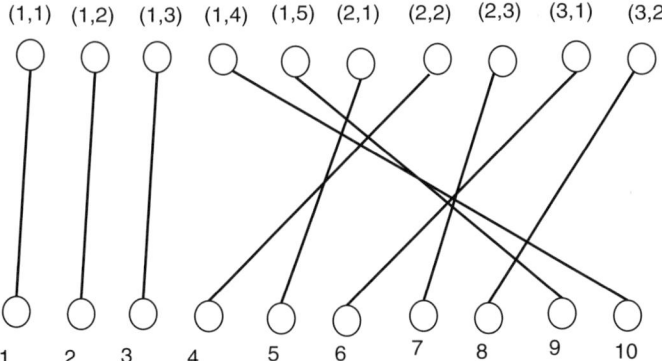

Fig. 4.2 The matching M being the result of the shuffle operation on the graph G from Fig. 4.1

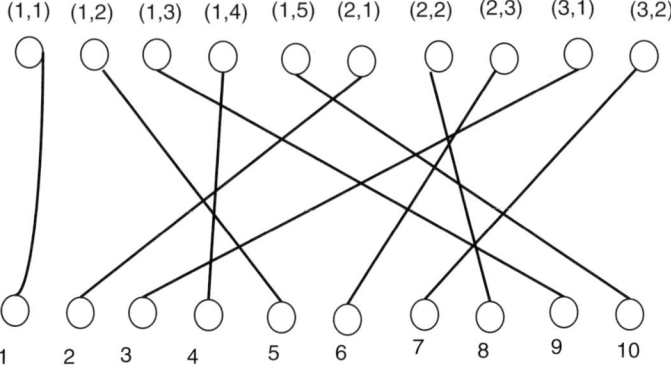

Fig. 4.3 The matching M^c being the result of the shuffle operation on the graph G from Fig. 4.1

$$s = \{(i,j,l) : y^i_{jl} = 1\}. \tag{4.10}$$

The composition, denoted by **FSU**, of folding, shuffling and unfolding, in this order, is defined as a transformation of s. The folding is defined as follows:

4.3.1 The Folding

For $(i,j,l) \in s$, we define

$$
\mathbf{F}(i,j,l) =
\begin{cases}
(i,j,l) & \text{if } j \le \frac{d_i}{2} \quad \text{and} \quad l \le \frac{D}{2}, \\[2mm]
(i,d_i+1-j,l) & \text{if } j > \frac{d_i}{2} \quad \text{and} \quad l \le \frac{D}{2}, \\[2mm]
(i,j,D+1-l) & \text{if } j \le \frac{d_i}{2} \quad \text{and} \quad l > \frac{D}{2}, \\[2mm]
(i,d_i+1-j,D+1-l) & \text{if } j > \frac{d_i}{2} \quad \text{and} \quad l > \frac{D}{2},
\end{cases}
$$

Let us define the multiset

$$C(s) = \{((i',j'),l') : \mathbf{F}(i,j,l) = ((i',j'),l') \text{ for some } (i,j,l) \in s\}.$$

We observe the following.

Lemma 4.8. *Let* $t(i,j) = \{l : ((i,j),l) \in C(s)\}$ *for* $i = 1,\ldots,n$ *and* $j = 1,\ldots,\frac{d_i}{2}$. *We have the following cardinality of* $t(i,j)$

$$|t(i,j)| = 2.$$

Proof. We observe that $\mathbf{F}(i,j,l)$ and $\mathbf{F}(i,d_i+1-j,l')$ for $i = 1,\ldots,n$ and $j = 1,\ldots,\frac{d_i}{2}$ are the only two triples in $t(i,j)$. It is worth noticing that if $l = D+1-l'$, then $\mathbf{F}(i,j,l) = \mathbf{F}(i,d_i+1-j,l')$. $\square$

The shuffling is defined as follows:

4.3.2 The Shuffling

Define a bipartite multigraph $G = (V_1 \cup V_2, E)$ as follows:

$$V_1 = \left\{ (i,j) \ : \ i = 1,\ldots,n; \ j = 1,\ldots,\frac{d_i}{2} \right\},$$

$$V_2 = \left\{ l \ : \ l = 1,\ldots,\frac{D}{2} \right\},$$

and
$$E = C(s).$$

We have the following observation about this graph.

Lemma 4.9. *There are two disjoint matchings M and M^c in the graph $G = (V_1 \cup V_2, E)$ such that $M \cup M^c = E$ and*

$$|M| = |M^c| = \frac{D}{2}.$$

Proof. The folding operation ensures that each node degree in the bipartite multigraph $G = (V_1 \cup V_2, E)$ equals 2, see Lemma 4.8. Thus the graph is 2−regular. This, however, implies that there are two disjoint perfect matchings M and M^c, see Bondy and Murty [?], such that
$$M \cup M^c = E.$$

However, $|E| = D$. Thus, either of the two matchings has cardinality $\frac{D}{2}$. This proves the lemma. □

We also observe the following.

Lemma 4.10. *For any $i = 1, \ldots, n$, we have*

$$|\{l : ((i,j),l) \in M \text{ for some } j\}|$$
$$= |\{l : ((i,j),l) \in M^c \text{ for some } j\}| = \frac{d_i}{2}.$$

Proof. The lemma follows immediately from the fact that there are exactly $\frac{d_i}{2}$ nodes (i,j) in V_1. Each of these nodes has degree 2 by Lemma 4.8. By Lemma 4.9. $((i,j),l) \in M$ and $((i,j),l') \in M^c$ for some l an l'. Thus the lemma holds. □

Finally, the unfolding.

4.3.3 The Unfolding

Define
$$\mathbf{U}(i,j,l) = \begin{cases} (i,j,l) & \text{if } ((i,j),l) \in M \\ (i, d_i+1-j, D+1-l) & \text{if } ((i,j),l) \in M^c. \end{cases}$$

4.3.4 Folding, Shuffling and Unfolding Yield An Assignment

The transformation **FSU** produces an assignment **FSU**(s) for an initial assignment s.

Lemma 4.11. FSU(s) *is an assignment.*

Proof. Follows immediately from Lemmas 4.9 and 4.10. □

The one-to-one correspondences between the assignments $\{y_{jl}^i\}$ their corresponding index sets s, see (4.10), and their corresponding **d**−partitions π defined in Chap. 3 justify the notation $J_3(s)$ instead of $J_3(\pi)$ in the reminder of this section. We now show that the **FSU** yields an assignment with its cost not exceeding the cost of the original assignment s.

Lemma 4.12. *We have*

$$J_3(s) \geq J_3(\mathbf{FSU}(s)).$$

Proof. We consider the following four cases:

1. $(i,j,l) \in s$, where $j = 1,\ldots,\frac{d_i}{2}$ and $l = 1,\ldots,\frac{D}{2}$. Then $FSU(i,j,l)$ is either (i,j,l) or $(i,d_i+1-j,D+1-l)$. Thus, $FSU(i,j,l)$ contributes either C_{jl}^i or $C_{(d_i+1-j)(D+1-l)}^i$ to $J_3(\mathbf{FSU}(s))$. By Lemma 4.4, $C_{(d_i+1-j)(D+1-l)}^i = C_{jl}^i$. Hence, $\mathbf{FSU}(i,j,l)$ makes the same contribution to $J_3(\mathbf{FSU}(s))$ as (i,j,l) does to $J_3(s)$.
2. $(i,d_i+1-j,l) \in s$, where $j = 1,\ldots,\frac{d_i}{2}$ and $l = 1,\ldots,\frac{D}{2}$. Then $\mathbf{FSU}(i,d_i+1-j,l)$ is either (i,j,l) or $(i,d_i+1-j,D+1-l)$. Thus, $\mathbf{FSU}(i,d_i+1-j,l)$ contributes either C_{jl}^i or $C_{(d_i+1-j)(D+1-l)}^i$ to $J_3(\mathbf{FSU}(s))$. Whereas, (i,d_i+1-j,l) contributes $C_{(d_i+1-j)l}^i$ to $J_3(s)$. By Lemma 4.5, $C_{(d_i+1-j)l}^i = C_{j(D+1-l)}^i$, and by Lemma 4.6, $C_{j(D+1-l)}^i \geq C_{jl}^i$. However, by Lemma 4.4 $C_{(d_i+1-j)(D+1-l)}^i = C_{jl}^i$.
3. $(i,d_i+1-j,D+1-l) \in s$, where $j = 1,\ldots,\frac{d_i}{2}$ and $l = 1,\ldots,\frac{D}{2}$. Then, $\mathbf{FSU}(i,d_i+1-j,D+1-l)$ is either (i,j,l) or $(i,d_i+1-j,D+1-l)$. Thus, $\mathbf{FSU}(i,d_i+1-j,D+1-l)$ contributes either C_{jl}^i or $C_{(d_i+1-j)(D+1-l)}^i$ to $J_3(\mathbf{FSU}(s))$. On the other hand $(i,d_i+1-j,D+1-l)$ contributes $C_{(d_i+1-j)(D+1-l)}^i = C_{jl}^i$ to $J_3(s)$.
4. $(i,j,D+1-l) \in s$, where $j = 1,\ldots,\frac{d_i}{2}$ and $l = 1,\ldots,\frac{D}{2}$. Then $\mathbf{FSU}(i,j,D+1-l)$ is either (i,j,l) or $(i,d_i+1-j,D+1-l)$. Thus, $\mathbf{FSU}(i,j,D+1-l)$ contributes either C_{jl}^i or $C_{(d_i+1-j)(D+1-l)}^i$ to $J_3(\mathbf{FSU}(s))$. On the other hand, $(i,j,D+1-l)$ contributes $C_{j(D+1-l)}^i$ to $J_3(s)$. By Lemma 4.6, $C_{j(D+1-l)}^i \geq C_{jl}^i$, and by Lemma 4.4, $C_{jl}^i = C_{(d_i+1-j)(D+1-l)}^i$.

Therefore in each of the four cases the **FSU**(i,j,l) contributes to J_3 (**FSU**(s)) no more that $(i,j,l) \in s$ to $J_3(s)$. Therefore, the lemma holds. □

The folding and shuffling may produce **FSU**(s) which is not order preserving. However, by Lemma 3.4 an order preserving solution s' can be constructed for which

$$J_3(\mathbf{FSU}(s)) \geq J_3(s'). \tag{4.11}$$

4.4 Optimality of Cyclic Solutions for Total Deviation

The main results of the previous section can be summarized as follows.

Theorem 4.13. *Let*

$$S = s_1, \ldots, s_D, s_{D+1}, \ldots, s_{2D},$$

be a feasible sequence for $2\mathbf{d} = (2d_1, \ldots, 2d_n)$. *Then, a sequence*

$$T = t_1, \ldots, t_D, t_{D+1}, \ldots, t_{2D},$$

where $i = 1, \ldots, n$ *occurs* d_i *times in the first half* $t_1, \ldots, t_D$ *and* d_i *times in the second half* $t_{D+1}, \ldots, t_{2D}$ *can be constructed such that*

$$F(T) \leq F(S).$$

Proof. The **FSU** transformation of S produces a sequence T which by Lemma 4.10 is half-balanced. Furthermore, by Lemma 4.12 and Theorem 3.14 we get $F(T) \leq F(S)$ as required. □

Theorem 4.14. *Let ST be a concatenation of sequences S and T such that*

$$x_{i|S|} - |S| r_i = 0 \text{ for all } i, \tag{4.12}$$

then $F(ST) = F(S) + F(T)$.

Proof. We have

$$
\begin{aligned}
F(ST) &= \sum_{i=1}^{n} \sum_{k=1}^{|S|+|T|} F_i(x_{ik} - kr_i) \\
&= \sum_{i=1}^{n} \sum_{k=1}^{|S|} F_i(x_{ik} - kr_i) + \sum_{i=1}^{n} \sum_{k=1}^{|T|} F_i(x_{i|S|+k} - (|S| + k)r_i) \\
&= \sum_{i=1}^{n} \sum_{k=1}^{|S|} F_i(x_{ik} - kr_i) + \sum_{i=1}^{n} \sum_{k=1}^{|T|} F_i(x_{i|S|} + \Delta_{ik} - |S| r_i - kr_i) \\
&= \sum_{i=1}^{n} \sum_{k=1}^{|S|} F_i(x_{ik} - kr_i) + \sum_{i=1}^{n} \sum_{k=1}^{|T|} F_i(\Delta_{ik} - kr_i) \\
&= F(S) + F(T),
\end{aligned}
$$

where

$$x_{i|S|+k} = x_{i|S|} + \Delta_{ik}.$$

Notice that (4.12) implies

$$\frac{x_{i|S|}}{|S|} = \frac{\Delta_{i|T|}}{|T|} = r_i.$$

□

We are now ready to prove the main result of this section.

Theorem 4.15. *For the total deviation problem, let S be an optimal sequence for $d_1, \ldots, d_n$. Then S^m, $m \geq 1$, is optimal for $md_1, \ldots, md_n$.*

Proof. By induction on m. The theorem obviously holds for $m = 1$. Suppose that the theorem holds for any $1 \leq m \leq k$. We prove that it also holds for $m = k + 1$. Consider an optimal sequence

$$T = t_1 \cdots t_{mD}$$

for $md_1, \ldots, md_n$. If m is even, then by Theorem 4.13, this sequence can be transformed by **FSU** without cost increasing into a sequence

$$T' = t'_1 \cdots t'_{mD/2} t'_{1+mD/2} \cdots t'_{mD},$$

where i occurs $md_i/2$ times in each of the two halves of T'. Thus, by Theorem 4.14 each half must be optimal for $md_1/2, \ldots, md_n/2$. Therefore, by the inductive assumption, each half is the concatenation of $m/2$ copies of S, and the theorem holds for an even $m = k + 1$. If m is odd, then consider a sequence ST for $(m+1)d_1, \ldots, (m+1)d_n$. We have $F(ST) = F(S) + F(T)$ by Theorem 4.14. By Theorem 4.13, the sequence ST can be transformed by **FSU** without cost increasing into a sequence

$$T' = t'_1 \cdots t'_{(m+1)D/2} t'_{1+(m+1)D/2} \cdots t'_{(m+1)D}$$

where i occurs $(m+1)d_i/2$ times in each of the two halves of T'. Thus, by Theorem 4.14 each half must be optimal for $(m+1)d_1/2, \ldots, (m+1)d_n/2$. Therefore, by the inductive assumption, each half is the concatenation of $(m+1)/2$ copies of S, and $F(ST) = F(S) + F(T) \geq (m+1)F(S)$. Consequently, $F(T) \geq mF(S)$ which proves the theorem for odd $m = k + 1$. Thus the theorem holds for any m.   □

It is worth observing that the constructive folding, shuffling, and unfolding operations are used in this chapter to prove the existence of optimal cyclic sequences rather than to actually construct optimal sequences. The latter can be obtained by first calculating the greatest common divisor m of $d_1, \ldots, d_n$, then by using the algorithm given in Chap. 3 to obtain an optimal sequence for $d_1/m, \ldots, d_n/m$, and finally by concatenating the sequence m times to construct an optimal sequence for the original demands $d_1, \ldots, d_n$.

4.5 Optimality of Cyclic Solutions for Maximum Deviation

The folding, shuffling and unfolding of just-in-time sequences does not increase the maximum deviation either. That is the counterpart of Lemma 4.12 for the bottleneck deviation holds. To show this it is sufficient to prove the counterparts of Lemmas 4.4, 4.5, and 4.6. This will be now done in Lemmas 4.16, 4.17, and 4.18 respectively. Check Table 3.47 for an example.

Lemma 4.16. *We have*

$$B^i_{(d_i+1-j)(D+1-k)} = B^i_{jk} \tag{4.13}$$

for any $i = 1, \ldots, n$, $k = 1, \ldots, D$ and $j = 1, \ldots, d_i$.

Proof. By definition (3.45)

$$
\begin{aligned}
B^i_{(d_i+1-j)(D+1-k)} &= \max\{F_i((d_i+1-j)-1-(D+1-k-1)r_i), \\
&\quad F_i((d_i+1-j)-(D+1-k)r_i)\} \\
&= \max\{F_i(-j+kr_i), F_i(1-j-r_i+kr_i)\}
\end{aligned}
$$

thus by the symmetry of F_i we get

$$B^i_{(d_i+1-j)(D+1-k)} = \max\{F_i(j-kr_i), F_i(j-1-(k-1)r_i)\} = B^i_{jk}.$$

$\square$

By replacing k by $2D+1-k$ in (4.13) of Lemma 4.16 we obtain.

Lemma 4.17. *We have*

$$B^i_{(d_i+1-j)k} = B^i_{j(D+1-k)}$$

for any $i = 1, \ldots, n$, $k = 1, \ldots, D$ and $j = 1, \ldots, d_i$.

Finally, we have the counterpart of Lemma 4.6 for the maximum deviation problem.

Lemma 4.18. *We have*

$$B^i_{j(D+1-k)} \geq B^i_{jk}$$

for any $i = 1, \ldots, n$, $k = 1, \ldots, \lceil \frac{D}{2} \rceil$ and $j = 1, \ldots, \lfloor \frac{d_i}{2} \rfloor$.

Proof. By definition (3.45)

$$
\begin{aligned}
B^i_{j(D+1-k)} &= \max\{F_i(j-1-(D+1-k-1)r_i), F_i(j-(D+1-k)r_i)\} \\
&= \max\{F_i(j-1-d_i+kr_i), F_i(j-d_i-r_i+kr_i)\}
\end{aligned}
$$

thus by the symmetry of F_i we have

$$B^i_{j(D+1-k)} = \max\{F_i(d_i+1-j-kr_i), F_i(d_i-j-(k-1)r_i)\}.$$

On the other hand

$$B^i_{jk} = \max\{F_i(j-kr_i), F_i(j-1-(k-1)r_i)\}.$$

First, we observe that

$$d_i+1-j-kr_i \geq j-kr_i \tag{4.14}$$

and

$$d_i - j - (k-1)r_i \geq j - 1 - (k-1)r_i \tag{4.15}$$

for $j = 1, \ldots, \left\lfloor \frac{d_i}{2} \right\rfloor$. Second, we observe that

$$d_i + 1 - j - kr_i \geq d_i - j - (k-1)r_i \geq 0 \tag{4.16}$$

for $k = 1, \ldots, \left\lceil \frac{D}{2} \right\rceil$ and $j = 1, \ldots, \left\lfloor \frac{d_i}{2} \right\rfloor$, and

$$j - kr_i \geq j - 1 - (k-1)r_i. \tag{4.17}$$

Thus, we have the following three cases:

1. $j - 1 - (k-1)r_i \geq 0$. Then, by (4.15)

$$F_i(d_i - j - (k-1)r_i) \geq F_i(j - 1 - (k-1)r_i)$$

and by (4.14) and (4.17)

$$F_i(d_i + 1 - j - kr_i) \geq F_i(j - kr_i).$$

Consequently

$$B^i_{j(D+1-k)} \geq B^i_{jk}.$$

2. $j - kr_i < 0$. Then,

$$d_i + 1 - j - kr_i \geq 1 - j + (k-1)r_i \geq 0 \tag{4.18}$$

and

$$d_i - j - (k-1)r_i \geq -j + kr_i \geq 0$$

for $k = 1, \ldots, \left\lceil \frac{D}{2} \right\rceil$. Thus, by the symmetry of F_i

$$F_i(d_i + 1 - j - kr_i) \geq F_i(1 - j + (k-1)r_i) = F_i(j - 1 - (k-1)r_i)$$

and

$$F_i(d_i - j - (k-1)r_i) \geq F_i(-j + kr_i) = F_i(j - kr_i).$$

Consequently,

$$B^i_{j(D+1-k)} \geq B^i_{jk}.$$

3. $j - 1 - (k-1)r_i < 0$ and $j - kr_i \geq 0$. Then by (4.18) and the symmetry of F_i

$$F_i(d_i + 1 - j - kr_i) \geq F_i(1 - j + (k-1)r_i) = F_i(j - 1 - (k-1)r_i)$$

and by (4.14)

$$F_i(d_i + 1 - j - kr_i) \geq F_i(j - kr_i).$$

Thus, again

$$B^i_{j(D+1-k)} \geq B^i_{jk}.$$

This proves the lemma. □

We have the following counterpart of Theorem 4.14.

Theorem 4.19. *Let ST be a concatenation of sequences S and T such that*

$$x_{i|S|} - |S|r_i = 0 \text{ for all } i, \tag{4.19}$$

then $G(ST) = \max\{G(S), G(T)\}$.

Proof. We have

$$
\begin{aligned}
G(ST) &= \max_{i,k}\{F_i(x_{ik} - kr_i)\} \\
&= \max\{\max_{i,1\leq k\leq |S|}\{F_i(x_{ik} - kr_i)\}, \\
&\qquad \max_{i,1\leq k\leq T|}\{F_i(x_{i|S|+k} - (|S|+k)r_i)\}\} \\
&= \max\{\max_{i,1\leq k\leq |S|}\{F_i(x_{ik} - kr_i)\}, \\
&\qquad \max_{i,1\leq k\leq T|}\{F_i(x_{i|S|} + \Delta_{ik} - |S|r_i - kr_i)\}\} \\
&= \max\{\max_{i,1\leq k\leq |S|}\{F_i(x_{ik} - kr_i)\}, \\
&\qquad \max_{i,1\leq k\leq T|}\{F_i(\Delta_{ik} - kr_i)\}\} \\
&= \max\{G(S), G(T)\},
\end{aligned}
$$

where

$$x_{i|S|+k} = x_{i|S|} + \Delta_{ik}.$$

□

We are now ready to prove the main result of this section.

Theorem 4.20. *For the maximum deviation problem, let S be an optimal sequence for $d_1, \ldots, d_n$. Then S^m, $m \geq 1$, is optimal for $md_1, \ldots, md_n$.*

Proof. By induction on m. The theorem obviously holds for $m = 1$. Suppose that the theorem holds for any $1 \leq m \leq k$. We prove that it also holds for $m = k+1$. Consider an optimal sequence

$$T = t_1 \cdots t_{mD}$$

for $md_1, \ldots, md_n$. If m is even, then this sequence can be transformed by **FSU** without cost increasing into a sequence

$$T' = t'_1 \cdots t'_{mD/2} t'_{1+mD/2} \cdots t'_{mD},$$

where i occurs $md_i/2$ times in each of the two halves of T'. Thus, each half must be optimal for $md_1/2,\ldots,md_n/2$ by Theorem 4.19. Therefore, by the inductive assumption, each half is the concatenation of $m/2$ copies of S, and the theorem holds for an even $m = k+1$. If m is odd, then consider a sequence ST for $(m+1)d_1,\ldots,(m+1)d_n$. We have $G(ST) = \max\{G(S),G(T)\}$ by Theorem 4.19. The sequence ST can be transformed by **FSU** without cost increasing into a sequence

$$T' = t'_1 \cdots t'_{(m+1)D/2} t'_{1+(m+1)D/2} \cdots t'_{(m+1)D}$$

where i occurs $(m+1)d_i/2$ times in each of the two halves of T'. Therefore, by the inductive assumption, each half is the concatenation of $(m+1)/2$ copies of S, and $G(ST) = \max\{G(S),G(T)\} \geq G(S)$. Consequently, $G(T) \geq G(S)$ which proves the theorem for odd $m = k+1$. Thus the theorem holds for any m. $\square$

4.6 The Oneness

Let Ω be the set of all just-in-time sequences S for an instance $\mathbf{d} = (d_1,\ldots,d_n)$ with maximum absolute deviation, or the bottleneck, not exceeding 1. More precisely

$$\Omega = \{S : |x_{ik} - kr_i| \leq 1 \text{ for } i = 1,\ldots,n; k = 1,\ldots,D\}. \tag{4.20}$$

We show in Chap. 5 that $\Omega \neq \emptyset$ for any $\mathbf{d}$. We now prove that minimizing the total absolute deviation

$$A^* = \min_S\{A(S) = \sum_{i=1}^{n}\sum_{k=1}^{D}|x_{ik} - kr_i|\}, \tag{4.21}$$

and the total squared deviation

$$Q^* = \min_S\{Q(S) = \sum_{i=1}^{n}\sum_{k=1}^{D}(x_{ik} - kr_i)^2\}, \tag{4.22}$$

on the set Ω are equivalent which, however, is not necessarily the case on other sets.

The minimization on the set Ω has obvious advantages for the computational complexity as it considerably reduces the number of cost coefficients C^i_{jk} (and B^i_{jk}) necessary to calculate in the equivalent assignment problem, see Chap. 3 and the Exercise 3.23. We denote by

$$Opt_A = \{S : A(S) = A^*\} \text{ and } Opt_Q = \{S : Q(S) = Q^*\}$$

the sets of optimal sequences for the total absolute deviation and the total squared deviation respectively.

We have the following key relationship between the values of $A(S)$ and $Q(S)$.

Theorem 4.21. *We have*

$$A(S) = \begin{cases} Q(S) + C & \text{if } S \in \Omega, \\ Q(S) - E(S) + C & \text{otherwise,} \end{cases} \tag{4.23}$$

where $E(S) > 0$ and a constant $C > 0$.

Proof. We have

$$kr_i = \lfloor kr_i \rfloor + \varepsilon_{ik}, \tag{4.24}$$

where $0 \le \varepsilon_{ik} < 1$ for $i = 1, \ldots, n$ and $k = 1, \ldots, D$. On the other hand

$$x_{ik} = \begin{cases} \lfloor kr_i \rfloor + a_{ik} & \text{if } x_{ik} > \lfloor kr_i \rfloor, \\ \lfloor kr_i \rfloor - b_{ik} & \text{otherwise,} \end{cases} \tag{4.25}$$

where a_{ik} is a positive integer and b_{ik} is a non-negative integer. Thus, for any feasible sequence S we obtain by (4.21)–(4.25)

$$\begin{aligned}
A(S) - Q(S) &= \sum_{(i,k):x_{ik}>\lfloor kr_i \rfloor} ((a_{ik} - \varepsilon_{ik}) - (a_{ik} - \varepsilon_{ik})^2) \\
&+ \sum_{(i,k):x_{ik}\le\lfloor kr_i \rfloor} ((b_{ik} + \varepsilon_{ik}) - (b_{ik} + \varepsilon_{ik})^2) \\
&= \sum_{(i,k):x_{ik}>\lfloor kr_i \rfloor} [(\varepsilon_{ik} - \varepsilon_{ik}^2) - (a_{ik} - 1)(a_{ik} - 2\varepsilon_{ik})] \\
&+ \sum_{(i,k):x_{ik}\le\lfloor kr_i \rfloor} [(\varepsilon_{ik} - \varepsilon_{ik}^2) - b_{ik}(b_{ik} + 2\varepsilon_{ik} - 1)] \\
&= \sum_{(i,k)} (\varepsilon_{ik} - \varepsilon_{ik}^2) - \sum_{(i,k):x_{ik}>\lfloor kr_i \rfloor} (a_{ik} - 1)(a_{ik} - 2\varepsilon_{ik}) \\
&- \sum_{(i,k):x_{ik}\le\lfloor kr_i \rfloor} b_{ik}(b_{ik} + 2\varepsilon_{ik} - 1) \\
&= C - E(S)
\end{aligned}$$

where

$$C = \sum_{(i,k)} (\varepsilon_{ik} - \varepsilon_{ik}^2),$$

and

$$E(S) = \sum_{(i,k):x_{ik}>\lfloor kr_i \rfloor} (a_{ik} - 1)(a_{ik} - 2\varepsilon_{ik}) + \sum_{(i,k):x_{ik}\le\lfloor kr_i \rfloor} b_{ik}(b_{ik} + 2\varepsilon_{ik} - 1).$$

If $S \in \Omega$, then $x_{ik} > \lfloor kr_i \rfloor$ implies $a_{ik} = 1$, and $x_{ik} \le \lfloor kr_i \rfloor$ implies $b_{ik} + \varepsilon_{ik} \le 1$. However, the latter implies either $b_{ik} = 0$ or $b_{ik} = 1$ and $\varepsilon_{ik} = 0$. Therefore, $S \in \Omega$ implies $E(S) = 0$. Otherwise, there are i and k such that either $a_{ik} > 1$ or $b_{ik} = 1$ and $\varepsilon_{ik} > 0$ or $b_{ik} > 1$ which implies that $E(S) > 0$. Finally,

$$C = \sum_{(i,k)} (\varepsilon_{ik} - \varepsilon_{ik}^2) > 0$$

since $0 < \varepsilon_{i1} = r_i < 1$ as long as $n > 1$. $\quad\square$

An instance $\mathbf{d} = (d_1, \ldots, d_n)$ has the oneness property if

$$Opt_A \cap \Omega \neq \emptyset.$$

We now show that if $\mathbf{d}$ has the oneness property, then any optimal sequence in Opt_Q must have its bottleneck not exceeding 1.

Theorem 4.22. *If* $Opt_A \cap \Omega \neq \emptyset$, *then* $Opt_A \cap \Omega \subseteq Opt_Q \subseteq \Omega$.

Proof. If $S^* \in Opt_A \cap \Omega$, then $A(S^*) \leq A(S)$ for any other S. By (4.23)

$$Q(S^*) = A(S^*) - C \leq A(S) - C + E(S) = Q(S),$$

since $E(S) > 0$. Thus, $Opt_A \cap \Omega \subseteq Opt_Q$. Moreover, $Opt_Q \subseteq \Omega$ whenever $S^* \in Opt_A \cap \Omega$. Otherwise, there would be $S \in Opt_Q$ and $S \notin \Omega$ such that

$$Q(S^*) = Q(S),$$

which leads to a contradiction since by (4.23)

$$A(S) + E(S) = A(S^*),$$

and $E(S) > 0$, hence $S^* \notin Opt_A$. Therefore,

$$Opt_A \cap \Omega \subseteq Opt_Q \subseteq \Omega.$$

$\square$

The following example proves that $Opt_A \cap \Omega = \emptyset$ for some instances, that is there are instances that do not have the oneness property. The characterization of the instances not having the oneness property remains an open question.

Example 4.23. Consider the instance $d = (46, 46, 1, 1, 1, 1, 1, 1, 1, 1)$ with $n = 10$. The sequence S_Q with a single copy of each $i = 3, 4, \ldots, 10$ in positions

$$15, 26, 35, 46, 55, 66, 75, \quad \text{and } 86,$$

the copies of $i = 1$ and 2 alternating in the remaining 92 positions and with a copy of $i = 1$ in position 1 minimizes (4.22) and $Q(S_Q) = 146.40$. Moreover, $S_Q \notin \Omega$ since

$$x_{1,13} - 13 \times \frac{23}{50} = 7 - \frac{299}{50} = 1\frac{1}{50} \tag{4.26}$$

and

$$x_{2,87} - 87 \times \frac{23}{50} = 39 - \frac{2,001}{50} = -1\frac{1}{50}. \tag{4.27}$$

Thus, by Theorem 4.22 $Opt_A \cap \Omega = \emptyset$, in other words any sequence that minimizes (4.21) must have its bottleneck higher than 1.

The sequence S_Q gives $A(S_Q) = 312.96$ whereas the optimum is 312.76 and it is attained by the sequence S_A with a single copy of each $i = 3, 4, \ldots, 10$ in positions

$$17, 26, 35, 46, 55, 66, 75, \quad \text{and } 84,$$

the copies of $i = 1$ and 2 alternating in the remaining 92 positions and with a copy of $i = 1$ in position 1. However, the bottleneck of S_A is higher than the 1.02 for S_Q since

$$y_{1,15} - 15 \times \frac{23}{50} = 8 - \frac{69}{10} = 1.10$$

and

$$y_{2,85} - 85 \times \frac{23}{50} = 38 - \frac{391}{10} = -1.10.$$

4.7 Exercises

Exercise 4.24. Prove the counterpart of Theorem 4.13 for the maximum deviation problem.

Exercise 4.25. Consider an instance with $n = 10$ products and demand vector

$$\mathbf{d} = (24, 24, 28, 28, 42, 42, 42, 42, 48, 16)$$

Thus, $D = 336$ and

$$r_1 = r_2 = \frac{1}{14}, \ r_3 = r_4 = \frac{1}{12}, r_5 = r_6 = r_7 = r_8 = \frac{1}{8}, r_9 = \frac{1}{7}, r_{10} = \frac{1}{21},$$

The deviation cost functions

$$\begin{aligned}
F_1(y) &= F_2(y) = a_1 |y|, \ F_3(y) = F_4(y) = a_2 |y|, \\
F_5(y) &= F_6(y) = F_7(y) = F_8(y) = a_3 |y|, \\
F_9(y) &= a_4 |y|, \ F_{10}(y) = a_5 |y|,
\end{aligned}$$

where

$$\begin{aligned}
a_1 &= 168^2 \cdot 8 \cdot 7 \cdot a_2, \\
a_2 &- 168^2 \cdot 6 \cdot 6 \cdot a_3, \\
a_3 &= 168^2 \cdot 4 \cdot 2 \cdot a_4, \\
a_4 &= 168^2 \cdot 7 \cdot a_5, \\
a_5 &= 1.
\end{aligned}$$

Prove that for any optimal sequence of the above instance, there is an i and time k such that

$$|x_{ik} - kr_i| > 1.$$

4.8 Comments and References

Bautista et al. [36] show that there always is cyclic optimal solution for $F_i = F$ for $i = 1,\ldots,n$ in the case of the total deviation problem. Kubiak [37] shows this for a general case of possibly different F_i. The proof presented in this chapter follows the prove given by Kubiak [37]. The proof that the same holds for the bottleneck deviation is new. Corominas and Moreno [38] were first to present an instance with any optimal solution minimizing total absolute deviation having maximum deviation higher than 1. Kovalyov et al. [39] present results of extensive computational experiments where they randomly generated 100,000 instances and were not able to find a single one which would not have the oneness property. Theorems 4.21 and 4.22 were shown by Corominas and Moreno [38]. Lebacque et al. [40] confirm that no simultaneous optimization of the absolute and squared deviations is possible. They provide a number of small instances for which this holds.

Chapter 5
Bottleneck Minimization

5.1 Introduction

We presented a solution to the maximum deviation problem in Chap. 3. We now take a closer look at the bottleneck deviation problem where the function to be minimized over all just-in-time sequences S for a given vector $\mathbf{d} = (d_1, \ldots, d_n)$ is defined as follows:

$$H(S) = \max_{i,k} |x_{ik} - r_i k|. \tag{5.1}$$

We keep the same notation here as in Chap. 3. More precisely, we deal with the following minimization problem referred to in this chapter as simply the *bottleneck* problem.

$$B^* = \min_{S} \{ H(S) = \max_{i,k} |x_{ik} - r_i k| \} \tag{5.2}$$

Subject to

$$
\begin{array}{ll}
x_{ik} \leq x_{ik+1} & \text{for } i = 1, \ldots, n \text{ and } k = 1, \ldots, D-1 \\
\sum_{i=1}^{n} x_{ik} = k & \text{for } k = 1, \ldots, D \\
x_{ik} \text{ non-negative integers} & \text{for } i = 1, \ldots, n \text{ and } k = 1, \ldots, D.
\end{array}
\tag{5.3}
$$

We leave it to the reader to show that this formulation is equivalent to the formulation (3.37–3.40) for $G(S) = H(S)$, see also Exercise 5.25.

We consider this particular bottleneck problem due to its generic nature, properties, shown in Sects. 5.3 and 5.5, and applications that will be discussed in this and subsequent chapters. We begin by showing in Sect. 5.2 a different solution to this bottleneck problem than the bottleneck assignment approach given in Chap. 3. This solution determines a consecutive interval of positions, called a *position window*, for each copy of each model so as not to violate a given bound on the absolute deviation, or the bottleneck, and then tries to sequence the copies in their position windows if possible. This approach results in a more efficient algorithm for the bottleneck problem. Moreover, it provides simple formulas for calculating the position window ends in case of the absolute deviation bottleneck.

W. Kubiak, *Proportional Optimization and Fairness,* International Series in Operations Research & Management Science 127, DOI 10.1007/978-0-387-87719-8_5,
© Springer Science+Business Media LLC 2009

Next, Sect. 5.4 shows the upper and lower bounds on the bottleneck, in particular it proves that any optimal solution to the bottleneck problem respects quota that is its deviation is *less* than 1. Finally, Sect. 5.6 addresses the case of two models, $n = 2$. It proves that the optimal solutions for $n = 2$ are in fact the just-in-time sequences obtained by the Webster's method of apportionment discussed in Chap. 2. Hence, it actually proves that the bottleneck of the latter sequences does not exceed $\frac{1}{2}$ for $n = 2$. This result implies that then the Webster's sequences are the most *regular words*. Section 5.6.4 applies these results by showing that the Webster's sequences maximize utilization of a resource shared by two cyclic processes.

5.2 The Position Window Based Algorithm

The idea of the solution to the problem (5.2) relies directly on the level curves introduced in Chap. 3 and it was given by Steiner and Yeomans [41]. It is explained here with the help of an example in Fig. 5.1, where the four level curves for a model i with demand $d_i = 3$ are shown. Suppose that one wishes to test if there exists a just-in-time sequence S with maximum deviation not exceeding B, that is $H(S) \leq B$, where B is a given upper bound imposed a priori on the maximum deviation. Then, one could draw a horizontal line at the distance B above the horizontal axis as in Fig. 5.1, and then learn from the level curve graphs how far the three copies of model i are allowed to deviate from their ideal positions in order *not* to violate the

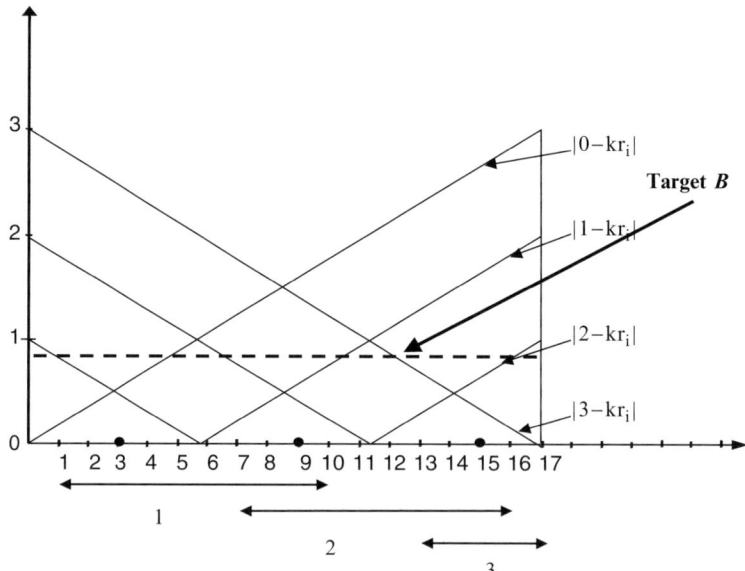

Fig. 5.1 The computation of the earliest and the latest positions for a model i with three copies

bound B imposed on the deviation. Figure 5.1 shows that model i will not violate B as long as its first copy is somewhere between 1 and 10 inclusive, its second copy somewhere between 7 and 16 inclusive, and its third copy somewhere between 13 and 17 inclusive.

On the other hand, copy 2 sequenced either before 7 or after 16 would result in the deviation being above B along the graph of the level curve $f_2^i(k) = |2 - r_i k|$. Consequently, the two crossing points of the horizontal line B and the graph of $f_j^i(k) = |j - r_i k|$ determine the *earliest* and the *latest* positions for copy j to be sequenced. These two positions are simply the ends of a position window for j. We assume that the earliest position is 1 and the latest is D if the corresponding crossing points do not exist in the interval $[1, D]$. These positions are defined generally as follows.

For $i = 1, \ldots, n$ and $j = 1, \ldots, d_i$ and the bottleneck $B \geq 0$ let us define the *earliest start position*

$$E(i, j) = \lceil \frac{j - B}{r_i} \rceil, \tag{5.4}$$

and the *latest finish position*

$$L(i, j) = \lfloor \frac{j - 1 + B}{r_i} + 1 \rfloor, \tag{5.5}$$

of the position windows for the copy j of model i.

We begin with the following result.

Theorem 5.1. *A just-in-time sequence S with its bottleneck deviation not exceeding B exists if and only if copy j of i, $i = 1, \ldots, n$ and $j = 1, \ldots, d_i$, occupies position k in S such that*

$$E(i, j) \leq k \leq L(i, j).$$

Proof. Let S be a sequence with its bottleneck deviation not exceeding B. We show that copy j of i falls into the interval $[E(i, j), L(i, j)]$ of S. The proof is by contradiction. Suppose the copy is in position $k < E(i, j)$ which implies $kr_i < j - B < j$ and $x_{ik} = j$. Then

$$|x_{ik} - kr_i| = x_{ik} - kr_i > j - j + B = B,$$

and we obtain a contradiction since the bottleneck deviation of S does not exceed B. Now, suppose the copy is in position $k > L(i, j)$ which implies $(k - 1)r_i > j - 1 + B$ and $x_{ik-1} = j - 1$. Then

$$|x_{ik-1} - (k - 1)r_i| = (k - 1)r_i - x_{ik-1} > j - 1 + B - j + 1 = B,$$

and we obtain a contradiction again since the bottleneck deviation of S does not exceed B. Consequently, copy j of i falls into $[E(i, j), L(i, j)]$, since S includes d_i copies of i, and thus the necessary condition holds.

Now, let us assume that copy j of i, $i = 1, \ldots, n$ and $j = 1, \ldots, d_i$, occupies a position k in S such that

$$E(i, j) \leq k \leq L(i, j).$$

Consider a given i. Let its d_i copies be in positions $1 \leq k_1, \ldots, k_{d_i} \leq D$ of S, where by assumption $\lceil \frac{j-B}{r_i} \rceil \leq k_j \leq \lfloor \frac{j-1+B}{r_i} + 1 \rfloor$. Then, $x_{ik} = j$, for $k_j \leq k < k_{j+1}$ and $1 \leq j < d_i - 1, x_{ik} = 0$, for $1 \leq k < k_1$, and $x_{ik} = d_i$, for $k_{d_i} < k \leq D$. Consider, the interval $k_j \leq k < k_{j+1}$. There, we have

$$B \geq j - \frac{j-B}{r_i} r_i \geq j - E(i,j)r_i \geq j - k_j r_i \geq j - k r_i$$

for any k. Moreover,

$$k r_i - j \leq (k_{j+1} - 1)r_i - j \leq (L(i, j+1) - 1)r_i - j \leq (\frac{j+B}{r_i} + 1 - 1)r_i - j = B$$

for any k. Thus, we have

$$|x_{ik} - k r_i| \leq B$$

for any $k_1 \leq k < k_{d_i}$. It remains to show that

$$k r_i \leq B \text{ for } 1 \leq k < k_1 \tag{5.6}$$

and

$$d_i - k r_i \leq B \text{ for } k_{d_i} < k \leq D. \tag{5.7}$$

The condition (5.6) holds since

$$(k_1 - 1)r_i \leq (L(i, 1) - 1)r_i \leq (\frac{B}{r_i} + 1 - 1)r_i \leq B,$$

and condition (5.7) holds since

$$0 \leq d_i - k_{d_i} r_i \leq d_i - E(i, d_i)r_i \leq d_i - (\frac{d_i - B}{r_i})r_i = B.$$

This proves

$$|x_{ik} - k r_i| \leq B \text{ for all } k.$$

Since our choice of i was arbitrary, then the sufficient condition holds and so does the theorem. $\quad\square$

To find a just-in-time sequence for $\mathbf{d} = (d_1, \ldots, d_n)$ with its bottleneck deviation not exceeding a given B we construct a bipartite graph $G = (V_1 \cup V_2, \mathcal{E})$, where $V_1 = \{1, \ldots, D\}$ is the set of the sequence positions and $V_2 = \{(i, j)|i = 1, \ldots, n; j = 1, \ldots, d_i\}$ is the set of copies to sequence. The edge $(k, (i, j)) \in \mathcal{E}$ if and only if $k \in [E(i, j), L(i, j)]$. This bipartite graph is V_1−convex since for each $(i, j) \in V_2$ if $(k, (i, j)) \in \mathcal{E}$ and $(m, (i, j)) \in \mathcal{E}$ with $k \leq m$, then $(l, (i, j)) \in \mathcal{E}$ for all $k \leq l \leq m$.

Any perfect matching M in G can be turned into a just-in-time sequence S^M for $(d_1, \ldots, d_n)$ by placing i in the position k of the sequence if $(k, (i, j)) \in M$ for some j, that is $S_k^M = i$. Conversely, any just-in-time sequence S for $(d_1, \ldots, d_n)$ can be turned into a perfect matching M^S as follows, if $S_k = i$, then $(k, (i, j)) \in M^S$, where

j is the number of copies of i in the k−prefix $S_1 \cdots S_k$ of S. Thus, the matching M^S is in fact an *order preserving perfect matching*, that is if $(k,(i,j)) \in M^S$ and $(k',(i,j+1)) \in M^S$, then $k < k'$ for all i and $j = 1, \ldots, d_i - 1$.

Actually, any perfect matching M in G can be turned into an order preserving perfect matching since if $(k,(i,j)) \in M$ and $(k',(i,j+1)) \in M$ and $k' < k$, then

$$E(i,j) \le E(i,j+1) \le k' < k \le L(i,j) \le L(i,j+1).$$

Thus, $(k',(i,j)) \in \mathcal{E}$ and $(k,(i,j+1)) \in \mathcal{E}$ so the positions of copies j and $j+1$ be exchanged since G is V_1−convex. Therefore, we have just shown the following theorem on which the solution to the problem (5.2) is based.

Theorem 5.2. *A just-in-time sequence for* $\mathbf{d} = (d_1, \ldots, d_n)$ *with its bottleneck deviation not exceeding given B exists if and only if the bipartite graph* $G = (V_1 \cup V_2, \mathcal{E})$ *has an order preserving perfect matching.*

The following example illustrates this algorithm.

Example 5.3. Consider $\mathbf{d} = (5,3,2)$ and $B = 0.5$. The formulas in (5.4) and (5.5) give the following earliest start positions and latest finish positions for this instance.

j	$E(1,j)$	$L(1,j)$	$E(2,j)$	$L(2,j)$	$E(3,j)$	$L(3,j)$
1	1	2	2	2	3	3
2	3	4	5	6	8	8
3	5	6	9	9		
4	7	8				
5	9	10				

The graph G for this instance is shown in Fig. 5.2, where the edges marked by the thick lines must be included in any perfect matching in G.

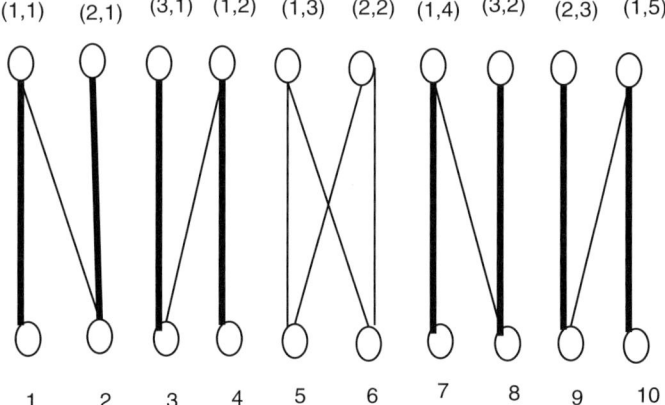

Fig. 5.2 The bipartite graph for $\mathbf{d} = (5, 3, 2)$ and $B = 0.5$

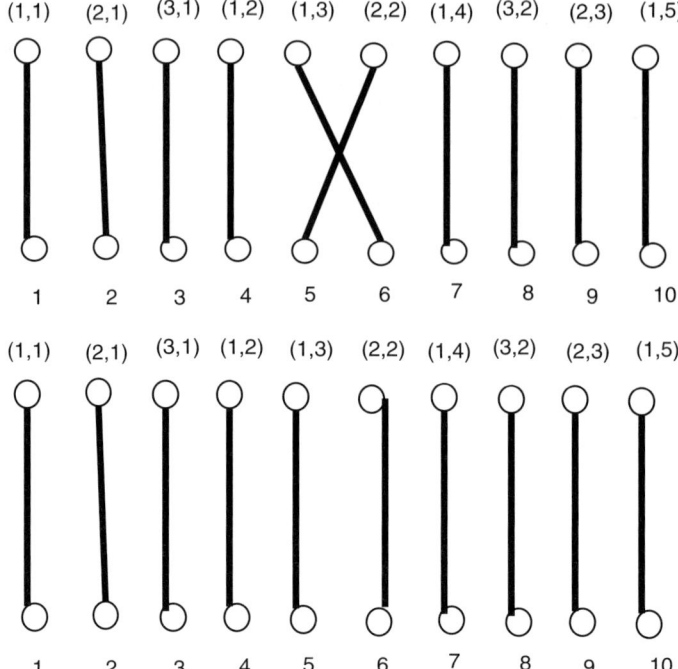

Fig. 5.3 The only two possible perfect matchings in graph G from Fig. 5.2

Therefore, there are exactly two perfect matchings in G, both shown in Fig. 5.3. The the top one gives the sequence

$$1 \rightarrow 2 \rightarrow 3 \rightarrow 1 \rightarrow 2 \rightarrow 1 \rightarrow 1 \rightarrow 3 \rightarrow 2 \rightarrow 1$$

whereas the bottom

$$1 \rightarrow 2 \rightarrow 3 \rightarrow 1 \rightarrow 1 \rightarrow 2 \rightarrow 1 \rightarrow 3 \rightarrow 2 \rightarrow 1. \qquad \square$$

A perfect matching in a V_1−convex graph G, if any exists, can be found by the algorithm that assigns position k to the copy (i, j) with the smallest value of $L(i, j)$ among all the available copies with $(k, (i, j)) \in \mathcal{E}$, if such exists. Otherwise, no perfect matching exists. Observe that for each i

$$E(i, j) \le E(i, j+1) \text{ and } L(i, j) \le L(i, j+1)$$

for $j = 1, \ldots, d_i$, moreover merging of any two ordered lists of numbers into a single ordered list of numbers can be done in time linear with respect the number of numbers on both lists. Thus, the algorithm can be implemented to run in $O(D)$ time. The search for an optimal B^* is based on the following observation.

Lemma 5.4. *The product $D \cdot B^*$ is an integer.*

Proof. Let B^* be an optimal bottleneck for $\mathbf{d} = (d_1, \ldots, d_n)$. Then, there are i and k such that

$$|x_{ik} - kr_i| = B^*.$$

Thus, $D \cdot B^*$ is an integer since

$$D |x_{ik} - kr_i| = |Dx_{ik} - kd_i|$$

is an integer. □

By Lemma 5.4 and Theorems 5.6, 5.9, and 5.10 given in Sect. 5.4, the optimal bottleneck B^* takes on one of the values $\frac{D-1}{2D}, \ldots, \frac{D-1}{D}$. Therefore, a binary search needs to test $O(\log D)$ possible values to find the optimal B^*. Thus, it takes $O(D \log D)$ to find the optimum bottleneck in the problem (5.2) subject to (5.3).

On the other hand, the general approach to the bottleneck deviation problem based on the reduction to the bottleneck assignment problem (3.42) presented in Chap. 3 requires calculating the D^3 entries of the cost matrix $[B_{jk}^i]$. However, by Theorem 5.6 and Exercise 3.23 the number of required entries can be reduced to $O(nD)$ as all entries with value 1 and higher can be eliminated for the absolute value functions F_i. Moreover, these entries are as follows

$$B_{jk}^i = \max\{|j - 1 - (k - 1)r_i|, |j - kr_i|\} \tag{5.8}$$

for the absolute value functions F_i. An open question then remains whether the bottleneck assignment problem (3.42) with costs (5.8) can be solved in time $O(nD)$, that is in time linear with respect to the required number of entries. Observe that this solution to (5.2) subject to (5.3), if it exists, would avoid the binary search for the optimal B^* as well as would offer the time complexity $O(nD)$, which is better than $O(D \log D)$ for large D.

5.3 The Complexity

The succinct, binary input encoding of the input vector $\mathbf{d} = (d_1, \ldots, d_n)$ requires $O(\sum \log(d_i + 1))$ bits therefore the optimization algorithm shown in Sect. 5.2 to run in time $O(D \log D)$ is in fact exponential with respect to this input encoding. The decision problem consisting in checking if there is a solution with its bottleneck not exceeding given B can only be solved in $O(D)$ time in Sect. 5.2, again exponential. Clearly, any certificate that requires the just-in-time sequence itself is too long to be of polynomial length with respect to the binary input encoding. Though, a polynomial certificate remains to be shown to exist so that the problem can be shown to belong to *NP*, a polynomial certificate proving that the problem is in the class *co-NP* does exist. The certificate is based on the Hall's Theorem for bipartite convex graphs of the just-in-time sequences and it results in the following feasibility test proven by Brauner and Crama [42].

Theorem 5.5. *A just-in-time sequence S with its bottleneck not exceeding B exists if and only if for each pair of integers x and y such that $0 \leq x < y \leq D$ the following two inequalities are satisfied simultaneously*

$$\sum_{i=1}^{n} \max(0, \lfloor yr_i + B \rfloor - \lceil xr_i - B \rceil) \geq y - x \tag{5.9}$$

and

$$\sum_{i=1}^{n} \max(0, \lceil yr_i - B \rceil - \lfloor xr_i + B \rfloor) \leq y - x. \tag{5.10}$$

The theorem allows checking the feasibility of B for a given instance $\mathbf{d} = (d_1, \ldots, d_n)$ in time (nD^2) which is actually less efficient than the $O(D)$ test proposed in Sect. 5.2. However, the *infeasibility* certificate is made up of a pair (x, y), $0 \leq x < y \leq D$, for which either (5.9) or (5.10) or both fail. Clearly, the test can be done in $O(n)$ time for the (x, y), and either number requires only $O(\log D)$ bits.

5.4 Bounds on the Bottleneck

5.4.1 The Upper Bounds

The following upper bound follows from Theorem 5.5 and was shown by Brauner and Crama [42], see also Steiner and Yeomans [41].

Theorem 5.6. *There always exists a just-in-time sequence S with its bottleneck not exceeding $1 - \frac{1}{D}$.*

Proof. We need to show that inequalities (5.9) and (5.10) of Theorem 5.5 are always satisfied for $B = 1 - \frac{1}{D}$. We begin with the inequality (5.9). Consider any pair of integers $0 \leq x < y \leq D$. We have

$$\lfloor yr_i + B \rfloor + \varepsilon_i = yr_i + B$$

and

$$\lceil xr_i - B \rceil = xr_i - B + \lambda_i$$

where $0 \leq \varepsilon_i, \lambda_i < 1$, for all i. Moreover, since $B = 1 - \frac{1}{D}$, then

$$\varepsilon_i, \lambda_i \in \{0, \frac{1}{D}, \ldots, 1 - \frac{1}{D}\}.$$

Consequently,

$$\lfloor yr_i + B \rfloor - \lceil xr_i - B \rceil = (y-x)r_i + 2B - (\varepsilon_i + \lambda_i) \geq (y-x)r_i > 0$$

and the inequality (5.9) holds for x and y.

Consider the inequality (5.10) now. We have

$$\lceil yr_i - B \rceil = yr_i - B + \omega_i$$

and

$$\lfloor xr_i + B \rfloor = xr_i + B - \tau_i$$

where $0 \leq \omega_i, \tau_i < 1$, for all i. Consequently,

$$\lceil yr_i - B \rceil - \lfloor xr_i + B \rfloor = (y-x)r_i - 2B + (\omega_i + \tau_i). \tag{5.11}$$

Let I be the set of all i for which the value of the expression in (5.11) is positive. We obtain

$$\sum_{i=1}^n \max(0, \lceil yr_i - B \rceil - \lfloor xr_i + B \rfloor) \leq (y-x) + \sum_{i \in I}(\omega_i + \tau_i - 2B) \leq (y-x)$$

since $B = 1 - \frac{1}{D}$ and

$$\tau_i, \omega_i \in \{0, \frac{1}{D}, \ldots, 1 - \frac{1}{D}\}.$$

This proves that the inequality (5.10) holds for for x and y. Therefore, the theorem holds by Theorem 5.5 since our choice of x and y was arbitrary. $\square$

The upper bound can be improved by the following result shown by Tijdeman [43].

Theorem 5.7. *Let λ_{ij} be a double sequence of non-negative numbers such that*

$$\sum_{1 \leq i \leq n} \lambda_{ik} = 1 \text{ for } k = 1, \ldots$$

For an infinite sequence S on the alphabet $\{1, \ldots, n\}$, $n \geq 2$, let x_{ik} be the number of is in the k-prefix of S. Then there exists a sequence S on $\{1, \ldots, n\}$ such that

$$\max_{i,k} | \sum_{1 \leq j \leq k} \lambda_{ij} - x_{ik}| \leq 1 - \frac{1}{2(n-1)}.$$

Let us define $\lambda_{ik} = r_i = \frac{d_i}{D}$ for $k = 1, \ldots$ and $i = 1, \ldots, n$. Then, this theorem ensures the existence of an infinite sequence S such that

$$\max_{i,k} |kr_i - x_{ik}| \leq 1 - \frac{1}{2(n-1)}. \tag{5.12}$$

We can ensure the required number d_i of copies of model i in the D-prefix of sequence S on the alphabet $\{1, \ldots, n\}$ as follows. Consider the D-prefix of S and suppose that there is i with $x_{iD} > d_i$. Then, there is j with $x_{jD} < d_j$. It can be easily checked that replacing the last i in the D-prefix by j does not increase the absolute maximum deviation for the D-prefix. Therefore, we can readily obtain a D-prefix where each i occurs exactly d_i times and with maximum deviation not exceeding $1 - \frac{1}{2(n-1)}$.

The sequence satisfying the bound in (5.12) is built as follows, see Tijdeman [43]. Let $L_k, k = 1, \ldots, D$ be the set of models satisfying the following condition at k:

$$\sigma_i = kr_i - x_{i,k-1} \geq \frac{1}{2n-2}, \tag{5.13}$$

where, as usual, the $x_{i,k-1}$ is the cumulative number of units of model i sequenced between 1 and $k - 1$. Apportion k to model i from the set L_k with the minimum value of

$$\frac{1 - \frac{1}{2n-2} - \sigma_i}{r_i}. \tag{5.14}$$

This is equivalent to apportioning k to model i from the set L_k with the maximum value of

$$\frac{d_i}{x_{i,k-1} + 1 - \frac{1}{2(n-1)}}, \tag{5.15}$$

which is the divisor method with the divisor function

$$d(a) = a + 1 - \frac{1}{2(n-1)}$$

applied to the models in L_k. Observe that this divisor function actually depends on the number n of models, the equivalent of the number of states s. Moreover, the condition (5.13) ensures that the model getting the position k does not violate the lower bound $\frac{1}{2(n-1)} - 1$ on its deviation imposed by (5.12).

By Theorem 5.7, we consequently have the following stronger than in Theorem 5.6 upper bound.

Theorem 5.8. *The optimal value B^* satisfies the following inequality*

$$B^* \leq 1 - \max\{\frac{1}{D}, \frac{1}{2(n-1)}\}.$$

Though, $D \geq 2(n-1)$ most often. It is obviously possible that $D < 2(n-1)$, for instance when $d_i = 1$ for all i and $n > 2$.

5.4.2 The Lower Bounds

The following lower bounds are shown in Chap. 6.

Theorem 5.9. *For $n > 2$, a standard instance $\mathbf{d} = (d_1, \ldots, d_n)$ of the bottleneck deviation problem defined in (5.2) has value, B, less than $\frac{1}{2}$ if and only if $d_i = 2^{i-1}$ for $i = 1, 2, \ldots, n$. Then,*

$$B = \frac{2^{n-1} - 1}{2^n - 1} = \frac{1}{2} - \frac{1}{2(2^n - 1)} = \frac{1}{2} - \frac{1}{2D}.$$

Recall that for the standard instance $\gcd(d_1, \ldots, d_n) = 1$. While, by Theorem 5.9, for any $n \geq 3$ there is only one standard instance with bottleneck less than $\frac{1}{2}$, the number of standard instances with bottleneck less than $\frac{1}{2}$ for $n = 2$ is *infinite*. We have the following result of Brauner and Crama [42], and Kubiak [44].

Theorem 5.10. *For $n = 2$, $B < \frac{1}{2}$ if and only if one of demands d_1 or d_2 is odd and the other even. Moreover, then $B = \frac{1}{2} - \frac{1}{2D}$.*

5.5 Main Properties

Theorem 4.20 shows optimality of cyclic solutions for the maximum deviation problem. We now present a simpler proof for the absolute value bottleneck which is essentially based on Theorem 5.6. Let $g > 1$ be a positive integer and S a just-in-time sequence for $\mathbf{d} = (d_1, \ldots, d_n)$. The concatenation $S^g = S \cdots S$ repeats S exactly g times and thus it is a just-in-time sequence for $g\mathbf{d} = (gd_1, \ldots, gd_n)$. We have

Theorem 5.11. *The sequence S^g is optimal for $(gd_1, \ldots, gd_n)$ as long as S is optimal for $(d_1, \ldots, d_n)$. Moreover their bottlenecks are equal.*

Proof. Consider the concatenation S^g, $g \geq 1$, of a sequence S for $(d_1, \ldots, d_n)$. We have

$$x_{i,\alpha D+l} - (\alpha D + l)r_i = \alpha d_i + x_{il} - \alpha d_i - lr_i = x_{il} - lr_i$$

for any $0 \leq \alpha < g$ and $1 \leq l < D$ in S^g. Thus, both S^g, $g \geq 1$, and S have the same bottleneck. Now, let S' be an optimal sequence for $(gd_1, \ldots, gd_n)$ with its bottleneck B^*. By Theorem 5.6, $B^* \leq 1 - \frac{1}{D}$. Thus, there are exactly d_i copies of i in the D−prefix of S'. Otherwise, there would be an i such that $x_{iD} > d_i$ but $Dr_i = d_i$. Consequently, $x_{iD} - Dr_i \geq 1$, which leads to a contradiction. Therefore, $S' = TW$ is a concatenation of a just-in-time sequence T for $(d_1, \ldots, d_n)$ and a just-in-time sequence W, perhaps empty if $g = 1$, for $((g-1)d_1, \ldots, (g-1)d_n)$. Let B' be a bottleneck of T. We have $\max\{B, B'\} \leq B^*$, where B is the bottleneck of an optimal sequence S for $(d_1, \ldots, d_n)$. However, the bottleneck of S^g equals B as well. Thus S^g is optimal for $(gd_1, \ldots, gd_n)$. $\square$

Finally, a just-in-time sequence and its mirror reflection have the same bottleneck.

Theorem 5.12. *The mirror reflection S^R of sequence S for $(d_1, \ldots, d_n)$ has the same bottleneck as S itself.*

Proof. By definition

$$S_{D+1-k} = S_k^R \text{ for } k = 1, \ldots, D.$$

Thus

$$y_{ik} = d_i - x_{iD-k} \text{ for all } i \text{ and } k = 1, \ldots, D$$

where $x_{i0} = 0$ for all i and cumulative x_{ik} and y_{ik} for S and S^R respectively. Let $k' = D + 1 - k$, $k = 1, \ldots, D$ be the one-to-one correspondence between k' and k. Then,

$$
\begin{aligned}
\left| y_{ik'} - k' r_i \right| &= \left| y_{i,D+1-k} - (D+1-k)r_i \right| \\
&= \left| d_i - x_{i,D-(D+1-k)} - d_i + (k-1)r_i \right| \\
&= \left| x_{i,k-1} - (k-1)r_i \right|.
\end{aligned}
$$

Since $\left| x_{iD} - Dr_i \right| = 0$ for all i, then the theorem holds. $\square$

5.6 The Absence of Competition

The minimization of maximum deviation, and the total deviation as well for that matter, can be viewed as simply a way of dealing with the competition for the same ideal positions, see (3.11) in Chap. 3 for the definition of ideal positions, by allocating them so that the maximum, or the total deviation respectively, of a just-in-time sequence is minimized, unfortunately that also means that as a result the bottleneck itself must grow to $\frac{1}{2}$ and most likely higher. However, there is a class of instances which are competition-free. These are those with demands being the consecutive powers of two. We prove in Chap. 6 that these are the only competition-free instances for $n \geq 3$. Moreover, all instances with $n = 2$ are virtually competition-free and if there is a competition for an ideal position, then its resolution keeps the value of the bottleneck unchanged and equal $\frac{1}{2}$. In fact any optimal solution then is a standard two state solution obtained by the Webster's method of apportionment discussed in Chap. 2. Besides its being as fair as it can be according to the apportionment theory the solution for $n = 2$ simultaneously minimizes the bottleneck deviation, the total deviation and the utilization of a resource shared by two cyclical processes. Therefore, this case is unique as it is both fair and optimal. We will discuss it in this section.

We assume $n = 2$ and that furthermore the greatest common divisor of d_1, d_2, and $D = d_1 + d_2$ is 1, that is $\gcd\{d_1, d_2\} = 1$. Otherwise, the optimal solution for $\frac{d_1}{g}$ and $\frac{d_2}{g}$ can be repeated $g = \gcd\{d_1, d_2\}$ times resulting into an optimal solution for the original instance with demands d_1 and d_2, by Theorem 5.11.

5.6.1 Optimal Solutions for $n = 2$

We now show that it is essentially optimal for $n = 2$ to place the copies in their *ideal* positions. The ideal position for copy j, $j = 1, \ldots, d_i$, of model i, $i = 1, 2$, has been defined in Chap. 3 in (3.11) as

$$\lceil \frac{2j-1}{2r_i} \rceil.$$

We first precisely define all solutions for $n = 2$ with their bottlenecks being less or equal $\frac{1}{2}$. It turns out that such solutions always exist. Moreover, if one of the demands d_1 or d_2 is even and the other odd, then there is exactly one solution with its bottleneck not exceeding $\frac{1}{2}$. This solution has in fact its bottleneck *less* than $\frac{1}{2}$ and it is thus the only optimal. Moreover, if both d_1 and d_2 are odd, then there are exactly two solutions with their bottlenecks not exceeding $\frac{1}{2}$. Either of these two solutions has its bottleneck *equal* $\frac{1}{2}$ and thus it is optimal. We begin with the definition of these solutions in Lemmas 5.13–5.16.

Lemma 5.13. *Consider the ideal vertices* $a_j = \frac{2j-1}{2r_1}$ *for* $j = 1, \ldots, d_1$ *and* $b_k = \frac{2k-1}{2r_2}$ *for* $k = 1, \ldots, d_2$. *If* $\ell - 1 < a_j \leq \ell$ *and* $\ell - 1 < b_k \leq \ell$ *for some* $\ell = 1, \ldots, D$, *then* $a_j = b_k = \ell$.

Proof. If $\ell - 1 < a_j \leq \ell$ and $\ell - 1 < b_k \leq \ell$ for some $\ell = 1, \ldots, D$, then $a_j = \frac{2j-1}{2r_1} = (\ell - 1) + f_1$ and $b_k = \frac{2k-1}{2r_2} = (\ell - 1) + f_2$, with $0 < f_i \leq 1$ for $i = 1, 2$. Thus, we have

$$2j - 1 = 2r_1(\ell - 1) + 2r_1 f_1$$

$$2k - 1 = 2r_2(\ell - 1) + 2r_2 f_2$$

and, since $r_1 + r_2 = 1$, then

$$j + k - 1 = (\ell - 1) + r_1 f_1 + r_2 f_2. \tag{5.16}$$

Now, the left-hand side of the equation (5.16) and the $(\ell - 1)$ are integers, thus $r_1 f_1 + r_2 f_2$ must be an integer. However, $0 < r_1 f_1 + r_2 f_2 \leq 1$, since $r_1 f_1 + r_2 f_2 \leq (r_1 + r_2) \max(f_1, f_2) = \max(f_1, f_2) \leq 1$. Thus, $r_1 f_1 + r_2 f_2 = 1$. However, $r_1 f_1 + r_2 f_2 = 1$ if and only if $f_1 = f_2 = 1$, which proves the lemma. $\square$

The following lemma shows that the instances with one of d_1 and d_2 being odd and the other even are competition-free. Thus, all copies can fall in their ideal positions.

Lemma 5.14. *If one of the demands,* d_1 *and* d_2, *is odd and the other even, then* $\alpha_j = \lceil a_j \rceil$ *for* $j = 1, \ldots, d_1$ *and* $\beta_k = \lceil b_k \rceil$ *for* $k = 1, \ldots, d_2$ *are pairwise different.*

Proof. By Lemma 5.13, if $\alpha_j = \lceil a_j \rceil = \beta_k = \lceil b_k \rceil$ for some $j = 1, \ldots, d_1$ and $k = 1, \ldots, d_2$, then $a_j = b_k$ or simply $(2j - 1)d_2 = (2k - 1)d_1$. However, this equality is impossible since its one side is odd and the other even. Moreover, $a_k - a_j = \frac{(k-j)D}{d_1} > 1$ and $b_m - b_\ell = \frac{(m-\ell)D}{d_2} > 1$ for $j < k$ and $\ell < m$. Therefore,

$\alpha_j = \lceil a_j \rceil \neq \alpha_k = \lceil a_k \rceil$ for $j,k = 1,\ldots,d_1$ and $k \neq j$ as well as $\beta_\ell = \lceil b_\ell \rceil \neq \beta_m = \lceil b_m \rceil$ for $\ell,m = 1,\ldots,d_2$ and $\ell \neq m$, which proves the lemma. $\square$

However, if both d_1 and d_2 are odd, then both models compete for the middle position $\frac{D}{2}$.

Lemma 5.15. *If both d_1 and d_2 are odd, then none of the numbers $a_j = \frac{2j-1}{2r_1}$ for $j = 1,\ldots,\frac{d_1+1}{2} - 1, \frac{d_1+1}{2} + 1, \ldots, d_1$ and $b_k = \frac{2k-1}{2r_2}$ for $k = 1,\ldots,\frac{d_2+1}{2} - 1, \frac{d_2+1}{2} + 1, \ldots, d_2$ is integer. Moreover, $a_{\frac{d_1+1}{2}} = b_{\frac{d_2+1}{2}} = \frac{D}{2}$ is an integer.*

Proof. If d_1 and d_2 are odd, then $\frac{D}{2}$ is an integer. Moreover, $\frac{D}{2}$ and d_i are relatively prime for $i = 1,2$. Otherwise, $\gcd\{D,d_1,d_2\} > 1$. Therefore, if d_i divides $(2j-1)\frac{D}{2}$ for $j = 1,\ldots,d_i$ and $i = 1,2$, then d_i must divide $(2j-1)$. This can only happen for $j = \frac{d_i+1}{2}$. Therefore, $a_{\frac{d_1+1}{2}} = b_{\frac{d_2+1}{2}} = \frac{D}{2}$ and none of the numbers $\frac{(2j-1)D}{2d_i} = \frac{2j-1}{2r_i}$ for $j = 1,\ldots,\frac{d_i+1}{2} - 1, \frac{d_i+1}{2} + 1, \ldots, d_i$ and $i = 1,2$ is an integer. $\square$

Therefore, the competition for the middle position $\frac{D}{2}$ can be settled by moving either competitor to position $\frac{D}{2} + 1$, which is free.

Lemma 5.16. *If both d_1 and d_2 are odd, then $\alpha_j = \lceil a_j \rceil$ for $j = 1,\ldots,d_1$, and $\beta_k = \lceil b_k \rceil$ for $k = 1,\ldots,\frac{d_2+1}{2} - 1, \frac{d_2+1}{2} + 1, \ldots, d_2$, and $\beta_{\frac{d_2+1}{2}} = \lceil b_{\frac{d_2+1}{2}} \rceil + 1 = \frac{D}{2} + 1$ are pairwise different, and so are $\beta_j = \lceil b_j \rceil$ for $j = 1,\ldots,d_2$, and $\alpha_k = \lceil a_k \rceil$ for $k = 1,\ldots,\frac{d_1+1}{2} - 1, \frac{d_1+1}{2} + 1, \ldots, d_1$, and $\alpha_{\frac{d_1+1}{2}} = \lceil a_{\frac{d_1+1}{2}} \rceil + 1 = \frac{D}{2} + 1$.*

Proof. Follows immediately from Lemmas 5.13 and 5.15, and the fact that $\lceil a_j \rceil \neq \frac{D}{2} + 1$ for $j = 1,\ldots,d_1$, and $\lceil b_k \rceil \neq \frac{D}{2} + 1$ for $k = 1,\ldots,d_2$. The latter holds since $\lceil \frac{2j-1}{2r_i} \rceil \geq \frac{D}{2} + 2$ for $j = \frac{d_i+1}{2} + 1$ and $i = 1,2$. $\square$

Consider α_j, $j = 1,\ldots,d_1$ and β_j for $j = 1,\ldots,d_2$ and define the sets $X_k = \{j : \alpha_j \leq k\}$ and $Y_k = \{j : \beta_j \leq k\}$ for $k = 1,\ldots D$. By Lemmas 5.14 and 5.16 the solution

$$x_{1,k} = |X_k| \text{ and } x_{2,k} = |Y_k| \qquad (5.17)$$

for $k = 1,\ldots D$ meets the constraints (5.3). We now show that its bottleneck does not exceed $\frac{1}{2}$.

Theorem 5.17. *The bottleneck of the solution (5.17) does not exceed $\frac{1}{2}$.*

Proof. For each $j \in X_k$ we have

$$j \leq \left\lfloor kr_1 + \frac{1}{2} \right\rfloor,$$

thus

$$x_{1k} = |X_k| = \left\lfloor kr_1 + \frac{1}{2} \right\rfloor.$$

Therefore,

$$-\frac{1}{2} \le x_{1k} - kr_1 = \left\lfloor kr_1 + \frac{1}{2} \right\rfloor - kr_1 \le \frac{1}{2}, \tag{5.18}$$

or

$$|x_{1k} - kr_1| \le \frac{1}{2}.$$

By Lemmas 5.14 and 5.16

$$k - |X_k| = |Y_k| = x_{2k}.$$

Thus, by (5.18)

$$-\frac{1}{2} \le x_{2k} - kr_2 = k - \left\lfloor kr_1 + \frac{1}{2} \right\rfloor - kr_2 = -\left\lfloor kr_1 + \frac{1}{2} \right\rfloor + kr_1 \le \frac{1}{2}.$$

$\square$

The sequence with 1 in positions $\alpha_j = \lceil a_j \rceil$ for $j = 1, \ldots, d_1$ and 2 in positions $\beta_k = \lceil b_k \rceil$ for $k = 1, \ldots, d_2$ whenever one of d_1 or d_2 is odd and the other even is the only one with bottleneck not exceeding $\frac{1}{2}$ as shown in the following theorem.

Theorem 5.18. *If one of the demands, d_1 and d_2, is odd and the other even, then the solution defined in Lemma 5.14 is unique solution with bottleneck not exceeding $\frac{1}{2}$.*

Proof. By Theorem 5.1, (5.4) and (5.5) copy (i, j) for $j = 1, \ldots, d_i$ and $i = 1, 2$ must be in position k such that

$$\left\lceil \frac{j - \frac{1}{2}}{r_i} \right\rceil \le k \le \left\lfloor \frac{j - \frac{1}{2}}{r_i} \right\rfloor + 1 \tag{5.19}$$

for $B = \frac{1}{2}$. However,

$$\frac{j - \frac{1}{2}}{r_i} = \frac{(2j - 1)D}{2d_i}$$

is not an integer since $(2j - 1)D$ is odd and $2d_i$ is even. Therefore,

$$\left\lfloor \frac{j - \frac{1}{2}}{r_i} \right\rfloor + 1 = \left\lceil \frac{j - \frac{1}{2}}{r_i} \right\rceil,$$

and thus, there is only one k that satisfies (5.19). Therefore, if there is any solution it must be unique. However, the solution with 1 in positions $\alpha_j = \lceil a_j \rceil$ for $j = 1, \ldots, d_1$ and 2 in positions $\beta_k = \lceil b_k \rceil$ for $k = 1, \ldots, d_2$ does not exceed $\frac{1}{2}$ by Theorem 5.17. Thus, it is the unique solution. $\square$

Finally, there are exactly two solutions with their bottlenecks not exceeding $\frac{1}{2}$ whenever both d_1 and d_2 are odd.

Theorem 5.19. *If both demands d_1 and d_2 are odd, then the two solutions defined in Lemma 5.16 are the only solutions with bottleneck not exceeding $\frac{1}{2}$.*

Proof. As in the proof of Theorem 5.18 copy (i, j) for $j = 1, \ldots, d_i$ and $i = 1, 2$ must be in position k such that the condition (5.19) is satisfied for $B = \frac{1}{2}$. However,

$$\frac{j - \frac{1}{2}}{r_i}$$

becomes integer for $j = \frac{d_i + 1}{2}$, and it remains fractional for any other j. Therefore, copy $j = \frac{d_i + 1}{2}$ can be in either of two positions $\frac{D}{2}$ or $\frac{D}{2} + 1$. The solution from Lemma 5.16 either gives the position $\frac{D}{2}$ to 1 and the position $\frac{D}{2} + 1$ to 2 or gives the position $\frac{D}{2}$ to 2 and the position $\frac{D}{2} + 1$ to 1. Therefore, these are the only two solutions possible for $B = \frac{1}{2}$. $\square$

5.6.2 The Bottleneck and the Webster's Method Are One for $n = 2$

We proved in Theorem 2.5 of Chap. 2 that every divisor method stays within the quota for all 2-state problems. However, what distinguishes the Webster's method from all other divisor methods for $n = 2$ is that it then actually minimizes the bottleneck as well. This will be now shown. Let us order the ideal vertices

$$a_j = \frac{2j - 1}{2r_1} \text{ for } j = 1, \ldots, d_1 \text{ and } b_k = \frac{2k - 1}{2r_2} \text{ for } k = 1, \ldots, d_2$$

in non-decreasing order. Clearly, if we replace

$$a_j \text{ by } \frac{d_1}{j - \frac{1}{2}}, \text{ and } b_k \text{ by } \frac{d_2}{k - \frac{1}{2}}$$

we get the list ordered in non-increasing order. Actually, the order will be decreasing if one of the demands, d_1 and d_2, is odd and the other even, and there will be a single tie if both d_1 and d_2 are odd. Now, consider the number

$$\frac{d_i}{\ell - \frac{1}{2}} \tag{5.20}$$

in the position $h + 1$ of this order. Then, clearly $x_{i,h+1} = \ell$ and $x_{i,h} = \ell - 1$ in the solution (5.17), thus

$$\frac{d_i}{\ell - \frac{1}{2}} = \frac{d_i}{x_{i,h} + \frac{1}{2}}.$$

Therefore, position $h + 1$ is apportioned to i that satisfies the following equation

$$\frac{d_i}{x_{i,h} + \frac{1}{2}} = \max_{k=1,2} \frac{d_k}{x_{k,h} + \frac{1}{2}}$$

which is exactly what the Webster's method does. Therefore, we just shown that the bottleneck solutions for $n = 2$ with $B = \frac{1}{2}$ are exactly those obtained by the Webster method of apportionment. The following example shows that neither Adams's nor Jefferson's parametric methods minimize the bottleneck.

Example 5.20. Consider $d_1 = 10$ and $d_2 = 7$. By Theorem 5.18 the unique optimal sequence for this instance is

$$1 \to 2 \to 1 \to 2 \to \boxed{1 \to 1} \to 2 \to 1 \to 2 \to 1 \to 2 \to \boxed{1 \to 1} \to 2 \to 1 \to 2 \to 1$$

which is the Webster's sequence. The Adams's divisor method would use

$$\frac{d_i}{\ell - 1}$$

instead of (5.20) to obtain the following sequence for the instance

$$1 \to 2 \to 1 \to 2 \to 1 \to 2 \to \boxed{1 \to 1} \to 2 \to 1 \to 2 \to \boxed{1 \to 1} \to 2 \to 1 \to 2 \to 1 \quad (5.21)$$

and the Jefferson's method would use

$$\frac{d_i}{\ell}$$

instead of (5.20) to obtain the sequence

$$1 \to 2 \to 1 \to 2 \to \boxed{1 \to 1} \to 2 \to 1 \to 2 \to \boxed{1 \to 1} \to 2 \to 1 \to 2 \to 1 \to 2 \longleftrightarrow 1. \quad (5.22)$$

The positions where the sequences differ are framed. Moreover, the Jefferson's sequence is not unique since the copies in the last two positions can be interchanged, which is indicated by $\longleftrightarrow$. Thus, neither the Adams's sequence nor the Jefferson's sequence are optimal for the instance. We could double check this statement by observing that $x_{1,6} = 3$ for (5.21) and thus

$$\left| x_{1,6} - 6r_1 \right| = \left| 3 - 6 \times \frac{10}{17} \right| = \frac{9}{17} > \frac{1}{2},$$

and $x_{1,11} = 7$ for (5.22), hence

$$\left| x_{1,11} - 11r_1 \right| = \left| 7 - 11 \times \frac{10}{17} \right| = \frac{9}{17} > \frac{1}{2}.$$

As pointed out earlier, by Theorem 2.5 all divisor methods are within the quota for $n = 2$. Thus any divisor method gives a solution $x_{i,k}$ equals either $\lfloor kr_i \rfloor$ or $\lceil kr_i \rceil$ and consequently

$$\left| x_{i,k} - kr_i \right| < 1.$$

Therefore, though all divisor methods ensure the bottleneck less that 1 for $n = 2$ only the Webster's method ensures that then the bottleneck does not exceed $\frac{1}{2}$. We now show that it is this small bottleneck that makes the Webster's method to provide optimal utilization of a resource shared by two cyclical processes. We begin with the definition of the most regular words.

5.6.3 The Most Regular Words

The *most regular* words are well-known in formal language theory, discrete geometry and optimal resource allocation, see Gaujal [45] and Vuillon [46] for review. This last application will be discussed later in this section. We now show that the most regular words are exactly the same as the sequences generated by the Webster's method for $n = 2$. We begin by precisely defining the most regular words.

Let (c_1, c_2) be a point in $\mathbb{R}^2$ with both coordinates c_1 and c_2 being non-negative integers. The cell centered at (c_1, c_2), or just a cell $\mathbb{C}(c_1, c_2)$, is the square

$$\mathbb{C}(c_1, c_2) = \left\{ (x, y) \in \mathbb{R}^2 : |x - c_1| \leq \frac{1}{2} \text{ and } |y - c_2| \leq \frac{1}{2} \right\}. \qquad (5.23)$$

Consider the line segment L

$$\frac{d_2}{d_1} x$$

in $\mathbb{R}^2$ between the points $(0, 0)$ and (d_1, d_2). Here, d_1 and d_2 are demands for models 1 and 2, respectively. That is the set

$$L(d_1, d_2) = \left\{ (x, y) \in \mathbb{R}^2 : y = \frac{d_2}{d_1} x \text{ and } 0 \leq x \leq d_1 \right\}. \qquad (5.24)$$

The $L(d_1, d_2)$-diagonal is the set of all cells that have non-empty intersection with the segment $L(d_1, d_2)$, that is

$$\mathbb{D}(d_1, d_2) = \{ \mathbb{C}(c_1, c_2) : \mathbb{C}(c_1, c_2) \cap L(d_1, d_2) \neq \emptyset \}.$$

Example 5.21. The example in Fig. 5.4 shows the line segment $L(4, 3)$ for $d_1 = 4$ and $d_2 = 3$ along with its diagonal cells $\mathbb{C}(0, 0), \mathbb{C}(1, 0), \mathbb{C}(1, 1), \mathbb{C}(2, 1), \mathbb{C}(2, 2), \mathbb{C}(3, 2), \mathbb{C}(3, 3),$ and $\mathbb{C}(4, 3)$. $\square$

Define a directed graph $\mathbb{W}(d_1, d_2) = (\mathbb{D}(d_1, d_2), \mathbb{A}(d_1, d_2))$, with its nodes being the cells of the diagonal $\mathbb{D}(d_1, d_2)$ and its set of arcs $\mathbb{A}(d_1, d_2)$ including a pair $(\mathbb{C}(c_1, c_2), \mathbb{C}(c_1', c_2'))$ if and only if either $c_1' = c_1 + 1$ and $c_2' = c_2$ or $c_1' = c_1$ and $c_2' = c_2 + 1$. That is, in the former case the cell $\mathbb{C}(c_1', c_2')$ is the *horizontal* translation of the cell $\mathbb{C}(c_1, c_2)$ by one unit along the horizontal axis, in the latter the cell $\mathbb{C}(c_1', c_2')$ is the *vertical* translation of the cell $\mathbb{C}(c_1, c_2)$ by one unit along the vertical axis.

Directed paths from the cell $\mathbb{C}(0, 0)$ to the cell $\mathbb{C}(d_1, d_2)$ in $\mathbb{W}(d_1, d_2)$ generate words on the alphabet $\{1, 2\}$ as follows. Let us mark the arc $(\mathbb{C}(c_1, c_2), \mathbb{C}(c_1', c_2'))$ with the letter 1 if the $\mathbb{C}(c_1', c_2')$ is the horizontal translation of $\mathbb{C}(c_1, c_2)$, and let

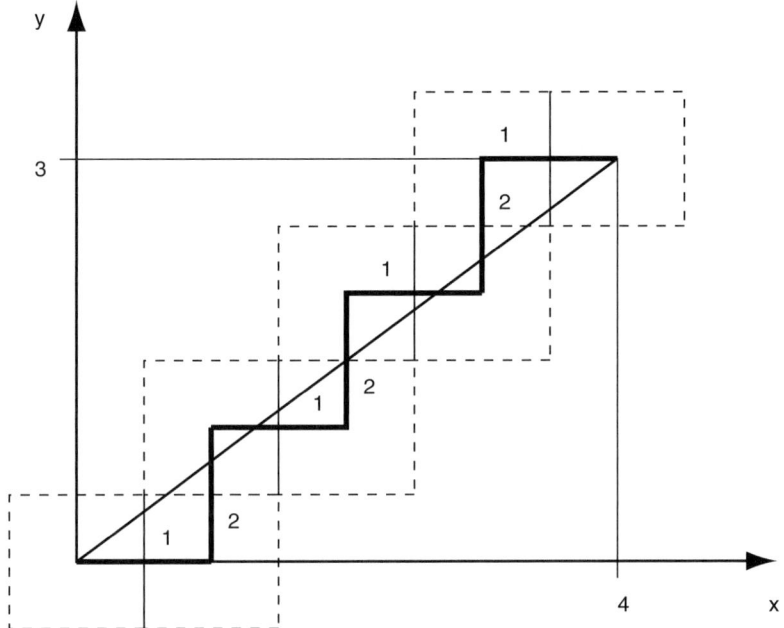

Fig. 5.4 The line segment $L(4,3)$ and its diagonal cells

us mark the arc $(\mathbb{C}(c_1,c_2),\mathbb{C}(c_1',c_2'))$ with the letter 2 if the $\mathbb{C}(c_1',c_2')$ is the vertical translation of $\mathbb{C}(c_1,c_2)$. For a directed path from the cell $\mathbb{C}(0,0)$ to the cell $\mathbb{C}(d_1,d_2)$ in $\mathbb{W}(d_1,d_2)$ let w be a word on the alphabet $\{1,2\}$ made up by the concatenation of the letters marking the consecutive arcs on the path from $\mathbb{C}(0,0)$ to $\mathbb{C}(d_1,d_2)$. The word w is called the *most regular word* for d_1 and d_2, and obviously is made up of d_1 letters 1 and d_2 letters 2. Figure 5.4 shows the only regular word 1212121 for $d_1 = 4$ and $d_2 = 3$. The set of all regular words for d_1 and d_2 will be denoted by $\Gamma(d_1,d_2)$.

We now show that the set of most regular words $\Gamma(d_1,d_2)$ is the same as the set of solutions to the bottleneck problem (5.2) with the bottleneck $B \le \frac{1}{2}$, which in turn is the same as the set of the apportionment sequences generated by the Webster's method of apportionment, for d_1 and d_2.

Let x_{ik}, $i = 1,2$ and $k = 1,\ldots,D$ be the solution to the bottleneck problem with $B \le \frac{1}{2}$. Define the points

$$(x_{1,1},x_{2,1}), (x_{1,2},x_{2,2}), \ldots, (x_{1,D},x_{2,D}).$$

The cells centered in these points belong to the diagonal $\mathbb{D}(d_1,d_2)$ which is shown in the following lemma.

Lemma 5.22. *The cells* $\mathbb{C}(0,0), \mathbb{C}(x_{1,1},x_{2,1}), \mathbb{C}(x_{1,2},x_{2,2}), \ldots, \mathbb{C}(x_{1,D},x_{2,D}) \in \mathbb{D}$ (d_1,d_2).

Proof. The points

$$(r_1, r_2), (2r_1, 2r_2), \ldots, (Dr_1, Dr_2) \tag{5.25}$$

are in $L(d_1, d_2)$. Moreover, for any $k = 1, \ldots, D$

$$\left| x_{1,k} - kr_1 \right| \leq \frac{1}{2} \text{ and } \left| x_{2,k} - kr_2 \right| \leq \frac{1}{2},$$

since the solution has the bottleneck $B \leq \frac{1}{2}$. Thus

$$\mathbb{C}(x_{1k}, x_{2k}) \cap L(d_1, d_2) \neq \emptyset,$$

and $\mathbb{C}(x_{1k}, x_{2k}) \in \mathbb{D}(d_1, d_2)$. $\square$

On the other hand we have.

Lemma 5.23. *If the cells* $\mathbb{C}(0,0), \mathbb{C}(t_1, w_1), \mathbb{C}(t_2, w_2), \ldots, \mathbb{C}(t_D, w_D)$ *make up a directed path in* $\mathbb{W}(d_1, d_2)$ *from* $\mathbb{C}(0,0)$ *to* $\mathbb{C}(d_1, d_2)$, *then the solution*

$$x_{1k} = t_k \text{ and } x_{2,k} = w_t \tag{5.26}$$

has the bottleneck $B \leq \frac{1}{2}$.

Proof. The solution (5.26) meets the constraints (5.3), thus it is feasible. Moreover,

$$(kr_1, kr_2) \in \mathbb{C}(t_k, w_k), \tag{5.27}$$

for $k = 1, \ldots, D$. Otherwise, there would be $\ell = 1, \ldots, D$ such that $(kr_1, kr_2) \in \mathbb{C}(t_k, w_k)$ for $k = 1, \ldots, \ell - 1$ and $(\ell r_1, \ell r_2) \notin \mathbb{C}(t_\ell, w_\ell)$. We assume $t_0 = 0$ and $w_0 = 0$. Then, either $(\ell r_1, \ell r_2) \in \mathbb{C}(t_{\ell-1}, w_{\ell-1})$ or $(\ell r_1, \ell r_2) \in \mathbb{C}(t_{\ell+1}, w_{\ell+1})$. This holds since $(\ell r_1, \ell r_2)$ must belong to some cell $\mathbb{C}(t_k, w_k)$ for $k = 1, \ldots, D$ and $(\ell r_1, \ell r_2) - ((\ell - 1)r_1, (\ell - 1)r_2) = (r_1, r_2) < (1, 1)$. Therefore, we have either

$$\left| t_{\ell-1} - \ell r_1 \right| \leq \frac{1}{2} \text{ and } \left| w_{\ell-1} - \ell r_2 \right| \leq \frac{1}{2}, \tag{5.28}$$

or

$$\left| t_{\ell+1} - \ell r_1 \right| \leq \frac{1}{2} \text{ and } \left| w_{\ell+1} - \ell r_2 \right| \leq \frac{1}{2}. \tag{5.29}$$

However,

$$\left| t_\ell - \ell r_1 \right| > \frac{1}{2} \text{ or } \left| w_\ell - \ell r_2 \right| > \frac{1}{2}$$

since by assumption $(\ell r_1, \ell r_2) \notin C(t_\ell, w_\ell)$, and

$$t_\ell - \ell r_1 + w_\ell - \ell r_2 = 0$$

since $t_\ell + w_\ell = \ell$ and $r_1 + r_2 = 1$. Therefore

$$\left| t_\ell - \ell r_1 \right| = \left| w_\ell - \ell r_2 \right|,$$

thus

$$|t_\ell - \ell r_1| > \frac{1}{2} \text{ and } |w_\ell - \ell r_2| > \frac{1}{2}.$$

This leads to a contradiction with either (5.28) or (5.29) since either

$$t_{\ell-1} = t_\ell \text{ or } w_{\ell-1} = w_\ell$$

and either

$$t_{\ell+1} = t_\ell \text{ or } w_{\ell+1} = w_\ell.$$

Thus, by (5.27) the solution in (5.26) has the bottleneck $B \leq \frac{1}{2}$. □

5.6.4 Two Cyclic Processes Sharing a Resource

We now show that the just-in-time sequence obtained for $n = 2$ by the Webster's method maximizes utilization of a resource shared by two cyclic processes as well, or equivalently the minimization of bottleneck deviation (5.2) for $n = 2$ implies the maximization of utilization. To our knowledge this result is a very rare example of fair allocation of a resource leading to its best utilization.

Following Gaujal [45] we consider two cyclic processes $P1$ and $P2$ that share a common resource R. The resource can be used by at most one process at a time. Either process passes through a cycle including a single activity that does not require the resource and a single nonpreemptive activity that requires the resource. The activities are denoted by $A_{\overline{R}}$ and A_R, respectively, for the process $P1$, and by $B_{\overline{R}}$ and B_R for the process $P2$. Each activity has a fixed duration denoted by $a_{\overline{R}}, a_R, b_{\overline{R}}$ and b_R for $A_{\overline{R}}, A_R, B_{\overline{R}}$ and B_R respectively. Figure 5.5 shows a temporized free-choice Petri net modeling the two processes and the resource. The system models a typical manufacturing work-cell where two Computer Numerical Control (CNC) machine tools work independently on manufacturing two kinds of parts, A and B, but rely on a common robot R for loading and unloading the parts, see Gaujal et al. [47]. Another example is a set of jobs with time lags to be processed on a single machine, see Wenci Yu [48] for a review of problems with time lags. The set includes two types of jobs those with processing time a_R and the time lag $a_{\overline{R}}$, and those with processing time b_R and the time lag $b_{\overline{R}}$. Thus a job of the first type must wait at least $a_{\overline{R}}$ time units between its consecutive executions, and a job of the second type must wait at least $b_{\overline{R}}$ time units between its consecutive executions. The machine is the resource R shared by the two types of jobs.

Let us assume that the two processes are required to be performed at certain rates. For instance out of a given number of D parts produced by the robotic cell daily d_A should be of type A and d_B of type B. The question then is how to sequence the robot allocations so as to complete the production of D parts in the shortest possible time C. Clearly, the robot is occupied by A during $a_R d_A$ time units and by B during $b_R d_B$ time units, thus it remains idle no less than $C - a_R d_A - b_R d_B$ time

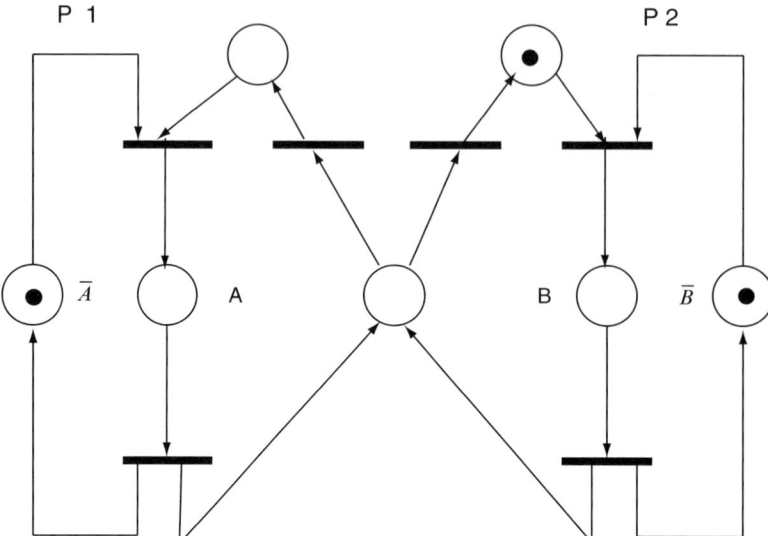

Fig. 5.5 Petri net modeling two cyclic processes A and B sharing a common resource R

units. Therefore, the minimization of C is equivalent to the maximization of the robot utilization (or the minimization of the robot idle time).

Formally, we look for a sequence S of length D over the two-letter alphabet A and B, see Fig. 5.5. The A in the sequence means that the token available in place R has been routed to place A regardless of the state of the system, that is the token has been allocated to the process $P1$, the B in the sequence means that the token in place R has been routed to place B, that is it has been allocated to the process $P2$. For any given sequence S of the resource allocations the free-choice Petri net of Fig. 5.5 can be unraveled as a decision-free Petri net, or a marked graph, where each place has exactly one input arc and exactly one output arc, see Gaujal [45] and Murata [49] for details. A circuit in a marked graph is a sequence of places and transitions $P_1 t_1 P_2 t_2 \cdots P_n$ where transition t_i is both the output transition of place P_i and the input transition of place P_{i+1}, $i = 1, \ldots, n-1$, $P_n = P_1$, and where neither a transition nor a place, except the place P_1, occurs twice. The circuit duration is the sum of durations of all the circuit places. The makespan of the marked graph for a given sequence S is the duration of the longest circuit of the marked graph, Ramamoorthy and Ho [50]. We have the following result.

Theorem 5.24. *The Webster's allocation sequences maximizes utilization of the common resource for any d_A and d_B, and independently of durations $a_{\overline{R}}$, $a_R, b_{\overline{R}}$ and b_R. Moreover, no other allocation sequence maximizes the utilization.*

Proof. The proof is based on a key result of Gaujal [45] who proves that the allocation sequence that maximizes utilization of the common resource for any d_A and d_B, and independently of durations $a_{\overline{R}}$, $a_R, b_{\overline{R}}$ and b_R is the most regular word in

$\Gamma(d_A,d_B)$. Lemmas 5.22 and 5.23 prove that the set $\Gamma(d_A,d_B)$ is exactly the set of solutions to the bottleneck deviation problem (d_A,d_B) with the bottleneck $B \leq \frac{1}{2}$, which in turn, see Sect. 5.6.2, is the same as the set of the apportionment sequences generated by the Webster's method of apportionment, for d_A and d_B. □

Gaujal [45] points out that in the case of $n \geq 3$ processes sharing a resource the optimal allocation sequence *depends* on the durations the processes require the resource for as well as on their time lags, see Exercise 5.29. The maximization of resource utilization problem then becomes *NP*-hard, see Ramamoorthy and Ho [50].

5.7 Exercises

Exercise 5.25. Show that the formulation (5.2–5.3) is equivalent to the formulation (3.37–3.40) for $G(S) = H(S)$.

Exercise 5.26. Show that the Steiner–Yeomans method is a quota-divisor method for any given T, see Józefowska et al. [26].

Exercise 5.27. Prove that the optimal bottleneck for the instance defined as follows $d_1 = 1$, $d_2 = 2$, $d_3 = 9$ and $d_i = \sum_{j=1}^{i-1} d_j = 2^{i-4} \times 12$ for $i = 4,\ldots,n$ equals $\frac{1}{2}$. Find other classes of instances with the same optimal bottleneck. Hint: See Brauner and Crama [42].

Exercise 5.28. Consider a job with processing time $a_R = 3$ and the time lag $a_{\overline{R}} = 5$, and a job with processing time $b_R = 4$ and the time lag $b_{\overline{R}} = 2$. Assume that the former is to be repeated 7 times and the latter 11 times daily. What are the optimal sequences of doing the jobs daily? What is the sequence makespan? What are the Jefferson's and Adams's sequence makespans?

Exercise 5.29. Show that the optimal allocation sequence for $n \geq 3$ depends on the timings of the n processes. Hint: See Gaujal [45].

Exercise 5.30. Show that any algorithm that produces just-in-time sequences with bottleneck $B < 1$ defines a house monotone, quota satisfying apportionment method and vice versa, that is any house monotone, quota satisfying apportionment method defines an algorithm producing just-in-time sequences with bottleneck $B < 1$.

5.8 Comments and References

The Theorem 5.1 belongs to Steiner and Yeomans [41], see also Brauner and Crama [42] for further refinements. For matching algorithms in bipartite convex graphs see Glover [51], Lipski and Preparata [52], Frederickson [53], Gallo [54], and Gabow and Tarjan [55]. The proof that the bottleneck problem is in *co-NP* based on Theorem 5.5 is given in Brauner and Crama [42]. The upper bound given in Theorem 5.6 comes from Brauner and Crama [42] and Theorem 5.7 was given by Tijdeman [43], see also Meijer [56], in 1980. For most regular words see Gaujal [45] and Vuillon [46]. The two cyclic processes sharing a common resource have been analyzed by Gaujal [45] and Gaujal et al. [47]. Murata [49] gives a good review of Petri Nets.

Chapter 6
Competition-Free Instances, The Fraenkel's Conjecture, and Optimal Admission Sequences

6.1 Introduction

A number of applications, we have seen some of them already in Chap. 3, deal with sequences over a finite alphabet where each letter of the alphabet is required to occur with a pre-specified rate r. The sequences are modeled generally as infinite however for the rational rates the sequences become cyclic and then we can limit ourselves to studying finite cycles. The sequence projection on a particular letter results in an isomorphic zero-one valued sequence with the ones in the positions occupied by the letter in the original sequence and zeros elsewhere. Thus, for each letter we can consider the zero-one valued sequences and search for an optimal one for a given letter regardless of all other letters. It turns out that the objective functions for a single letter are often minimized by sequences with the letter being in positions defined be the following formula

$$\left\lceil \frac{j}{r} - \frac{\theta}{r} \right\rceil \tag{6.1}$$

where $j = 1, 2, \ldots$ for some phase θ, which may be letter-dependent, and $0 \leq \theta < 1$. This was the case for the just-in-time sequences minimizing the total and maximum deviation, see Theorem 3.1. There, $\theta = \frac{1}{2}$. It also holds for a class of multimodular functions as proven by Hajek [6]. The multimodular functions were first studied by Hajek [6], see Sect. 6.9 for their definition, and later by Altman et al. [7] as discrete counterparts of continuos convex functions. Their most prominent application thus far is to the load balancing problem in queueing networks, where the sequences are designed to implement an admission policy, see Sect. 6.10 for details of this policy. Hajek [6], and Altman et al. [7] prove that the expected queue sizes and more generally expected travel times in queuing networks represented by stochastic event graphs are multimodular functions.

   The just-in-time sequences have the same goal of leveling the workloads throughout the supply chain. There, the admission sequence is simply the order in which the

W. Kubiak, *Proportional Optimization and Fairness,* International Series in Operations Research & Management Science 127, DOI 10.1007/978-0-387-87719-8_6,
© Springer Science+Business Media LLC 2009

models enter the assembly line. The model rates are determined by model forecasts relative to the total demand for all models.

The admission sequences defined by (6.1) spread the letter evenly throughout the sequence but do not require equal distances between any two consecutive occurrences of the letter. These distances may differ by at *most* one. Thus the sequence (6.1) is a relaxation of the exact covering sequences, also referred to as the constant gap sequences, which require equal distances between any two consecutive occurrences of the same letter. The sequences (6.1) are also known as Beatty sequences, Beatty [57], see Sect. 6.8.

Though an optimal solution for a single letter is provided by (6.1) and it can be easily calculated, the real problem begins with composing the individual optimal sequences for each letter into a sequence for the whole alphabet for clearly the positions of letters in optimal sequences for individual letters may overlap. Thus, a question of how the overlap conflicts should be resolved arises. We have seen in Chap. 3 that the conflict for just-in-time sequence can be resolved to optimality so that we can efficiently find sequences that optimize various functions of deviations – total or maximum. To our knowledge there is no such efficient, that is polynomial time, algorithm known for the multimodular functions so the problem there remains open.

An important related question also arises, namely, what are the letter rates for which the simple composition of individual letter optimal sequences defined by (6.1) leads to a feasible sequence for the whole alphabet? In other words, can we find the letter phase θ in (6.1) so that the individual letter sequences do not overlap? This simple composition is always possible for $n = 2$, we have seen this in Lemma 5.14 for one demand odd and the other even, we then have $\theta = \frac{1}{2}$ for either sequence, but it also holds for both demands being odd. Then, we can just take $\theta_1 = \frac{1}{2} + \varepsilon$ and $\theta_2 = \frac{1}{2} - \varepsilon$ for sufficiently small $\varepsilon > 0$. The case with the irrational rates is dealt with by the Beatty theorem [57], see Theorem 6.32.

However, the case $n = 2$ does not capture the complexity of the problem that remains open and leads to an intriguing and challenging conjecture. This conjecture referred to as the Fraenkel's Conjecture, see Sect. 6.8 for its details, claims that if the rates are requested to be pairwise distinct, then the simple composition without overlap is only possible if the demands are powers-of-two, thus the rates are as follows

$$\frac{1}{2^n - 1}, \frac{2}{2^n - 1}, \cdots, \frac{2^{n-1}}{2^n - 1}.$$

The conjecture was proven for $n = 3$ by Morikawa [58, 59], for $n = 4$ by Altman et al. [7], and for $n = 5$ and 6 by Tijdeman [5, 60]. As well, it was shown for the case with a letter having rate at least $\frac{2}{3}$ by Simpson [61]. Thus, the relaxation from the constant gap sequences to the sequences (6.1) admits a unique instance for distinct rates, if the conjecture holds. A well known result obtained independently by Mirsky et al. [62] shows that there are *no* constant gap sequences for pairwise distinct rates. Such can only be obtained if at least two letters have the same rates.

In this chapter we show that the competition-free instances introduced in Chap. 5 for the just-in-time sequence optimization define a special case of Fraenkel's

Conjecture that we refer to as the symmetric Fraenkel's Conjecture, see Sect. 6.2 for details. The competition-free instances admit solutions that sequence all copies in their ideal positions and thus minimize all the total deviation (3.1) and the maximum deviation (3.37) objective functions with the symmetric F_i, for $i = 1, \ldots, n$, at the same time.

We then show that the composition-free instances must be power-of-two instances for $n \geq 3$. This will be done in Sects. 6.3–6.7. We also prove in Sect. 6.8 that if the phases in (6.1) are chosen to be all equal, then the only instances for which the solutions can be *possibly* composed from optimal single letter solutions without overlapping are the power-of-two instances. All sequences obtained according to the parametric apportionment methods result in all phases being equal. However, we conjecture that the only parametric method for which the power-of-two is competition-free is the Webster's method. Finally, we discuss the multimodular functions and some of their applications in Sects. 6.9–6.11.

6.2 The Competition-Free and the Power-of-Two Instances

We now turn to the discussion of competition-free instances of the just-in-time problem, that is the total and maximum deviation minimization, that we alluded to in Chaps. 3 and 5. A *competition-free* instance $\mathbf{d} = (d_1, \ldots, d_n)$ is any instance that has all its ideal positions

$$\left\lceil \frac{2j - 1}{2r_i} \right\rceil \tag{6.2}$$

for $i = 1, \ldots, n$ and $j = 1, \ldots, d_i$ pairwise different. We show in Chap. 5 that any instance $\mathbf{d} = (d_1, d_2)$ with one demand being even and the other odd is competition-free. We now consider the competition-free instances for $n \geq 3$. We show that the move from $n = 2$ to $n \geq 3$ is rather a quantum leap that results in a unique competition-free instance for any $n \geq 3$. This unique instance is the *power-of-two instance* where $d_i = 2^{i-1}$ for $i = 1, 2, \ldots, n$.

Without loss of generality we consider the standard instances only. Recall from Chap. 1 that an instance is *standard* if $0 < d_1 \leq d_2 \leq \cdots \leq d_n$, $n \geq 2$, and $\gcd(d_1, \ldots, d_n) = 1$. Moreover, recall from Chap. 4 that there always exist cyclic optimal solutions. Without loss of generality we assume that no two demands are equal, that is $0 < d_1 < d_2 < \ldots < d_n$. Otherwise, clearly an instance would not be competition-free since all ideal positions for models with equal demands would overlap. Our ultimate goal in this chapter is to prove the following theorem:

Theorem 6.1. *For $n \geq 3$, an instance is competition-free if and only if it is power-of-two.*

We prove this theorem by showing that for the competition-free instances any demand d_i either divides the highest demand d_n or its complement $D - d_n$, this is proven in Sect. 6.4. The consequence of this is that the highest demand d_n must be even in competition-free instances, this is shown in Sect. 6.5. Our proof of Theorem 6.1 is by induction, thus Sect. 6.6 shows that the theorem holds for $n = 3$. Though we could have used the existing proof of Fraenkel's conjecture for $n = 3$, 4, 5, and 6 instead, see Sect. 6.8 for details of Fraenkel's Conjecture, we decided to prove the case $n = 3$ differently to show that this can be done without the apparatus developed for the special cases Fraenkel's Conjecture in the literature, see Tijdeman [5]. The proof is presented in Sect. 6.7 which puts all the arguments together. Before moving to the details of the proof we show that the competition-free instances attain the absolute minimum of the bottleneck deviation which no other instances are capable of attaining. This is shown in the next section.

6.3 Bottleneck of the Competition-Free Instances

We prove that the competition-free instances are exactly those that result in maximum absolute (or bottleneck) deviation being less than $\frac{1}{2}$. That is the competition-free instances satisfy the following inequality

$$B^* = \min_S \{ H(S) = \max_{i,k} |x_{ik} - kr_i| \} < \frac{1}{2} \tag{6.3}$$

subject to constraints (5.3) given in Chap. 5 and these are the only instances that do this.

The following lemma implies that the instances with bottleneck *less* than $\frac{1}{2}$ must be competition-free.

Lemma 6.2. *If $B^* < \frac{1}{2}$, then each copy is sequenced in its ideal position.*

Proof. By contradiction. Consider copy j of i sequenced in position $\lceil \frac{2j-1}{2r_i} \rceil + \Delta$, where $\Delta \geq 1$ and integer. Then $x_{i, \lceil \frac{2j-1}{2r_i} \rceil} \leq j - 1$. Moreover, $\lceil \frac{2j-1}{2r_i} \rceil r_i \geq \frac{2j-1}{2} = j - \frac{1}{2}$. Therefore, the deviation $|x_{i, \lceil \frac{2j-1}{2r_i} \rceil} - \lceil \frac{2j-1}{2r_i} \rceil r_i|$ at point $k = \lceil \frac{2j-1}{2r_i} \rceil$ is at least $\frac{1}{2}$, which contradicts $B^* < \frac{1}{2}$. Now, assume that copy j is sequenced in position $\lceil \frac{2j-1}{2r_i} \rceil - \Delta$, where again $\Delta \geq 1$ and integer. Then, $x_{i, \lceil \frac{2j-1}{2r_i} \rceil - 1} \geq j$. Moreover, $(\lceil \frac{2j-1}{2r_i} \rceil - 1)r_i \leq j - \frac{1}{2}$, which means that the deviation $|x_{i, \lceil \frac{2j-1}{2r_i} \rceil - 1} - (\lceil \frac{2j-1}{2r_i} \rceil - 1)r_i|$ is at least $\frac{1}{2}$. This again contradicts $B^* < \frac{1}{2}$ and proves the lemma.   □

Moreover, no ideal vertex of the competition-free instances is integral.

Lemma 6.3. *If $B^* < \frac{1}{2}$, then no ideal vertex is integral.*

Proof. By contradiction. Let the ideal vertex of copy j of i be integral. Then, $\lceil \frac{2j-1}{2r_i} \rceil = \frac{2j-1}{2r_i}$. By Lemma 6.2, $x_{i, \lceil \frac{2j-1}{2r_i} \rceil} = j$. Consequently, $|x_{i, \lceil \frac{2j-1}{2r_i} \rceil} - \lceil \frac{2j-1}{2r_i} \rceil r_i| = |j - (j - \frac{1}{2})| = \frac{1}{2}$, which contradicts $B^* < \frac{1}{2}$ and proves the lemma.   □

We have the following two important characteristics of the competition-free instances.

Lemma 6.4. *D must be odd for the competition-free instances.*

Proof. By contradiction. Suppose that D is even for some competition-free instance. Then, there must be a positive even number of odd demands d_i for the instance is standard. Let d_i and d_j, $i \neq j$, be two odd demands in the instance. Then, the middle copies $\frac{d_i+1}{2}$ and $\frac{d_j+1}{2}$ of i and j respectively occupy the same ideal position $\frac{D}{2}$. Thus, the instance is not competition-free which leads to a contradiction. $\square$

Similarly, we show that exactly one d_i of a competition-free instance may be odd.

Lemma 6.5. *For competition-free instances exactly one d_i, $i = 1, \ldots, n$, is odd .*

Proof. Consider i with an odd d_i. Then, $d_i = 2k + 1$ for some integer $k \geq 0$, and the ideal vertex of copy $k + 1 \leq 2k + 1$ of i is $\frac{2(k+1)-1}{2r_i} = \frac{D}{2}$. Consequently, each i with an odd d_i has one of its ideal positions at $\lceil \frac{D}{2} \rceil$, thus there is no more than one odd d_i in a competition-free instance. However, at least one d_i must be odd, otherwise the instance would not be standard since $\gcd(d_1, \ldots, d_n, D) \geq 2$. $\square$

We are ready to show that the instances with small bottleneck that meet the condition (6.3) are the same as the competition-free instances.

Theorem 6.6. *The (6.3) holds subject to constraints (5.3) for an instance $\mathbf{d} = (d_1, \ldots, d_n)$ if and only if $\mathbf{d} = (d_1, \ldots, d_n)$ is competition-free.*

Proof. Lemma 6.2 implies that all D ideal positions $\lceil \frac{2j-1}{2r_i} \rceil$ are pairwise different for $B^* < \frac{1}{2}$. Thus, the instance is competition-free. Now, by contradiction, assume that all D ideal positions $Z_j^i = \lceil \frac{2j-1}{2r_i} \rceil$ are pairwise different and $B^* \geq \frac{1}{2}$. Consider a solution that places all copies in their ideal positions. We have

$$|x_{ik} - kr_i| \geq \frac{1}{2},\qquad(6.4)$$

for some some i and position k in this solution. Without loss of generality assume that k is the *earliest* such position, that is for all i and $k' < k$ the deviation $|x_{ik'} - k'r_i|$ is less than $\frac{1}{2}$.

If k is between the ideal positions of copies j and $j+1$ of i, that is

$$Z_j^i \leq k < Z_{j+1}^i,\qquad(6.5)$$

for some copy $j = 1, \ldots, d_i - 1$ of i, then $x_{ik} = j$ and either

$$j - kr_i \geq \frac{1}{2} \quad \text{or} \quad j - kr_i \leq -\frac{1}{2}.$$

Then, however, we get either

$$Z_j^i \geq \frac{2j-1}{2r_i} \geq k \text{ or } \frac{2j+1}{2r_i} \leq Z_{j+1}^i \leq k,$$

which contradicts (6.5) as long as

$$\frac{2j-1}{2r_i} > k.$$

Thus, consider

$$\frac{2j-1}{2r_i} = k.$$

Then, for $k > 1$, we have $x_{i,k-1} = j - 1$. Thus,

$$j-1+(k-1)r_i = j - kr_i - 1 - r_i = -\frac{1}{2} - r_i < -\frac{1}{2}$$

which contradicts our assumption that k is the earliest position where (6.4) holds. However, $k > 1$. Otherwise, for $k = 1$, we have

$$r_i = \frac{1}{2},$$

thus $D = 2d_i$ is even. This leads to a contradiction since by Lemma 6.4 D must be odd for any competition-free instance.

If k is before the ideal position of the first copy of i, that is

$$k < Z_1^i, \tag{6.6}$$

then

$$|0 - kr_i| \geq \frac{1}{2}.$$

Thus

$$k \geq Z_1^i \geq \frac{1}{2r_i}, \tag{6.7}$$

holds since (6.7) contradicts (6.6). Finally, if k is after the ideal position of the last copy of i, that is

$$k > Z_{d_i}^i, \tag{6.8}$$

then

$$|d_i - kr_i| \geq \frac{1}{2}.$$

Thus

$$Z_{d_i}^i \geq \frac{2d_i - 1}{2r_i} \geq k, \tag{6.9}$$

holds since (6.9) contradicts (6.8). Therefore, we reach a contradiction for any possible case for k, which proves that if all D ideal positions $Z_j^i = \lceil \frac{2j-1}{2r_i} \rceil$ are pairwise different, then $B^* < \frac{1}{2}$. $\square$

For $n = 2$, the instance is competition-free if and only if one of demands d_1 or d_2 is odd and the other even which follows from Lemmas 5.13 and 5.15. Moreover, by Theorem 5.10

$$B^* < \frac{1}{2}$$

if and only if one of demands d_1 or d_2 is odd and the other even. These two also imply the following theorem.

Theorem 6.7. $B^* < \frac{1}{2}$ *if and only if one of demands d_1 or d_2 is odd and the other even.*

6.4 Polygons of the Competition-Free Instances

We now show further characteristics of the competition-free instances for $n \geq 3$. The main one is that any d_i, for $i = 1, \ldots, n-1$, divides either the highest demand d_n or its complement $D - d_n$ in a competition-free instance. Our proof is geometric, it relies on natural symmetries embedded in n regular polygons inscribed in a circle of circumference D and defined by the ideal vertices

$$\frac{2j-1}{2r_i}, \tag{6.10}$$

for $i = 1, \ldots, n$ and $j = 1, \ldots, d_i$.

We begin with introducing the essential definitions and terminology. Take a circle of circumference D, that is of the diameter $\frac{D}{\pi}$, and wrap the interval $[0, D]$ around it so that the ends 0 and D meet at the *North Pole*, the highest point 0 in Fig. 6.1. Call the resulting circle a *D-circle* . The circle is split into D unit arcs $[0, 1], \ldots, [D-1, 0]$ by the *unit marks* $0, \ldots, D-1$. The points $j + \frac{1}{2}$, for $j = 0, \ldots, D-1$ will be called the *half-points*. The lowest point on the D-circle opposite the North Pole will be called the *South Pole*. The South Pole is always at $\frac{D}{2}$, equal $7\frac{1}{2}$ in Fig. 6.1, since D must be odd for any competition-free instance by Lemma 6.9.

The ideal vertices $\frac{2j-1}{2r_i}$, $j = 1, \ldots, d_i$, of i define a polygon $P_i^{\frac{1}{2}}$ inscribed in D-circle, see Fig. 6.2. The superscript $\frac{1}{2}$ in our notation for the polygons will become clear later in the chapter, namely in Sect. 6.8. It suffices to say now that the polygons correspond to the Webster's parametric method of apportionment that uses parameter $\delta = \frac{1}{2}$ in calculating apportionments. The polygon $P_i^{\frac{1}{2}}$ is a regular d_i-gone with the side of length $\frac{D}{\pi} \sin \frac{180°}{d_i}$ as observed in the following lemma.

Lemma 6.8. *The ideal vertices for any i are equally spaced in $[0, D]$ with the first vertex at $\frac{D}{2d_i} = \frac{1}{2r_i}$, each next at a distance $\frac{D}{d_i} = \frac{1}{r_i}$ from its immediate predecessor, and the last one at $\frac{D}{2d_i} = \frac{1}{2r_i}$ from D.*

Fig. 6.1 D-circle, $D = 15$

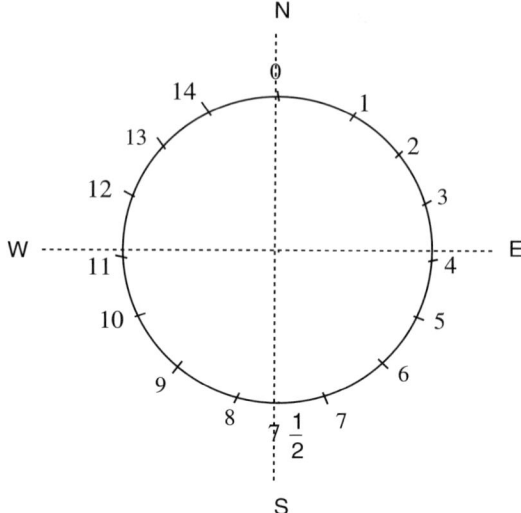

Fig. 6.2 The d-gone $P^{\frac{1}{2}}$ inscribed in D-circle, $D = 15$ and $d = 8$

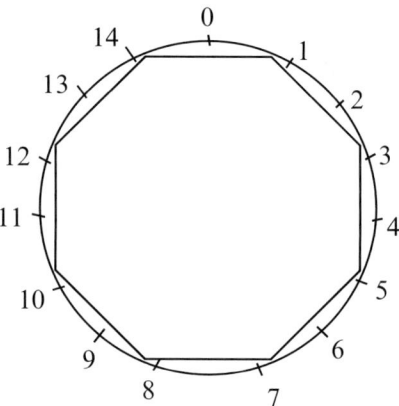

Proof. Follows immediately from the definition of ideal vertices, see (6.10). Namely, the distance between vertex $j+1$ and vertex j equals $\frac{2(j+1)-1}{2r_i} - \frac{2j-1}{2r_i} = \frac{D}{d_i}$, for $j = 1,\ldots,d_i - 1$. The first vertex is at $\frac{2-1}{2r_i} = \frac{D}{2d_i}$, and the last at $\frac{2d_i-1}{2r_i} = D - \frac{D}{2d_i}$. □

The vertices of different polygons never fall in the same unit arcs for the competition-free instances.

Lemma 6.9. *For the competition-free instances exactly one ideal vertex falls in the open interval $(i-1,i)$, for $i = 1,\ldots,D$.*

Proof. For competition-free instances all D ideal positions $\lceil \frac{2j-1}{2r_i} \rceil$ for $i = 1,\ldots,n$ and $j = 1,\ldots,d_i$ are pairwise different. By Lemma 6.3 and Theorem 6.6, all ideal vertices $\frac{2j-1}{2r_i}$, $i = 1,\ldots,n$ and $j = 1,\ldots,d_i$ are fractional which proves the lemma. $\square$

Finally, we have the following useful observation.

Lemma 6.10. *For the competition-free instances,* $1 < \frac{D}{d_n} < 2$ *and* $\frac{D}{d_i} > 2$ *for* $i = 1,\ldots,n-1$.

Proof. Obviously, $1 < \frac{D}{d_n}$ since $n \geq 3$ and all demands are positive integers. We show that $\frac{D}{d_n} < 2$ by contradiction. Suppose that $\frac{D}{d_n} \geq 2$, then $\frac{D}{2d_n} \geq 1$ and consequently by Lemmas 6.8 and 6.9 the first and last ideal vertices of product n do not fall inside $[0,1]$ and $[D-1,D]$ respectively. However, $\frac{D}{d_i} \geq \frac{D}{d_n}$, for $i = 1,\ldots,n-1$ and standard instances, thus, *no* ideal vertex falls in either $[0,1]$ or $[D-1,D]$ which contradicts Lemma 6.9 and proves that $\frac{D}{d_n} < 2$. We now show that $\frac{D}{d_i} > 2$ for $i = 1,\ldots,n-1$. By contradiction, suppose that $\frac{D}{d_i} \leq 2$ for some $i = 1,\ldots,n-1$. Then, $\frac{D}{2d_i} \leq 1$ and consequently, i shares both $[0,1]$ and $[D-1,D]$ with n, which again contradicts Lemma 6.9 and proves that $\frac{D}{d_i} > 2$ for $i = 1,\ldots,n-1$. $\square$

For each polygon $P_i^{\frac{1}{2}}$ with $d_i > 1$, let us call its two closest to the North Pole vertices (one to the left and one to the right of the North Pole) the *North Pole vertices* of $P_i^{\frac{1}{2}}$. By Lemma 6.8, if $d_i > 1$, then the side that connects the North Pole vertices of $P_i^{\frac{1}{2}}$ is *horizontal*. The arc between the *North Pole vertices* will be referred to as the *North arc* of $P_i^{\frac{1}{2}}$.

For an even d_i, by the symmetry with respect to the WE axis, there are two vertices of $P_i^{\frac{1}{2}}$ at the bottom of the D-circle opposite the North Pole vertices of $P_i^{\frac{1}{2}}$, we shall call them the *South Pole vertices* of $P_i^{\frac{1}{2}}$. Obviously, the side connecting them is horizontal. The arc between the *South Pole vertices* will be referred to as the *South arc* of $P_i^{\frac{1}{2}}$. Notice that the North Pole vertices are the same as the South Pole vertices for $d_i = 2$.

We are now ready to prove the main result of this section.

Lemma 6.11. *Consider a d_i-gone $P_i^{\frac{1}{2}}$ with $d_i > 1$, $i = 1,\ldots,n-1$. Then, either*

$$d_n = \alpha_i d_i$$

for some even α_i or

$$D - d_n - d_i = \beta_i d_i$$

for some even $\beta_i \geq 2$.

Proof. Consider an arbitrary arc $\overarc{XY}$ of D-circle between two adjacent vertices X and Y of $P_i^{\frac{1}{2}}$, $d_i > 1$, $i = 1,\ldots,n-1$, see Fig. 6.3. By Lemma 6.9, X does not coincide

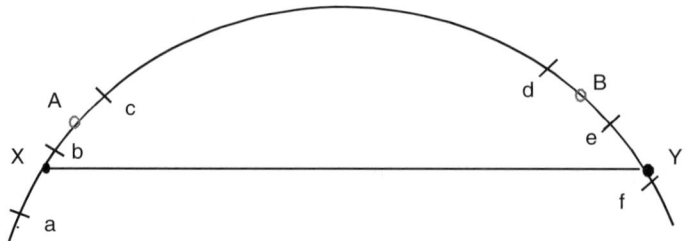

Fig. 6.3 An arc $\widehat{XY}$ of D-circle bounded by the adjacent vertices X and Y of d_i-gone $P_i^{\frac{1}{2}}$

with unit marks a or b, and Y does not coincide with unit marks e or f. By Lemma 6.10, the length $\overline{be}$ of the arc $\widehat{be}$ is at least 1. Also by Lemma 6.10, a vertex, say A, of $P_n^{\frac{1}{2}}$ must be somewhere between unit marks b and c, and another vertex, say B, of $P_n^{\frac{1}{2}}$ must be somewhere between unit marks d and e, see Fig. 6.3. Observe, that $\overline{be} = 1$ implies $A = B, c = e$, and $d = b$.

Let α be the number of $P_n^{\frac{1}{2}}$ vertices between b and e. We have $\alpha \geq 1$ and

$$(\alpha - 1) \times \frac{1}{r_n} < \overline{be}.$$

On the other hand,

$$\alpha \times \frac{1}{r_n} > \overline{be}$$

since $\frac{1}{r_n} > 1 + \max\{\overline{bA}, \overline{Be}\} > \overline{bA} + \overline{Be}$, where $\overline{bA}$ and $\overline{Be}$ are the lengths of the arcs $\widehat{bA}$ and $\widehat{Be}$ respectively. Therefore, the α satisfies the following inequalities

$$r_n \overline{be} < \alpha < r_n \overline{be} + 1 \tag{6.11}$$

and thus it is a unique integer. However, $\overline{be}$ can be either $\left\lfloor \frac{1}{r_i} \right\rfloor$ or $\left\lfloor \frac{1}{r_i} \right\rfloor - 1$. Let β be the unique α satisfying the inequalities (6.11) for $\overline{be} = \left\lfloor \frac{1}{r_i} \right\rfloor$ and γ be the unique α satisfying the inequalities (6.11) for $\overline{be} = \left\lfloor \frac{1}{r_i} \right\rfloor - 1$. We have

$$0 \leq \beta - \gamma < 1 + r_n. \tag{6.12}$$

Thus, for integer β and γ, the inequalities (6.12) imply either $\beta = \gamma$ or $\beta = \gamma + 1$. In the former case, we have $d_n = \beta d_i$, and since there is an even number of vertices of $P_n^{\frac{1}{2}}$ on the North arc of $P_i^{\frac{1}{2}}$, then β is even. Thus, the lemma holds. In the latter case, there are $d_i > a \geq 1$, $\widehat{XY}$-arcs between adjacent vertices of $P_i^{\frac{1}{2}}$ with $\overline{be} = \left\lfloor \frac{1}{r_i} \right\rfloor$ and $d_i - a$, $\widehat{XY}$-arcs between adjacent vertices of $P_i^{\frac{1}{2}}$ with $\overline{be} = \left\lfloor \frac{1}{r_i} \right\rfloor - 1$. The former have $\gamma + 1$ vertices of $P_n^{\frac{1}{2}}$ each, the latter have γ vertices of $P_n^{\frac{1}{2}}$ each. Thus, there are

$\left\lfloor \dfrac{1}{r_i} \right\rfloor - \gamma - 1$ vertices of polygons $P_k^{\frac{1}{2}}$ for $k \neq i$ and $k = 1, \ldots, n-1$ on each arc $\widehat{XY}$. Consequently,

$$D - d_n - d_i = (\left\lfloor \dfrac{1}{r_i} \right\rfloor - \gamma - 1)d_i = \beta_i d_i. \tag{6.13}$$

This completes the proof since by (6.11) and the definition of γ we have

$$\gamma - 1 < r_n(\left\lfloor \dfrac{1}{r_i} \right\rfloor - 1) < \left\lfloor \dfrac{1}{r_i} \right\rfloor - 1 < \left\lfloor \dfrac{1}{r_i} \right\rfloor$$

and thus $\beta_i \geq 2$. Finally, observe that since there is an even number of vertices of polygons $P_k^{\frac{1}{2}}$ for $k \neq i$ and $k = 1, \ldots, n-1$ on the North arc of $P_i^{\frac{1}{2}}$, then

$$\beta_i = \left\lfloor \dfrac{1}{r_i} \right\rfloor - \gamma - 1$$

is even.   □

6.5 Characteristics of Competition-Free Instances

The consequence of Lemma 6.11 and the well-known result of Mirsky et al. [62] concerning the exact covering sequences is that any competition-free instance must have d_n even. The proof of this claim will be the main result of this section.

We now recall an important result obtained independently by Mirsky, Newman, Davenport and Rado and concerning the exact covering sequences. For nonnegative integer numbers a and b denote by (a,b) the sequence $\{jb + a : j = 0,1,2,\ldots\}$. A finite set $\{(a_i,b_i) : 1 \leq i \leq n\}$ is called an *exact covering sequence* if every nonnegative integer occurs in exactly one (a_i,b_i). We have the following result.

Lemma 6.12. *If a set of pairs (a_i,b_i) for $i = 1,\ldots,n$ is an exact covering sequence, then there are k and l such that $k \neq l$ and $b_k = b_l = \max_i\{b_i\}$*

Proof. Suppose (a_i,b_i) for $i = 1,\ldots,n$ is an exact covering sequence. Then in the series

$$\sum_{i=1}^{n} \sum_{k \geq 0} x^{a_i + kb_i}$$

the x^n for each nonnegative integer n occurs exactly once. Therefore, we have

$$\sum_{i=1}^{n} \sum_{k \geq 0} x^{a_i + kb_i} = \sum_{i=1}^{n} \frac{x^{a_i}}{1 - x^{b_i}} = \frac{1}{1 - x}. \tag{6.14}$$

Let $\omega = e^{2i\pi/r}$ be a primitive rth root of unity for some integer $r > 1$, multiply both sides of (6.14) by $\omega - x$ to get

$$\sum_{i=1}^{n} \frac{(\omega - x)x^{a_i}}{1 - x^{b_i}} = \frac{\omega - x}{1 - x}.$$

Thus

$$\sum_{r \text{ divides } b_i} \frac{(\omega - x)x^{a_i}}{1 - x^{b_i}} + \sum_{r \text{ does not divide } b_i} \frac{(\omega - x)x^{a_i}}{1 - x^{b_i}} = \frac{\omega - x}{1 - x}.$$

By letting $x \to \omega$ we get

$$\sum_{r \text{ does not divide } b_i} \frac{(\omega - x)x^{a_i}}{1 - x^{b_i}} \to 0 \text{ and } \frac{\omega - x}{1 - x} \to 0$$

and

$$\sum_{r \text{ divides } b_i} \frac{(\omega - x)x^{a_i}}{1 - x^{b_i}} \to \sum_{r \text{ divides } b_i} \frac{-\omega^{a_i}}{-b_i \omega^{b_i-1}}.$$

Thus,

$$\sum_{r \text{ divides } b_i} \frac{-\omega^{a_i}}{-b_i \omega^{b_i-1}} = \sum_{r \text{ divides } b_i} \frac{\omega^{a_i+1}}{b_i \omega^{b_i}} = 0. \tag{6.15}$$

By definition $\omega^{b_i} = 1$ as long as r divides b_i, thus (6.15) becomes

$$\sum_{r \text{ divides } b_i} \frac{\omega^{a_i}}{b_i} = 0.$$

If we take $r = \max\{b_i\}$, then

$$\sum_{r \text{ divides } b_i} \frac{\omega^{a_i}}{b_i} = \sum_{r = b_i} \frac{\omega^{a_i}}{r} = 0.$$

However, $\omega \neq 0$ and thus the equation

$$\sum_{r = b_i} \omega^{a_i} = 0$$

can only hold if there are at least two different i and j such that $b_i = b_j = r$. $\quad\square$

We are now ready to show the main result of this section.

Lemma 6.13. *For the competition-free instances d_n must be even.*

Proof. By contradiction. Suppose that d_n is odd. Then, by Lemma 6.5 all d_i are even for $i = 1, \ldots, n-1$. Therefore, by Lemma 6.11, there are even $\lambda_1, \ldots, \lambda_{n-1}$ such that

$$D - d_n = (\lambda_i + 1)d_i, \tag{6.16}$$

for $i = 1, \ldots, n-1$. Consider the vertices of polygons $P_1^{\frac{1}{2}}, \ldots, P_{n-1}^{\frac{1}{2}}$. By (6.16) they are equal

$$\frac{(2j-1)(\lambda_i+1)D}{2(D-d_n)} \tag{6.17}$$

for $i = 1, \ldots, n-1$, and the vertices of polygon $P_n^{\frac{1}{2}}$ are equal

$$\frac{(2j-1)D}{2d_n}. \tag{6.18}$$

The polygon $\overline{P}_n^{\frac{1}{2}}$ with its vertices equal

$$\frac{(2j-1)D}{2(D-d_n)}, \tag{6.19}$$

has these vertices in different unit arcs than $P_n^{\frac{1}{2}}$ which follows from Theorems 6.6 and 6.7 since d_n is odd and $D - d_n$ is even. Thus, the set of unit arcs with the vertices of polygons $P_1^{\frac{1}{2}}, \ldots, P_{n-1}^{\frac{1}{2}}$ given in (6.17) coincides with the set of unit arcs with the vertices of the polygon $\overline{P}_n^{\frac{1}{2}}$ in (6.19). Therefore, the sets $\{(2j-1)(\lambda_i+1) : j = 1, 2, \ldots\}$ for $i = 1, \ldots, n-1$ are disjoint and their union covers the set $\{(2j-1) : j = 1, 2, \ldots\}$ of odd positive integers. Thus, the sets $\{(\lambda_i+1)j - \frac{\lambda_i}{2} : j = 1, 2, \ldots\}$ for $i = 1, \ldots, n$ are disjoint and their union covers $\mathbb{N}$. Then, however, by the result of Mirsky, Newman, Davenport and Rado given in Lemma 6.12, we must have

$$\lambda_l = \lambda_k$$

for some $l \neq k$. and consequently

$$d_l = d_k$$

for some $l \neq k$ by (6.16), which leads to a contradiction. This proves that d_n must be even in a competition-free instance. $\square$

Finally, we have the following observation.

Lemma 6.14. *For the competition-free instances, if d_{n-1} is even, then $\alpha_{n-1} = 2$.*

Proof. By Lemma 6.13 d_n must be even in a competition-free instance. Thus, if d_{n-1} is even as well, then we have $d_n = \alpha_{n-1}d_{n-1}$ for some $\alpha_{n-1} \geq 2$ by Lemma 6.11. We have $\frac{D}{d_n}(\alpha_{n-1} - 1) < \alpha_{n-1}$ since only vertices of $P_n^{\frac{1}{2}}$ may occur between the North Pole vertices of $P_{n-1}^{\frac{1}{2}}$, and by Lemma 6.9 each of them occupies a different unit arc, hence $\frac{D}{d_n} < \frac{\alpha_{n-1}}{\alpha_{n-1}-1}$. Let us now consider the South Pole vertices of an even d_{n-1}-gone $P_{n-1}^{\frac{1}{2}}$. Again, the distance between the first and the last of these vertices of $P_n^{\frac{1}{2}}$ is $\frac{D}{d_n}(\alpha_{n-1} - 1)$. Extend these sequence of vertices by adding one vertex of $P_n^{\frac{1}{2}}$ to each end. The distance between the new end vertices then becomes $\frac{D}{d_n}(\alpha_{n-1} + 1)$. On the other hand there are at least $\alpha_{n-1} + 3$ unit arcs between these two end vertices:

α_{n-1} occupied by the vertices of $P_n^{\frac{1}{2}}$, two occupied by the vertices of $P_{n-1}^{\frac{1}{2}}$, and one occupied by an odd d-gone vertex, see Lemma 6.5. The d-gone is neither $P_n^{\frac{1}{2}}$ nor $P_{n-1}^{\frac{1}{2}}$ since both d_n and d_{n-1} are even. Consequently, $\frac{D}{d_n}(\alpha_{n-1}+1) > \alpha_{n-1}+3$, and thus $\frac{\alpha_{n-1}+3}{\alpha_{n-1}+1} < \frac{D}{d_n}$. Therefore, we have shown that $\frac{\alpha_{n-1}+3}{\alpha_{n-1}+1} < \frac{D}{d_n} < \frac{\alpha_{n-1}}{\alpha_{n-1}-1}$, and thus, $\frac{2}{\alpha_{n-1}+1} < \frac{1}{\alpha_{n-1}-1}$. This implies, $\alpha_{n-1} < 3$, hence $\alpha_{n-1} = 2$, which proves the lemma. $\square$

6.6 The Competition-Free Instances for $n = 3$

This section shows that there is only one competition-free instance for $n = 3$. This instance is $d_1 = 1, d_2 = 2, d_3 = 4$. This is not a straightforward extension of the $n = 2$ case where the number of instances with small deviations was shown infinite. The case $n = 3$ needs to be considered separately since our proof of Theorem 6.1 will be done by induction on n beginning with $n = 3$. We begin by showing the following.

Lemma 6.15. *We have either $d_1 = 1, d_2 = \alpha, d_3 = 2\alpha$ for some even α or $d_3 = \beta d_1 d_2$, where $d_1 > 1$, and integer $\beta \geq 1$.*

Proof. If the unique odd d_i, where i is either 1 or 2, does not divide d_3, then by Lemma 6.11

$$d_{1+(i \bmod 2)} = D - d_3 - d_i = \alpha_i d_i.$$

Since $d_{1+(i \bmod 2)}$ is even then again by Lemma 6.11

$$d_3 = \alpha_{1+(i \bmod 2)} d_{1+(i \bmod 2)} = \alpha_{1+(i \bmod 2)} \alpha_i d_i.$$

Thus, $d_i \leq \gcd(d_i, d_{1+(i \bmod 2)}, d_3) = 1$, and the lemma follows from Lemma 6.14.

Now, let the odd d_i, where i is either 1 or 2, divides d_3. Then by Lemma 6.11

$$d_3 = \alpha_2 d_2 \text{ and } d_3 = \alpha_1 d_1.$$

Thus,

$$d_2 = \frac{\alpha_1 d_1}{\alpha_2}$$

which means that α_2 divides $\alpha_1 d_1$. If α_2 and d_1 are relatively prime, then α_2 divides α_1 and thus $d_1 \leq \gcd(d_1, d_2, d_3) = 1$, and again and the lemma follows from Lemma 6.14. Otherwise, $d_1 = kx$ and $\alpha_2 = lx$, for some integer $x \geq 2$, and k and l being relatively prime. Then

$$d_3 = kx\alpha_1,$$
$$d_1 = kx$$

and

$$d_2 = \frac{\alpha_1 k}{l}.$$

Thus, l divides α_1, that is

$$\alpha_1 = ml \text{ and } d_2 = mk.$$

Consequently, $k \le \gcd(d_1, d_2, d_3) = 1$ and

$$d_3 = xml$$
$$d_1 = x$$
$$d_2 = m.$$

This ends the proof. $\square$

For the former case in Lemma 6.15 we have.

Lemma 6.16. *The instances with $d_1 = 1, d_2 = \alpha, d_3 = 2\alpha$, where $\alpha \ge 4$ and even, are not competition-free.*

Proof. Consider copy $\alpha - k$ of 2 and copy $2(\alpha - k)$ of 3, where $0 \le k \le \alpha - 1$. The ideal vertex of the former is

$$\frac{(2\alpha - 2k - 1)}{2\alpha}(3\alpha + 1) = (3\alpha - k) - \frac{1}{2} - \frac{k}{\alpha} - \frac{1}{2\alpha} \tag{6.20}$$

and the ideal vertex of the latter equals

$$(3\alpha - k) + \frac{1}{4} - \frac{k}{\alpha} - \frac{1}{4\alpha}. \tag{6.21}$$

Let us chose k so that

$$\frac{\alpha}{4} < k \le \frac{\alpha}{3} < \frac{\alpha}{2},$$

which is possible for $\alpha \ge 6$. Then

$$-1 < \frac{1}{4} - \frac{k}{\alpha} - \frac{1}{4\alpha} < 0$$

and

$$-1 < -\frac{1}{2} - \frac{k}{\alpha} - \frac{1}{2\alpha} < 0.$$

Therefore, the two vertices fall in the same unit interval $[(3\alpha - k) - 1, 3\alpha - k]$ and thus the instance is not competition free. For $\alpha = 4$ we easily check that (6.20) for $k = 1$ equals

$$(3\alpha - k) - \frac{1}{2} - \frac{k}{\alpha} - \frac{1}{2\alpha} = (3\alpha - 1) - \frac{7}{8}$$

whereas (6.21)

$$(3\alpha - k) + \frac{1}{4} - \frac{k}{\alpha} - \frac{1}{4\alpha} = (3\alpha - 1) - \frac{1}{16}.$$

Thus again the two fall in the same unit interval $[(3\alpha - k) - 1, 3\alpha - k]$ and thus again the instance is not competition free.   □

For the latter case in Lemma 6.15 we have.

Lemma 6.17. *The instances with $d_3 = \beta d_1 d_2$, where $d_1 > 1$, and integer $\beta \geq 1$, are not competition-free.*

Proof. If βd_1 is odd, then d_2 must be even by Lemma 6.5. Then, however by Lemma 6.14, $\beta d_1 = 2$ and we get a contradiction. Therefore, βd_1 must be even. If at the same time d_2 is even, then again by Lemma 6.14, $\beta d_1 = 2$ and thus $d_1 = 2$ which leads to a contradiction since then all three, d_1, d_2 and d_3 are even and thus $\gcd(d_1, d_2, d_3) \geq 2$. Hence d_2 must be odd. Now, consider

$$l = \frac{(2j-1)d_3}{2d_1} = \frac{(2j-1)\beta d_2}{2},$$

where $j = 1, \ldots, d_1$. If β is odd, then

$$k = l + \frac{1}{2}$$

is a positive integer less than d_3, and the ideal vertex for copy k of 3 is then

$$\frac{(2k-1)D}{2d_3} = \frac{lD}{d_3} = \frac{(2j-1)d_3}{2d_1} \times \frac{D}{d_3} = \frac{(2j-1)D}{2d_1}$$

the same as the ideal vertex for copy j of 1, thus the instance is not competition-free.

Finally, consider the case of even β. Suppose that $\left\lfloor \frac{d_2}{d_1} \right\rfloor$ is even, then there are a sides of d_1-gone with the odd number $\left\lceil \frac{d_2}{d_1} \right\rceil$ of d_2-gone vertices between the vertices of each of these sides, and by symmetry with respect to the NS axis the a is even. Observe that by symmetry with respect to the NS axis, there is an even number $\left\lfloor \frac{d_2}{d_1} \right\rfloor$ of d_2-gone vertices between the North Pole vertices of d_1-gone. Therefore, the remaining $(d_1 - a)$ sides of d_1-gone are with the even $\left\lfloor \frac{d_2}{d_1} \right\rfloor$ number of d_2-gone vertices between the vertices of each of these sides. Then, however

$$d_2 = a \left\lceil \frac{d_2}{d_1} \right\rceil + (d_1 - a) \left\lfloor \frac{d_2}{d_1} \right\rfloor$$

which is a contradiction since the right hand side is even and d_2 is odd. Thus, $\left\lfloor \frac{d_2}{d_1} \right\rfloor$ must be odd. Then, consider integers

$$l = \frac{\beta k d_1}{2} \text{ and } j = \frac{k+1}{2},$$

where $k = 1, \ldots, d_2$ and odd. The ideal vertex for copy j of 2 is

$$\frac{kD}{2d_2} = \frac{k\beta d_1}{2} + \left\lfloor \frac{k}{2} \right\rfloor + \frac{1}{2} + \frac{kd_1}{2d_2}$$

and the ideal vertex for copy l of 3 is

$$\frac{kD}{2d_2} - \frac{D}{2d_3} = \frac{k\beta d_1}{2} + \left\lfloor \frac{k}{2} \right\rfloor + \frac{kd_1}{2d_2} - \frac{d_1 + d_2}{2d_3}.$$

For $k = \left\lfloor \frac{d_2}{d_1} \right\rfloor$, we have

$$0 < \frac{1}{2} + \frac{kd_1}{2d_2} < 1.$$

Note that $\left\lfloor \frac{d_2}{d_1} \right\rfloor < \frac{d_2}{d_1}$ since odd d_2 is not divisible by even d_1. Thus, it remains to show that

$$0 < \frac{kd_1}{2d_2} - \frac{d_1 + d_2}{2d_3}$$

for the k. This is equivalent to showing that

$$1 + \frac{d_2}{d_1} < \left\lfloor \frac{d_2}{d_1} \right\rfloor \beta d_1$$

or

$$\frac{3}{2 \left\lfloor \frac{d_2}{d_1} \right\rfloor} + 1 \le \beta d_1$$

which holds since $\left\lfloor \frac{d_2}{d_1} \right\rfloor \ge 1$ and $\beta d_1 \ge 4$. Thus, ideal vertex for copy j of 2 and the ideal vertex for copy l of 3 both fall in the same interval $[\frac{k\beta d_1}{2} + \lfloor \frac{k}{2} \rfloor, \frac{k\beta d_1}{2} + \lfloor \frac{k}{2} \rfloor + 1]$ and the instance is not competition-free. This ends the proof. $\quad\square$

The main result of this section follows.

Theorem 6.18. *The optimal solution to (6.3) for a standard instance* $\mathbf{d} = (d_1, d_2, d_3)$ *has value* $B^* < \frac{1}{2}$ *if and only if* $d_1 = 1, d_2 = 2$ *and* $d_3 = 4$ *and* $B^* = \frac{3}{7}$.

Proof. Follows immediately from Theorem 6.21 and Lemmas 6.15–6.17. $\quad\square$

6.7 Putting it Together

We are now ready to prove Theorem 6.1. The induction is based on the following key lemma.

Lemma 6.19. *If an instance* $(d_1, \ldots, d_n)$ *is competition-free, then so is the instance* $(d_1, \ldots, d_{n-1})$, $n \ge 2$.

Proof. We have two cases to consider. First, suppose that the unique odd d_i in the competition-free instance $(d_1, \ldots, d_n)$ does not divide d_n. Consider the instance $(d_1, \ldots, d_{n-1})$ and

$$D' = \sum_{k=1}^{n-1} d_k = D - d_n.$$

We then have the ideal vertex of copy $j = 1, \ldots, d_i$ of i at

$$\frac{2j-1}{2r'_i} = \frac{(2j-1)D'}{2d_i} = \frac{(2j-1)(D-d_n)}{2d_i}. \tag{6.22}$$

By Lemma 6.11

$$D' = D - d_n = (\beta_i + 1)d_i \tag{6.23}$$

where $\beta_i + 1$ is odd. Thus from (6.22) and (6.23) we get

$$\frac{2j-1}{2r'_i} = \frac{(2j-1)(\beta_i+1)}{2}. \tag{6.24}$$

Since $(2j-1)(\beta_i+1)$ is odd, then the copies $j = 1, \ldots, d_i$ of i have all their ideal vertices at the half-points. The distance between any two adjacent vertices of i is thus $\beta_i + 1$, and the first vertex of i is at the half-point $\frac{\beta_i+1}{2}$. Moreover, for any $k \neq i$ and $k = 1, \ldots, n-1$, its ideal vertices are at

$$\frac{2j-1}{2r'_k} = \frac{2j-1}{2r_k} - (2j-1)\frac{\alpha_k}{2} \tag{6.25}$$

where $j = 1, \ldots, d_k$. Thus, any ideal vertex of k advances to the left by an integer $(2j-1)\frac{\alpha_k}{2}$, since α_k is even by Lemma 6.11, from its original vertex $\frac{2j-1}{2r_k}$. It remains to show that each of the vertices in (6.24) and (6.25) occupies a different unit interval $(l-1, l)$, $l = 1, \ldots, D'$, that is there is exactly one ideal vertex of $(d_1, \ldots, d_{n-1})$ that falls into $(l-1, l)$, and thus the instance is competition-free. The proof uses the D-circle introduced in Sect. 6.4. Let C_D be the D-circle for the competition-free instance $(d_1, \ldots, d_{n-1}, d_n)$. Delete all unit arcs from C_D with ideal vertices of d_n-gone to obtain a circle C. In this circle, each unit arc is occupied by exactly one vertex since it is the case in the original circle C_D. Moreover, by Lemma 6.11, exactly β_i unit arcs with the vertices of d_k-gones, $k \neq i$ and $k = 1, \ldots, n-1$, remain between any two adjacent vertices of d_i-gone after the deletion. Hence, the d_i-gone vertices are in the unit arcs

$$\left[\left\lfloor \frac{(2j-1)(\beta_i+1)}{2} \right\rfloor, \left\lceil \frac{(2j-1)(\beta_i+1)}{2} \right\rceil \right]. \tag{6.26}$$

of C. Observe, however, that these vertices may not necessarily coincide with the half-points in (6.24) Moreover, Lemma 6.11 guarantees that the removal of the unit arcs with the ideal vertices of d_n-gone from C_D advances ideal vertex j of $k \neq i$ and $k = 1, \ldots, n-1$ by

$$j\alpha_k - \frac{\alpha_k}{2}$$

with respect to the original position

$$\left\lceil \frac{2j-1}{2r_k} \right\rceil,$$

thus the vertices of all d_k-gones, $k \neq i$ and $k = 1,\ldots,n-1$ coincide with those in (6.25). Finally, the vertices of the d_i-gone in the C-circle can be moved to the half-points inside of their unit arcs (6.26) which results into a $C_{D'}$-circle with the vertices define by (6.24) and (6.25) for the instance $(d_1,\ldots,d_{n-1})$. Thus the instance $(d_1,\ldots,d_{n-1})$ is competition-free.

Now assume that the unique odd d_i in the instance $(d_1,\ldots,d_n)$ divides d_n We then have $\frac{D}{d_i} = \alpha_i + \frac{D'}{d_i}$, for $i = 1,\ldots,n-1$. Again, take the D-circle for the original instance and delete from it all unit arcs with the ideal vertices of the d_n-gone inside. By Lemma 6.11, exactly α_k unit arcs are deleted between any two adjacent vertices of d_i-gone for $i = 1,\ldots,n-1$. Consequently, the distance between any two adjacent vertices of d_k-gone becomes $\frac{D'}{d_k}$, and we can easily obtain a D'-circle for the instance $(d_1,\ldots,d_{n-1})$. with exactly one ideal vertex in each unit arc. Thus, the instance $(d_1,\ldots,d_{n-1})$ is competition-free. $\square$

This lemma is key in the following theorem.

Theorem 6.20. *If an instance $(d_1,\ldots,d_n)$ is competition-free, then $d_i = 2^{i-1}$ for $i = 1,\ldots,n$, $n \geq 3$.*

Proof. By induction on $n \geq 3$. The theorem holds for $n = 3$ by Theorem 6.18. Let us assume that the theorem holds for $n = k \geq 3$. We prove that it also holds for $n = k+1$. Consider a competition-free instance $(d_1,\ldots,d_n)$. By Lemma 6.19, the instance $(d_1,\ldots,d_{n-1})$ is competition-free. Thus, by the induction assumption $d_i = 2^{i-1}$ for $i = 1,\ldots,n-1$. Therefore, d_{n-1} is even, and by Lemma 6.14, $\alpha_{n-1} = 2$. Therefore, by Lemma 6.11 $d_n = 2d_{n-1} = 2^{n-1}$, hence $d_i = 2^{i-1}$ for $i = 1,\ldots,n$, and the theorem holds for $n = k+1$. This completes the proof since by induction the theorem holds for any $n \geq 3$. $\square$

Thus, it remains to show that the opposite also holds.

Theorem 6.21. *The instance $(d_1,\ldots,d_n)$ with $d_i = 2^{i-1}$ for $i = 1,\ldots,n$ has $B^* = \frac{2^{n-1}-1}{2^n-1} < \frac{1}{2}$ and so is competition-free.*

Proof. Consider the ideal positions for the instance $(d_1,\ldots,d_n)$ with $d_i = 2^{i-1}$ for $i = 1,\ldots,n$. We have the following ideal position for copy $j = 1,\ldots,2^{i-1}$ of i,

$$Z_j^i = \left\lceil \frac{2j-1}{2r_i} \right\rceil = (2j-1)2^{n-i} - \left\lfloor \frac{2j-1}{2^i} \right\rfloor = (2j-1)2^{n-i},$$

and thus observe that all ideal positions for this instance are pairwise different. Now, let us consider the earliest $E(i,j) = \left\lceil \frac{j-B}{r_i} \right\rceil$ and the latest $L(i,j) = \left\lfloor \frac{j-1+B}{r_i} + 1 \right\rfloor$ positions for copy j of i and the bottleneck

$$B = \frac{2^{n-1} - 1}{2^n - 1}.$$

We have

$$E(i,j) = \left\lceil \frac{j-B}{r_i} \right\rceil = (2j-1)2^{n-i} - \left\lfloor \frac{j-1}{2^{i-1}} \right\rfloor = (2j-1)2^{n-i} = Z_j^i$$

and

$$L(i,j) = \left\lfloor \frac{j-1+B}{r_i} + 1 \right\rfloor = (2j-1)2^{n-i} - \left\lceil \frac{2^{i-1} - j}{2^{i-1}} \right\rceil = (2j-1)2^{n-i} = Z_j^i.$$

Therefore, a feasible sequence for the instance $(d_1, \ldots, d_n)$ with $d_i = 2^{i-1}$ for $i = 1, \ldots, n$ and $B = \frac{2^{n-1}-1}{2^n-1} < \frac{1}{2}$ exists. The sequence simply has copy j of i in its ideal position Z_j^i. Finally, by Lemma 4.22 no other sequence for the instance $d_i = 2^{i-1}$ for $i = 1, \ldots, n$ with $B < \frac{1}{2}$ exists. Otherwise, the sequence would have some copy j of some i in a position different then Z_j^i, which would result in a bottleneck at least $\frac{1}{2}$. Therefore, $B^* = \frac{2^{n-1}-1}{2^n-1}$. $\square$

It is clear now that Theorem 6.1 follows immediately from Theorems 6.20 and 6.21.

6.8 Fraenkel's Conjecture and Competition-Free Instances

We now explore Theorem 6.1 in the context of the well-known Fraenkel's Conjecture which we now define. For rational numbers $\alpha \geq 1$ and β denote by $S(\alpha, \beta)$ the sequence $\{\lfloor j\alpha + \beta \rfloor : j = 1, 2, \ldots\}$. A finite set $\{S(\alpha_i, \beta_i) : 1 \leq i \leq n\}$ is called an *exact cover* if every positive integer occurs in exactly one $S(\alpha_i, \beta_i)$. We have the following conjecture of Fraenkel, see also Tijdeman [5].

Conjecture 6.22 (Fraenkel). If $\{S(\alpha_i, \beta_i) : 1 \leq i \leq n\}$ is an exact cover with $\alpha_1 > \ldots > \alpha_n$ and $n \geq 3$, then $\{\alpha_1, \ldots, \alpha_n\} = \{\frac{2^n-1}{2^{i-1}} : i = 1, \ldots, n\}$.

The Fraenkel's Conjecture remains open. However, the following results have been shown. Simpson [61] proves the following theorem.

Theorem 6.23. *The Fraenkel's Conjecture is vacuously true for $\alpha_n \leq \frac{3}{2}$ or equivalently for $r_n \geq \frac{2}{3}$.*

Morikawa [58, 59], and Altman et al. [7], and Tijdeman [5, 60] show the following.

Theorem 6.24. *The Fraenkel's Conjecture holds for $n = 3, 4, 5,$ and 6.*

Uspensky [63] shows that.

Theorem 6.25. *The Fraenkel's Conjecture is vacuously true whenever $\beta_i = 0$ for $i = 1, \ldots, n$, or $\beta_i = 1$ for $i = 1, \ldots, n$.*

Proof. Uspensky proves that the set $\{S(\alpha_i, 0) : 1 \leq i \leq n\}$ is never an eventual exact cover for real $\alpha_i \geq 1, i = 1, \ldots, n$ and $n \geq 3$. This also implies that no set $\{S(\alpha_i, 1) : 1 \leq i \leq n\}$ is an eventual exact cover for real $\alpha_i \geq 1, i = 1, \ldots, n$. These two imply the theorem. $\square$

We now show the following theorem that we refer to as the symmetric Fraenkel's Conjecture.

Theorem 6.26. *If $\{S(\alpha_i, -\frac{\alpha_i}{2}) : 1 \leq i \leq n\}$ is an exact cover with $\alpha_1 > \ldots > \alpha_n$ and $n \geq 3$, then $\{\alpha_1, \ldots, \alpha_n\} = \{\frac{2^n - 1}{2^{i-1}} : i = 1, \ldots, n\}$.*

Proof. If $\{S(\alpha_i, \beta_i) : 1 \leq i \leq n\}$ is an exact cover, then $\sum_{i=1}^{n} \frac{1}{\alpha_i} = 1$. Now, let D be the common denominator of $\frac{1}{\alpha_1}, \ldots, \frac{1}{\alpha_n}$. Then, we have $\frac{1}{\alpha_1} = \frac{d_1}{D}, \ldots, \frac{1}{\alpha_n} = \frac{d_n}{D}$, where $d_1, \ldots, d_n$ are some positive integers, and $D = \sum_{i=1}^{n} d_i$. Without loss of generality $\gcd(d_1, \ldots, d_n) = 1$. Consider, the interval $[0, D]$. Since $\{S(\alpha_i, -\frac{\alpha_i}{2}) : 1 \leq i \leq n\}$ is an exact cover, then all $j\alpha_i - \frac{\alpha_i}{2} = \frac{(2j-1)\alpha_j}{2} = \frac{(2j-1)}{2r_i}, j = 1, \ldots d_i$ are fractional.

This claim holds as hollows. Suppose that there is j and i such that $\frac{(2j-1)}{2r_i} = \frac{(2j-1)D}{2d_i}$ is integral. Hence, since $2j - 1$ is odd, then D must be even. Therefore, there must be an even number of odd numbers among $d_1, \ldots, d_n$. Otherwise, either all demands are even and then $\gcd(d_1, \ldots, d_n) \geq 2$ or else there is an odd number of odd demands and then D is odd, both cases lead to a contradiction. Now, let $d_a = 2k - 1$ and $d_b = 2l - 1$, for some positive integers k and l, be two odd demands and $a \neq b$. Obviously, $k \leq d_a$ and $l \leq d_b$ and thus $\frac{(2k-1)D}{2d_a} = \frac{D}{2}$ of a and $\frac{(2l-1)D}{2d_b} = \frac{D}{2}$ of b are equal and thus $\{S(\alpha_i, -\frac{\alpha_i}{2}) : 1 \leq i \leq n\}$ is *not* an exact cover, which leads to a contradiction.

Therefore, the instance $(d_1, \ldots, d_n)$ is competition-free. Thus, by Theorem 6.20, $d_i = 2^{i-1}$, $i = 1, \ldots, n$. Consequently the symmetric Fraenkel's Conjecture holds. $\square$

This theorem immediately implies the Fraenkel's Conjecture for the sets of the form $\{S(\alpha_i, -\frac{\alpha_i}{2} + a) : 1 \leq i \leq n\}$, where a is an integer, and $\{S(\alpha_i, -\frac{\alpha_i}{2} + \lambda_i) : 1 \leq i \leq n\}$, where $-\frac{1}{2^n} < \lambda_i < \frac{1}{2^n}$. What happens for other forms of β_is seems mostly open. However, we show that the conjecture holds for the parametric case as well.

For a rational numbers $\alpha \geq 1$ and $0 \leq \delta \leq 1$, define the parametric sequence as follows

$$j\alpha + (\delta - 1)\alpha \tag{6.27}$$

for $j = 1, 2, \ldots$ The parametric sequences are related to the parametric sequences of apportionment as follows. Consider the set of parametric sequences $\{S(\alpha_i, (\delta - 1)$

$\alpha_i) : 1 \le i \le n$}. If the set is an exact cover, then $\sum_{i=1}^{n} \frac{1}{\alpha_i} = 1$. Let D be the common denominator of $\frac{1}{\alpha_1}, \ldots, \frac{1}{\alpha_n}$. Then, $\frac{1}{\alpha_1} = \frac{d_1}{D}, \ldots, \frac{1}{\alpha_n} = \frac{d_n}{D}$, for some positive integers $d_1, \ldots, d_n$, and $D = \sum_{i=1}^{n} d_i$. Without loss of generality $\gcd(d_1, \ldots, d_n) = 1$. We then have

$$j\alpha_i + (\delta - 1)\alpha_i = (j + (\delta - 1))\alpha_i = \frac{j + (\delta - 1)}{r_i} \tag{6.28}$$

for $j = 1, \ldots d_i$. Thus the sequence corresponds to a δ-parametric sequence

$$\frac{d_i}{j + (\delta - 1)}.$$

The Adams's for $\delta = 0$, the Jefferson's for $\delta = 1$, and the Webster's for $\delta = \frac{1}{2}$. Therefore, by Theorem 6.25 the Fraenkel's conjecture is vacuously true for $\delta = 1$ or the Adam's sequence, and for $\delta = 0$ or the Jefferson's sequence. We have the following general theorem for the parametric sequences.

Theorem 6.27. *If $\{S(\alpha_i, (\delta - 1)\alpha_i) : 1 \le i \le n\}$ is an exact cover with $0 \le \delta \le 1$, $\alpha_1 > \ldots > \alpha_n$ and $n \ge 3$, then $\{\alpha_1, \ldots, \alpha_n\} = \{\frac{2^n - 1}{2^{i-1}} : i = 1, \ldots, n\}$.*

Proof. By Theorem 6.25 we may assume $0 < \delta < 1$. Rewrite (6.28) as follows

$$\frac{j + (\delta - 1)}{r_i} = \frac{2j - 1}{2r_i} + \frac{\varepsilon}{r_i},$$

where $-\frac{1}{2} < \varepsilon = \delta - \frac{1}{2} < \frac{1}{2}$. The vertices

$$\frac{2j - 1}{2r_i},$$

for $i = 1, \ldots, n$ and $j = 1, \ldots, d_i$ define a regular d_i-gone $P_i^{\frac{1}{2}}$ and the vertices

$$\frac{2j - 1}{2r_i} + \frac{\varepsilon}{r_i},$$

for $i = 1, \ldots, n$ and $j = 1, \ldots, d_i$ define a regular d_i-gone denoted by $P_i^{\frac{1}{2} + \varepsilon}$. The polygon $P_i^{\frac{1}{2} + \varepsilon}$ is obtained from $P_i^{\frac{1}{2}}$ by rotating the latter by the angle

$$\theta_i = \frac{2\pi|\varepsilon|}{d_i},$$

$i = 1, \ldots, n$, clockwise if $\varepsilon > 0$, or counter-clockwise otherwise. Assume $\varepsilon > 0$. Suppose that the polygons $P_i^{\frac{1}{2} + \varepsilon}$ make up an exact cover for $(d_1, \ldots, d_n)$ which is not the power-of-two instance. Rotate them by θ_i counter-clockwise to obtain the polygons $P_i^{\frac{1}{2}}$, $i = 1, \ldots, n$. Consider those polygons after the rotation now. By Theorem 6.20

the instance $(d_1, \ldots, d_n)$ is not competition-free, thus there is a unit arc $[k, k+1]$ of the D-circle, and by the NS-symmetry a unit arc $[D-(k+1), D-k]$, that includes vertices of at least two polygons $P_i^{\frac{1}{2}}$ and $P_j^{\frac{1}{2}}$, $i \neq j$, after the rotations. Without loss of generality let us assume that the k is smallest possible. Thus, each unit arc $[0, 1], \ldots, [k-1, k]$, as well as each unit arc $[D, D-1], \ldots, [D-(k-1), D-k]$ by the NS-symmetry, includes at most one vertex of the polygons $P_i^{\frac{1}{2}}$, $i = 1, \ldots, n$. Thus, there are $l \leq k$ vertices of polygons $P_i^{\frac{1}{2}}$, $i = 1, \ldots, n$ in each of the arcs $[0, k)$ and $[D-k, D)$, and $l+2$ in the arc $[D-(k+1), D)$. This, however, leads to a contradiction as follows. Consider the arcs $R = [0, k)$ and $L = [D-(k+1), D)$. If $l < k$, then the clockwise rotation of polygon $P_i^{\frac{1}{2}}$ by, see Fig. 6.4

$$\theta_i < \frac{\pi}{d_i}, \tag{6.29}$$

for $i = 1, \ldots, n$, results into at most l vertices of polygons $P_i^{\frac{1}{2}+\varepsilon}$, $i = 1, \ldots, n$, in the arc R which leads to a contradiction since the polygons make up an exact cover and thus there must be exactly k of then in R. Otherwise, that is if $l = k$, the same rotation results into at least $l+2$ vertices of polygons $P_i^{\frac{1}{2}+\varepsilon}$, $i = 1, \ldots, n$, in the arc L. This leads to a contradiction again since then $l+2 = k+2$, and there are only $k+1$ unit arcs in L. Therefore, the Fraenkel's Conjecture holds for all apportionment parametric sequences for $\varepsilon >$. However, a similar reasoning proves the $\varepsilon < 0$ case as well, and the $\varepsilon = \frac{1}{2}$ case is shown in Theorem 6.26. $\square$

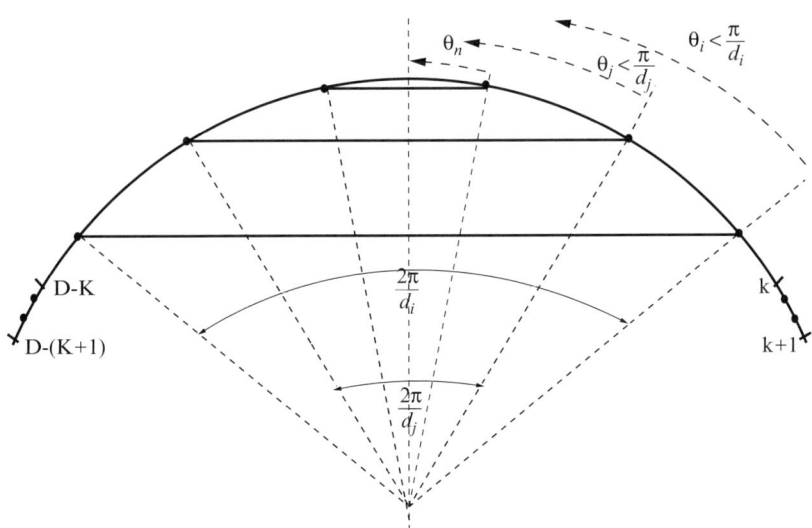

Fig. 6.4 The North arcs of polygons $P_n^{\frac{1}{2}}$, $P_i^{\frac{1}{2}}$, and $P_j^{\frac{1}{2}}$ in the rotations by θ_n, θ_i, and θ_j respectively

6.9 Regular Sequences and Multimodular Functions

We begin by introducing regular sequences (words). The main problem is to find, if any exists, a regular word for given rates with which letters should occur in the sequence. Though this problem remains open, we provide a summary of conditions under which such words exist. Our discussion relies on the results on the competition-free instances and the symmetric Fraenkel's Conjecture discussed in Sect. 6.8.

6.9.1 Regular Sequences

For the rate $0 < r < 1$ and the phase $0 \leq \theta < 1$ the zero-one sequence

$$\sigma(r, \theta) = \lfloor (j+1)r + \theta \rfloor - \lfloor jr + \theta \rfloor \tag{6.30}$$

$j \in \mathbb{Z}$ is called a *regular* sequence (or a *Sturmian* word if r is irrational). Let $\{a_1, \ldots, a_n\}$ be a finite alphabet. Let $\{a_1, \ldots, a_n\}^{\mathbb{Z}}$ be the set of infinite sequences on $\{a_1, \ldots, a_n\}$. For the letter a_i and sequence $s \in \{a_1, \ldots, a_n\}^{\mathbb{Z}}$ let $I(s, a_i) \in \{0, 1\}^{\mathbb{Z}}$ be the indicator in s of the letter a_i, that is $I(s, a_i)_j = 1$ if and only if $s_j = a_i$. An example of the concepts just introduced follows.

Example 6.28. Consider a periodic sequence s^{∞} with the period

$$s = abacaba.$$

The three indicators $I(s, a)$, $I(s, b)$, and $I(s, c)$ are

$$\ldots 0101010110101010 \ldots$$

$$\ldots 1001000100100010 \ldots$$

and

$$\ldots 01000000100000010 \ldots$$

with the periods $1010101, 0100010,$ and 0001000 respectively.

A sequence s is said to have asymptotic rate r of the letter a_i if

$$\lim_{N \to \infty} \frac{1}{N} \sum_{k=1}^{N} I(s, a_i)_k = r$$

Observe that s in Example 6.28 has asymptotic rate $\frac{4}{7}$ of the letter a, $\frac{2}{7}$ of the letter b and $\frac{1}{7}$ of the letter c.

For a zero-one sequence $s \in \{0, 1\}^{\mathbb{Z}}$, the support in s of 1 is the set of all positions in s with the value 1

$$S = \{j \in \mathbb{Z} : s_j = 1\}.$$

For the regular word $\sigma(r, \theta)$ the support is by (6.30) as follows

$$S = \left\{ \left\lceil \frac{j}{r} - \frac{\theta}{r} \right\rceil - 1 : j \in \mathbb{Z} \right\}.$$

The main problem considered in this section can be formulated as follows, see also Altman et al. [7].

Problem 6.29. Given the rates $r_1, \ldots, r_n$ that sum up to 1, is there a sequence $s \in \{a_1, \ldots, a_n\}^{\mathbb{Z}}$ such that

$$I(s, a_i) = \sigma(r_i, \theta_i) \tag{6.31}$$

for some $0 \le \theta_i < 1$ for all $i = 1, \ldots, n$?

Since the sequence in (6.31)

$$\sigma(r_i, \theta_i) = \lfloor (j+1)r_i + \theta_i \rfloor - \lfloor jr_i + \theta_i \rfloor \tag{6.32}$$

for $j = 1, 2, \ldots$ has its support

$$S_i = \left\{ \left\lceil \frac{j}{r_i} - \frac{\theta_i}{r_i} \right\rceil - 1 : j \in \mathbb{Z} \right\}$$

and all supports are shifted by -1 we can consider supports

$$S_i' = \left\{ \left\lceil \frac{j}{r_i} - \frac{\theta_i}{r_i} \right\rceil : j \in \mathbb{Z} \right\}$$

instead. Therefore, problem (6.29) is equivalent to the following problem.

Problem 6.30. Given the rates $r_1, \ldots, r_n$, are there phases $\theta_1, \ldots, \theta_n$ such that the sets

$$S_i' = \left\{ \left\lceil \frac{j}{r_i} - \frac{\theta_i}{r_i} \right\rceil : j \in \mathbb{Z} \right\}$$

for $i = 1, \ldots, n$ make up an exact cover of $\mathbb{Z}$, that is each integer is in exactly one set S_i'?

The problem is a challenging open problem. Graham [64] proves that if all the rates $r_1, \ldots, r_n$ are distinct, then all of them are rational. However, rational rates imply that the sequences (6.32) are cyclic. Therefore, the $\mathbb{Z}$ in the problem (6.30) can be replaced by the set $\mathbb{N}$ of positive integers. Then, however, the Fraenkel's conjecture claims that the only rates must be power-of-two rates. We show in Theorem 6.27 that if the sequence $s \in \{a_1, \ldots, a_n\}^{\mathbb{N}}$ is set up according to the parametric methods of apportionment, then this conjecture holds. It also holds unconditionally for $n = 3, 4, 5$, and 6 as well as vacuously for any instance with maximum rate $\max\{r_i\} \ge \frac{2}{3}$, see Sect. 6.8. Moreover, we observe that for this question to have an affirmative answer not all phases θ_i, $i = 1,\dots,n$, may be greater than $\frac{1}{2}$ and not all of them may be less than $\frac{1}{2}$. This follows from the same arguments as used in the proof of Theorem 6.27. Finally, by Lemmas 5.14 and 5.15 the regular word is always possible for $n = 2$, see also the proof of Lemma 6.31 below.

Although a complete characterization of all the rates for which a sequence $s \in \{a_1,\dots,a_n\}^{\mathbb{Z}}$ satisfying (6.31) exists is an open problem even if some rates are allowed to equal, we have the following result for two distinct rates, see also Altman et al. [7].

Lemma 6.31. *If the rates $r_1,\dots,r_n$ are made up of at most two distinct numbers, then they admit a regular sequence.*

Proof. Let $r_1 = \dots = r_k = p$ and $r_{k+1} = \dots = r_n = q$. Consider an instance with $n = 2$, $R_1 = kp$ and $R_2 = (n-k)q$. If p is rational, then $p = \frac{d_1}{D}$ and $q = \frac{d_2}{D}$ and $D = kd_1 + (n-k)d_2$, for some positive integers d_1 and d_2. Without loss of generality $\gcd(d_1, d_2) = 1$. Let $\gcd(k, n-k) = g$, then $k' = \frac{k}{g}$ and $m' = \frac{n-k}{g}$ and $D' = k'd_1 + m'd_2$. The instance $(k'd_1, m'd_2)$ is a standard instance such that $R_1 = \frac{k'd_1}{D'}$ and $R_2 = \frac{m'd_2}{D'}$. For one of $k'd_1$ and $m'd_2$ being odd and the other even, Lemma 5.14 shows a solution s which is a regular word with $\theta = \frac{1}{2}$ for either letter. The solution can be made to work for both $k'd_1$ and $m'd_2$ being odd as well. Then, we can just take $\theta_1 = \frac{1}{2} + \varepsilon$ and $\theta_2 = \frac{1}{2} - \varepsilon$ for the two letters respectively and for sufficiently small $\varepsilon > 0$. Thus the sequence s^g, that is s concatenated g times is a regular word for the instance $(kd_1, (n-k)d_2)$. In that sequence, we replace the lth, the $(k+l)$th, $\dots$, the $(k(d_1 - 1) + l)$th occurrence of the letter a by the letter l, $l = 1,\dots,k$, respectively, and we replace the lth, the mth, the $((n-k)+m)$th, $\dots$, the $((n-k)(d_2 - 1) + m)$th occurrence of the letter b by the letter m, $m = 1,\dots,n-k$, respectively. The resulting sequence is regular for the alphabet $\{1,\dots,n\}$ and the rates $r_1,\dots,r_k, r_{k+1},\dots,r_n$.

If p is irrational, then by the Beatty theorem, see Theorem 6.32 below, the sequences $\left\lfloor \frac{j}{kp} \right\rfloor$ and $\left\lfloor \frac{j}{(n-k)q} \right\rfloor$, $j = 1, 2,\dots$ partition $\mathbb{N}$ since $kp + (n-k)q = 1$. In that sequence replace the lth, the $(k+l)$th, $\dots$, the $(kj+l)$th, etc. occurrence of the letter a by the letter l, and replace the lth, the $((n-k)+m)$th, $\dots$, the $((n-k)j+m)$th, etc. occurrence of the letter b by the letter m. By the Beatty theorem the resulting sequences are regular. The sequences also have the asymptotic rates as requested. $\square$

The following theorem is referred to in the proof and it was first shown by Beatty [57].

Theorem 6.32 (Beatty). *The sequences $\{\lfloor \alpha j \rfloor : j = 1, 2,\dots\}$ and $\{\lfloor \beta j \rfloor : j = 1, 2,\dots\}$ partition $\mathbb{N}$ if and only if*

(i) $\frac{1}{\alpha} + \frac{1}{\beta} = 1$.
(ii) α *is irrational.*

Finally, the following lemma gives a relative way of generating regular word, see Altman et al. [7] for its proof, for given rates for which the regular word already exists.

Lemma 6.33. *If the rates $r_1, r_2, \ldots, r_n$ admit a regular sequence, then so do the rates* $\underbrace{\dfrac{r_1}{k}, \ldots, \dfrac{r_1}{k}}_{k-times}, r_2, \ldots, r_n$ *where the the letter a_1 with the rate r_1 is replaced by $k \geq 1$ distinct letters $a_1^1, \ldots, a_1^k$ with the rate $\frac{r_1}{k}$ for each.*

We finish this section with an example of regular words generated in the special cases discussed earlier in this section.

Example 6.34. The power-of-two instance $\mathbf{d} = (1, 2, 4, 8, 16)$ admits a regular sequence with period

$$abacabadabacabaeabacabadabacaba.$$

Thus, by Lemma 6.33 the instance $\mathbf{d}_1 = (1, 2, 4, 4, 4, 4, 4, 8)$ admits a regular sequence, its period is

$$xbyczbwdxbyczbwexbyczbwdxbyczbw.$$

The instance $\mathbf{d}' = (9, 10)$ admits a regular sequence since $n = 2$. The sequence has period

$$abababababababababa.$$

Thus, by Lemma 6.31 the instance with two different demands $\mathbf{d}' = (3, 3, 3, 10)$

$$axayazaxayazaxayaza.$$

is regular.

6.9.2 Multimodular Functions

The multimodular functions were introduced by Hajek [6] as follows. Define vectors $v_0, v_1, \ldots, v_m \in \mathbb{Z}^m$

$$v_0 = (-1, 0, \ldots, 0)$$
$$v_1 = (1, -1, \ldots, 0)$$
$$v_2 = (0, 1, -1, \ldots, 0)$$
$$\cdots$$
$$v_m = (0, 0, 0, \ldots, 1)$$

Let $V = \{v_0, v_1, \ldots, v_m\}$. A function J on $\mathbb{Z}^m$ is *multimodular* if for all $u \in \mathbb{Z}^m$,

$$J(u + v) + J(u + w) \geq J(u) + J(u + v + w)$$

for $v, w \in V$ and $v \neq w$. A sequence of integers

$$s = s_1 s_2 \cdots$$

is said to have an asymptotic mean r if

$$\lim_{n \to \infty} \frac{1}{n} \sum_{k=1}^{n} s_k = r.$$

We let the multimodular function J on $\mathbb{Z}^m$ to "slide" along an infinite sequence of integers s with a given asymptotic mean r,

$$\liminf_{n \to \infty} \frac{1}{n} \sum_{k=1}^{n} J(s_k, s_{k+1}, \ldots, s_{k+m-1}) \tag{6.33}$$

to calculate the asymptotic average of J along the s. The following theorem is due to Hajek [6]. It gives a lower bound for the asymptotic average (6.33), and it also shows that the bound is attained on regular words.

Theorem 6.35. *Let J be a multimodular function on $\mathbb{Z}^m$. If s is any integer sequence with asymptotic mean r then*

$$\liminf_{n \to \infty} \frac{1}{n} \sum_{k=1}^{n} J(s_k, s_{k+1}, \ldots, s_{k+m-1})$$

$$\geq \int_0^1 J((\lfloor r + \phi \rfloor, \lfloor 2r + \phi \rfloor - \lfloor r + \phi \rfloor, \ldots,$$

$$\lfloor mr + \phi \rfloor - \lfloor (m-1)r + \phi \rfloor)) d\phi$$

In case of a regular word $s = \sigma(r, \theta)$ for some $0 \leq \theta < 1$

$$\lim_{n \to \infty} \frac{1}{n} \sum_{k=1}^{n} J(\sigma(r, \theta)_k, \sigma(r, \theta)_{k+1}, \ldots, \sigma(r, \theta)_{k+m-1})$$

$$= \int_0^1 J((\lfloor r + \phi \rfloor, \lfloor 2r + \phi \rfloor - \lfloor r + \phi \rfloor, \ldots,$$

$$\lfloor mr + \phi \rfloor - \lfloor (m-1)r + \phi \rfloor)) d\phi.$$

Altman et al. [7, 65] extend this result as follows. Let $J^1, \ldots, J^n$ be multimodular functions on $\mathbb{Z}^m$ and let $g : \mathbb{R}^n \to \mathbb{R}$ be any increasing linear function. Consider sequences $s \in \{a_1, \ldots, a_n\}^{\mathbb{Z}}$ with asymptotic rates r_i for the letter a_i so that

$$r_1 + r_2 + \cdots + r_n = 1.$$

They prove the following theorem.

Theorem 6.36. *Let $J^1, \ldots, J^n$ be a multimodular functions on $\mathbb{Z}^m$. If $s \in \{a_1, \ldots, a_n\}^{\mathbb{Z}}$ with asymptotic rates r_i for the letter a_i, then*

$$\liminf_{N\to\infty} \frac{1}{N}\sum_{k=1}^{N} g(J^1(I(s,a_1)_k, I(s,a_1)_{k+1}, \ldots, I(s,a_1)_{k+m-1}), \ldots,$$

$$J^n(I(s,a_n)_k, I(s,a_n)_{k+1}, \ldots, I(s,a_n)_{k+m-1}))$$

$$\geq g\Big(\int_0^1 J^1((\lfloor r_1 + \phi\rfloor, \lfloor 2r_1 + \phi\rfloor - \lfloor r_1 + \phi\rfloor, \ldots, \lfloor mr_1 + \phi\rfloor$$

$$- \lfloor (m-1)r_1 + \phi\rfloor))d\phi,$$

$$\int_0^1 J^2((\lfloor r_2 + \phi\rfloor, \lfloor 2r_2 + \phi\rfloor - \lfloor r_2 + \phi\rfloor, \ldots, \lfloor mr_2 + \phi\rfloor$$

$$- \lfloor (m-1)r_2 + \phi\rfloor))d\phi, \ldots,$$

$$\int_0^1 J^n((\lfloor r_n + \phi\rfloor, \lfloor 2r_n + \phi\rfloor - \lfloor r_n + \phi\rfloor, \ldots, \lfloor mr_n + \phi\rfloor$$

$$- \lfloor (m-1)r_n + \phi\rfloor))d\phi\Big).$$

In case $I(s,a_i) = \sigma(r_i, \theta_i)$ for some $0 \leq \theta_i < 1$ for all $i = 1,\ldots,n$

$$\lim_{N\to\infty} \frac{1}{N}\sum_{k=1}^{N} g(J^1(\sigma(r_1,\theta_1)_k, \sigma(r_1,\theta_1)_{k+1}, \ldots, \sigma(r_1,\theta_1)_{k+m-1}), \ldots,$$

$$J^n(\sigma(r_n,\theta_n)_k, \sigma(r_n,\theta_n)_{k+1}, \ldots, \sigma(r_n,\theta_n)_{k+m-1}))$$

$$= g\Big(\int_0^1 J^1((\lfloor r_1 + \phi\rfloor, \lfloor 2r_1 + \phi\rfloor - \lfloor r_1 + \phi\rfloor, \ldots, \lfloor mr_1 + \phi\rfloor$$

$$- \lfloor (m-1)r_1 + \phi\rfloor))d\phi,$$

$$\int_0^1 J^2((\lfloor r_2 + \phi\rfloor, \lfloor 2r_2 + \phi\rfloor - \lfloor r_2 + \phi\rfloor, \ldots, \lfloor mr_2 + \phi\rfloor$$

$$- \lfloor (m-1)r_2 + \phi\rfloor))d\phi, \ldots,$$

$$\int_0^1 J^n((\lfloor r_n + \phi\rfloor, \lfloor 2r_n + \phi\rfloor - \lfloor r_n + \phi\rfloor, \ldots, \lfloor mr_n + \phi\rfloor$$

$$- \lfloor (m-1)r_n + \phi\rfloor))d\phi\Big).$$

Section 6.9.1 reviews known cases of the rates $r_1, r_2, \ldots, r_n$ for which it is possible to construct a sequence s such that all projections (indicators) $I(s,a_i)$ are regular, that is $I(s,a_i) = \sigma(r_i, \theta_i)$ for some $0 \leq \theta_i < 1$ for all $i = 1,\ldots,n$.

6.10 Optimal Admission of Arrivals

This motivating application comes from Hajek [6]. Consider a sequence of customers arriving at rate λ. Following Hajek we assume that the interarrival times $\{Y_k : k = 1,2,\ldots\}$ are independent, identically distributed random variables with finite mean $\frac{1}{\lambda}$. That is the arrival process is a renewal process, see Ross [66]. A prespecified admission sequence s over the alphabet $\{a_1,\ldots,a_n\}$ with given asymptotic rates r_i for the letter a_i is used to split the arrival stream between n exponential servers with mean service times $\frac{1}{\mu_1}, \ldots, \frac{1}{\mu_n}$. For each k, customer k is admitted to server i such that $s_k = a_i$. Thus, customers arrive at server i at the rate λr_i.

Define Q_k^i to be the number of customers in the queue of server i, including the one whose service is in progress, immediately prior to the arrival of customer k. Let X_k^i be the *potential* number of departures from the queue of server i between the arrival of customer k and customer $k+1$. We assume that the conditional probability distribution of X_k^i given that the time between the arrivals is t is Poisson with mean $\mu_i t$. Thus, the expected value of X_k^i is $\frac{\mu_i}{\lambda}$. The *actual* number of departures between the arrival of customer k and customer $k+1$ can not exceed $Q_k^i + I(s, a_i)_k$, which is the queue size just after customer k arrives. Thus, we have the following recursive formula

$$Q_{k+1}^i = \max\{Q_k^i + I(s, a_i)_k - X_k^i, 0\},$$

for $k = 1, 2, \ldots$ and

$$Q_1^i = 0.$$

Thus, by induction we obtain

$$Q_{k+1}^i = \max\left\{\sum_{j=1}^l (I(s, a_i)_{k+1-j} - X_{k+1-j}^i) : l = 0, \ldots, k\right\}$$

for $k = 1, 2, \ldots$, where by definition

$$\sum_{j=1}^0 (I(s, a_i)_{k+1-j} - X_{k+1-j}^i) = 0.$$

Therefore, the expected number of customers, or the expected queue size, in the queue of server i is

$$EQ_{k+1}^i = E\left(\max\left\{\sum_{j=1}^l I(s, a_i)_{k+1-j} - X_{k+1-j}^i : l = 0, \ldots, k\right\}\right)$$

$$= J_k^i(I(s, a_i)_1, \ldots, I(s, a_i)_k)$$

for $k = 1, 2, \ldots$ and the expected queue size is a multimodular function for each server.

Lemma 6.37. EQ_{k+1}^i *is a multimodular function on $\mathbb{Z}^k$.*

Hajek gives the following interpretation of this lemma. The difference $EQ_k^i(u) - EQ_k^i(u + v_j)$ represents the waiting time saving seen by customer k in the queue of server i if customer $j - 1$ were admitted to the server rather than customer j. This saving is non-negative since k cannot wait longer by having a previous customer admitted earlier to server i. Multimodularity of EQ_k^i ensures that for $j \neq l$ this time saving is no greater than $EQ_k^i(u + v_l) - EQ_k^i(u + v_j + v_l)$, which is the time saving that would result from admitting $j-1$ and $l-1$ rather than customer j and l respectively to server i.

Finally.

Theorem 6.38. *If s is any zero-one valued sequence with asymptotic mean r_i, then*

$$\liminf_{N\to\infty} \frac{1}{N}\sum_{i=1}^{N} EQ_k^i \geq J^i$$

If $s = \sigma(r_i, \theta_i)$, then

$$\lim_{N\to\infty} \frac{1}{N}\sum_{i=1}^{N} EQ_k^i = J^i,$$

where

$$J^i = \lim_{k\to\infty} \int_0^1 J_k^i((\lfloor r_i+\phi\rfloor, \lfloor 2r_i+\phi\rfloor - \lfloor r_i+\phi\rfloor, \ldots,$$
$$\lfloor kr_i+\phi\rfloor - \lfloor (k-1)r_i+\phi\rfloor))d\phi.$$

Therefore, regular admission sequences minimize expected queue sizes for each server. Altman et al. [7] extend this result to the expected workload and waiting time in an event graph (thus there is no inside choice for the route followed by the customers in such graph) that replaces a single server in Fig. 6.5. However, in either case regular sequences may not exist for some rates, after all the Fraenkel's Conjecture holds for $n = 3, 4, 5$, and 6, see Theorem 6.24, moreover, Simpson [61] shows the same for one rate being at least $\frac{2}{3}$, see Theorem 6.23. Therefore, the algorithm that finds an optimal admission sequence for the rates for which a regular sequence does not exist remains an important open question.

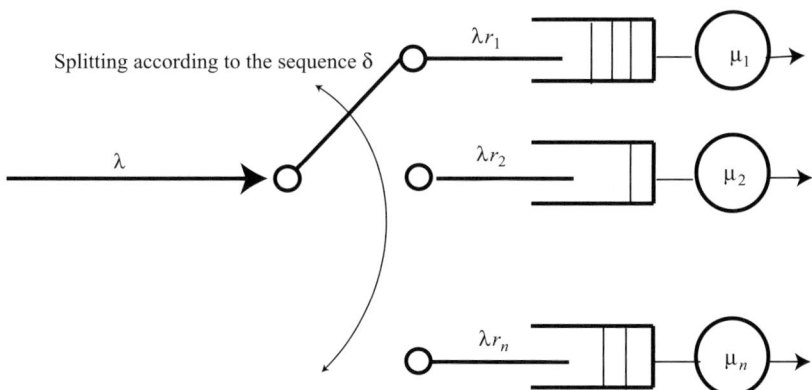

Fig. 6.5 Splitting arrival process according to the admission sequence s with asymptotic rates r_i for the letter a_i, $i = 1,\ldots,n$

6.11 Multimodular Function on Just-In-Time Sequences

We close this chapter with a proof that the function on just-in-time sequences considered in Chap. 3 is a multimodular function when limited to a set zero-one vectors. Let $A^i_{j,D}$ be a set of all vectors in $\{0,1\}^D$ with exactly $j = 0,\ldots,d_i$ ones, that is

$$A^i_{j,D} = \left\{ u : u \in \{0,1\}^D \text{ and } \sum_{k=1}^{D} u_k = j \right\}.$$

Let

$$A^i = \bigcup_{j=0}^{d_i} A^i_{j,D}.$$

For a vector $u \in A^i$ let $i_1,\ldots,i_j$ be all coordinates with value 1. We define a function g on A^i as follows

$$g(u) = \sum_{k=0}^{i_1-1} f_0(k) + \sum_{k=i_1}^{i_2-1} f_1(k) + \cdots + \sum_{k=i_j}^{D} f_j(k). \tag{6.34}$$

where the functions f_l are defined in (3.6) in Chap. 3, and $i_1 = D+1$ for $j = 0$. Notice that the function g on $A^i_{d_i,D}$ is exactly the function in (3.8) We have the following observation.

Lemma 6.39. *Function g is multimodular on A^i.*

Proof. We need to show that for any $u \in A^i$ and $v,w \in V$ with $v \neq w$

$$g(u+v) + g(u+w) \geq g(u) + g(u+v+w) \tag{6.35}$$

as long as all vectors $u, u+v, u+w$ and $u+v+w$ are in A^i. Suppose that $v = v_k$ and $w = v_l$, and without loss of generality $k < l$. If all four vectors are in $A^i_{j,D}$, for some j, then we have $0 < k$ and $l < D$. Thus, there are $i_a = k$ and $i_b = l$ such that $u_{k-1} = 0$, $u_k = 1$, and $u_{l-1} = 0$, $u_l = 1$. Consequently,

$$g(u+v) = g(u) + f_a(i_a - 1) - f_{a-1}(i_a - 1),$$

$$g(u+w) = g(u) + f_b(i_b - 1) - f_{b-1}(i_b - 1),$$

and

$$g(u+v+w) = g(u) + f_a(i_a - 1) - f_{a-1}(i_a - 1) + f_b(i_b - 1) - f_{b-1}(i_b - 1).$$

Thus, (6.35) holds as equality. For $v = v_D$ we get

$$g(u+v_D) = g(u) + f_{j+1}(D) - f_j(D) \tag{6.36}$$

and for $v = v_0$

$$g(u+v_0) = g(u) + \sum_{k=1}^{i_2-1} (f_0(k) - f_1(k)) + \cdots + \sum_{k=i_j}^{D} (f_{j-1}(k) - f_j(k)), \quad (6.37)$$

in particular for $u + v_0$ having all zeros, we have

$$g(u+v_0) = \sum_{k=0}^{D} f_0(k) = g(u) + \sum_{k=1}^{D} (f_0(k) - f_1(k)).$$

Thus,

$$g(u+v_0+v_D) = g(u+v_0) + f_j(D) - f_{j-1}(D)$$

and hence

$$g(u+v_D) + g(u+v_0) = g(u) + g(u+v_0+v_D) + f_{j+1}(D) - 2f_j(D) + f_{j-1}(D).$$

However,

$$f_{j+1}(D) - 2f_j(D) + f_{j-1}(D) \geq 0$$

by Lemma 4.1 where

$$f_j(D) = F(j - d_i)$$

by definition (3.6) with convex F. Therefore,

$$g(u+v_D) + g(u+v_0) \geq g(u) + g(u+v_0+v_D).$$

For $g(u+v_D)$ and $g(u+w)$ with $w \neq v_0$, we have

$$g(u+v_D+w) = g(u+w) + f_{j+1}(D) - f_j(D)$$

hence by (6.36)

$$g(u+v_D) + g(u+w) = g(u) + g(u+v_D+w).$$

Finally, for $w \neq v_D$ we have

$$g(u+v_0+w) = g(u+v_0) + f_{b-1}(i_b - 1) - f_{b-2}(i_b - 1),$$

thus

$$g(u+v_0) + g(u+w) = g(u) + g(u+v_0+w) + f_b(i_b - 1)$$
$$-2f_{b-1}(i_b - 1) + f_{b-2}(i_b - 1)$$

However,

$$f_b(i_b - 1) - 2f_{b-1}(i_b - 1) + f_{b-2}(i_b - 1) \geq 0$$

by Lemma 4.1 where

$$f_b(i_b - 1) = F(b - (i_b - 1)r_i)$$

by definition (3.6) with convex F. Therefore,

$$g(u + v_0) + g(u + w) \geq g(u) + g(u + v_0 + w).$$

$\square$

6.12 Exercises

Exercise 6.40. Prove that

$$\left\lfloor \frac{1}{r} \right\rfloor \leq \left\lceil \frac{j+1}{r} - \frac{\theta}{r} \right\rceil - \left\lceil \frac{j}{r} - \frac{\theta}{r} \right\rceil \leq \left\lceil \frac{1}{r} \right\rceil$$

for $j = 1, 2, \ldots$, $0 \leq \theta < 1$ and the rate r.

Exercise 6.41. Show that the rates $\frac{1}{11}, \frac{2}{11}, \frac{4}{11}$, and $\frac{4}{11}$ admit a regular sequence. Do the same for the rates $\frac{1}{11}, \frac{2}{11}, \frac{2}{11}$, and $\frac{6}{11}$ as well as for $\frac{1}{11}, \frac{1}{11}, \frac{3}{11}$, and $\frac{6}{11}$.

Exercise 6.42. Show that the Fraenkel's Conjecture is vacuously true for $\{S(\alpha_i, (\delta - 1)\alpha_i) : 1 \leq i \leq n\}$ with $\frac{3}{4} < \delta \leq 1$ or $0 \leq \delta < \frac{1}{4}$.

Exercise 6.43. Show that for any $\frac{1}{2} < \delta \leq \frac{3}{4}$ or $\frac{1}{4} \leq \delta < \frac{1}{2}$ there is an m so that for any $m \geq n$ the set $\{S(\alpha_i, (\delta - 1)\alpha_i) : 1 \leq i \leq n\}$ for $\{\alpha_1, \ldots, \alpha_n\} = \{\frac{2^n - 1}{2^{i-1}} : i = 1, \ldots, n\}$ is not an exact covering.

Exercise 6.44. Find the 25-prefix of the Beatty sequence for π do the same for $\sqrt{2}$.

Exercise 6.45. Consider admission into a D/D/1 queue with service times and interarrival times all deterministic. Assuming that the system is initially empty, derive the formula for the workload in the system immediately after the nth arrival. Prove that this formula defines multimodular function for any n. This application comes from Altman et al. [65], see also [67], where it is motivated by quality of service in ATM networks that should guarantee a given Cell Loss Ratio for a session.

6.13 Comments and References

Theorem 6.1 was stated as a conjecture by Brauner and Crama [42]. It was proven by Kubiak [44], another proof using balanced words, see Chap. 9 for definition of balanced words, was given by Brauner et al. [68]. Lemma 6.12 was independently shown by Mirsky, Newman, Davenport, and Rado, see Newman [62]. Our proof is based on the proofs in Wilf [69], and Altman et al. [7]. For more information on Fraenkel's Conjecture see Tideman [5,60]. Uspensky [63] shows Theorem 6.25.

The Fraenkel's conjecture was proved by Morikawa [58, 59]for $m = 3$, and Altman et al. [7] for $m = 4$, and by Tijdeman [5] for $m = 5$ and $m = 6$. Simpson [61] proved that the conjecture vacuously holds if the highest rate r_n is at least $\frac{2}{3}$, see Theorem 6.23. For Beatty sequences see Beatty [57], Stolarsky [70], and Tijdeman [60]. For some examples of Beatty sequences check [71].

Hajek defined multimodular functions in [6], and proved that regular sequences provide a lower bound for these functions. More on regular sequences and Sturmian words can be found in Chap. 2 by Berstel and Séébold [72]. Hajek went further showing the importance of multimodular functions and regular sequences to the problem of the expected queue size minimization in a system that admits customers to an exponential server according to a prespecified splitting sequence, see Sect. 6.10. Altman et al. in a couple of papers [65,73] prove that the expected traveling time in a system with multiple subsystems, each being a stochastic event graph, controlled by an admission sequence that routs customers entering the system to its sub-systems is an increasing multimodular function. The function's lower bound is obtained by assuming a regular admission sequence with a pre-specified asymptotic rate for each subsystem. The lower bound can be attained if the sequences do not overlap thus making an exact cover of $\mathbb{N}$. Altman et al. [7] give a number of characteristics of rates that guarantee that their individual sequences do not overlap. More on the multimodular optimization and regular sequences can be found in Altman et al. [67].

Chapter 7
Response Time Variability

7.1 Introduction

Most modern systems share their resources between different jobs. The jobs define a certain amount of work to be done, for instance the file size to be transmitted to or from a server or the number of cars of a particular model to be produced on a mixed-model assembly line. To ensure fair sharing of common resources between different jobs, this work is divided into atomic tasks, for instance data blocks or cars. These tasks, in turn, are required to be evenly distributed so that the time distance between any two consecutive tasks of the same job is as constant as possible. The following are some real-live examples.

The Asynchronous Transfer Mode (ATM) networks divide each application (voice, large data file, video) into *cells* of fixed size so that the application can be preempted after each cell. Furthermore, *isochronous* applications, for instance voice and video, require an inter-cell distance in a cell stream to be as close to being a constant as possible. For instance, multimedia systems avoid presenting video frames either too early or too late to avoid jagged motion perceptions. These applications may also require the inter-cell distance not to exceed some pre-specified distance, see Han et al. [74] and Altman et al. [7].

On a mixed-model, just-in-time assembly line a sequences of different models to produce is sought where each model is distributed as "evenly" as possible but appears a given number of times to satisfy demand for different models. Consequently, shortages, on one hand, and excessive inventories, on the other, are reduced, see Monden [22] and Chap. 3.

Another application is stride scheduling, see Chap. 10 for details, where each client is first issued a number of *tickets*. The resources are then allocated to the clients in discrete time slices called *quanta*. The client to be allocated resources in next quantum is selected through a certain function of the number of its past allocations and the number of its tickets, Waldspurger and Weihl [13].

W. Kubiak, *Proportional Optimization and Fairness,* International Series in Operations Research & Management Science 127, DOI 10.1007/978-0-387-87719-8_7,
© Springer Science+Business Media LLC 2009

Herrmann [75] presents a waste collection problem in a health-care facility where the time between two consecutive visits to the same waste collection point should be kept as constant as possible.

These problems are often considered as distance-constrained scheduling problems, where the temporal distance between any two consecutive executions of a task is not longer than a pre-specified distance, Han et al. [74]. See also Chap. 8. Sometimes even a stronger condition is imposed that the temporal distance is *equal* to the pre-specified distance. For instance, Altman et al. [7] study the constant gap words, and Wei and Liu [76], and Anily et al. [77] consider the Periodic Machine Maintenance problem with equal distances between consecutive services of the same machine.

The distance-constrained model, however, suffers from a serious practical disadvantage which is that there may not be a feasible solution that respects the distance constraints and at the same time ensures that tasks are done at given rates. In this chapter, we propose the total response time variability metric instead to avoid the feasibility problem but at the same time to preserve the main idea of having any two consecutive tasks at a time distance which remains as constant as the existing resources and other competing jobs permit. The total response time variability is also proposed as a main metric of the stride schedule by Waldspurger and Weihl [13]. We formulate the Response Time Variability (*RTV*) problem as follows.

Given n positive integers $d_1 \leq \cdots \leq d_n$, define $D = \sum_{i=1}^{n} d_i$ and the rates $r_i = \frac{d_i}{D}$ for $i = 1, \ldots, n$. Consider a just-in-time sequence $S = s_1 s_2 \ldots s_D$ of length D where i (a client, a model or a task; in this chapter we shall use the term model most often) occurs exactly d_i times. For any two consecutive occurrences (or copies) of i we define a distance α between them as the number of positions that separate them plus 1. This distance is often referred to as the end-to-end distance between the two occurrences. Since i occurs exactly d_i times in S, then there are exactly d_i distances $\alpha_1^i, \ldots, \alpha_{d_i}^i$ for i, where $\alpha_{d_i}^i$ is the distance between the last and the first occurrence of i in S. Obviously, the two are the same for $d_i = 1$. We shall assume that s_1 immediately follows s_D. Since

$$\alpha_1^i + \cdots + \alpha_{d_i}^i = D,$$

then the average distance $\overline{\alpha}_i$ between the is in S equals

$$\overline{\alpha}_i = \frac{D}{d_i} = \frac{1}{r_i},$$

and it is the same for each feasible sequence S. We define the response time variability for i as follows

$$RTV_i = \sum_{1 \leq j \leq d_i} (\alpha_j^i - \overline{\alpha}_i)^2,$$

and the total response time variability (*RTV*) as follows

$$RTV = \sum_{i=1}^{n} RTV_i = \sum_{i=1}^{n} \sum_{1 \leq j \leq d_i} (\alpha_j^i - \overline{\alpha}_i)^2.$$

Observe that the total response time variability is a weighted variance with the weight being equal to d_i for i, that is

$$RTV = \sum_{i=1}^{n} d_i Var_i,$$

where $Var_i = \frac{1}{d_i} \sum_{1 \leq j \leq d_i} (\alpha_j^i - \overline{\alpha}_i)^2$. By definition, $RTV_i = Var_i = 0$ for any i with $d_i = 1$.

The bottleneck metric (5.2) of Chap. 5 will be used as a secondary metric, with RTV being primary, and it will also be referred to as the Throughput Error (TE) in this chapter.

An input to the Response Time Variability problem is a list of n positive integers $d_1 \leq \cdots \leq d_n$. Their compact, for instance binary, encoding requires $O(\sum \log (d_i + 1))$ bits. Consequently, any algorithm, optimization or heuristic, for the problem that runs in time polynomial in D cannot be deemed polynomial with respect to the compact encoding of the input. Such algorithm polynomial with respect to the compact encoding of the input does not even exist if, as we assume in this chapter, a solution to the Response Time Variability problem is a sequence (or a word) over the alphabet made up of n different symbols, where model i occurs exactly d_i times. Obviously, producing any such sequence will take at least D steps. This assumption appears justified for, to our knowledge, there is in general no compact encoding of solutions (outputs) to the Response Time Variability problem that would limit their sizes by a polynomial of the compact input size. However, we show such a compact scheme in a special case of $n = 2$ in Sect. 7.2.3. We refer the reader to Grigoriev [33], and Brauner et al. [78] for a comprehensive discussion of the class of high multiplicity scheduling problems of which the Response Time Variability problem is a member.

The plan for the reminder of the chapter is as follows. Section 7.2 studies the optimization and the computational complexity of the Response Time Variability problem. It first introduces the number decomposition graphs as a useful tool for the analysis of the total response time variability. It then shows an optimization algorithm for the two model case. The algorithm minimizes both the total response time variability and the bottleneck at the same time. Next, the section shows a dynamic programming algorithm to prove polynomial in D solvability for a fixed number of models. Finally, the section proves that the problem is NP-hard. Section 7.3 presents a simple position exchange heuristic that exchanges positions of model copies in a sequence as long as the exchanges lead to a reduction in the value of RTV. The sequences subjected to this exchange heuristic are generated by various procedures: the bottleneck that solves the bottleneck minimization problem, the insertion based on a solution to the two model case, the Webster's and Jefferson's based on the well known parametric methods of the apportionment. Sections 7.4 and 7.5 present mathematical programming optimization algorithms for the Response Time Variability problem.

7.2 Optimization and Complexity

7.2.1 Number Decomposition Graphs and Response Time Variability

For i with $d_i \geq 2$, consider a vector $\alpha = (\alpha_1, \ldots, \alpha_{d_i})$ of d_i *positive* integers that sum up to D. Without loss of generality we assume that the coordinates of α are ordered in descending order, that is $\alpha_1 \geq \cdots \geq \alpha_{d_i}$. Any vector α that meets the above conditions will be referred to as the *decomposition* vector of D into d_i components. Now, let us define a *unit* exchange operation on the decomposition vector α as follows. Consider two components α_j and $\alpha_k > 1$, $j < k$, of α. Replace α_j by $\alpha_j + 1$ and α_k by $\alpha_k - 1$ and keep all other components of α unchanged. The new components, after possible ordering them, define another decomposition vector β. For instance, adding 1 to the second component and subtracting 1 from the third component of the vector $\alpha = (6, 6, 5)$ leads to the vector $\beta = (7, 6, 4)$. Let us consider a *weighted* directed graph $\mathcal{D}_i$ referred to as the *number decomposition graph* for d_i. This graph's set of nodes $\mathcal{V}_i$ includes all decomposition vectors of D into d_i components, and its set of arcs $\mathcal{A}_i$ includes all pairs of decomposition vectors (α, β) such that β is obtained from α by a single unit exchange between α_j and $\alpha_k > 1$, for some $j < k$, (and possible permutation of the resulting components to keep them in descending order). The weight of the arc (α, β) is defined as

$$2(\alpha_j - \alpha_k + 1). \tag{7.1}$$

We have the following straightforward properties of $\mathcal{D}_i$.

Lemma 7.1. *The following properties hold for* $\mathcal{D}_i$:

- *The graph $\mathcal{D}_i$ is acyclic with a single node without predecessors, called a top node, $\mathcal{N}_i$, and a single node without successors, $\mathcal{M}_i$.*
- $\mathcal{N}_i = (\underbrace{\lceil \frac{D}{d_i} \rceil, \ldots, \lceil \frac{D}{d_i} \rceil}_{D \bmod d_i}, \underbrace{\lfloor \frac{D}{d_i} \rfloor, \ldots, \lfloor \frac{D}{d_i} \rfloor}_{d_i - D \bmod d_i}).$
- $\mathcal{M}_i = (D - d_i + 1, \underbrace{1, \ldots, 1}_{d_i - 1}).$

Proof. If $(\alpha, \beta) \in \mathcal{A}_i$, then

$$\sum_{i=1}^{n} \alpha_i^2 < \sum_{i=1}^{n} \beta_i^2.$$

Therefore, $\mathcal{D}_i$ must be acyclic. The node $\mathcal{N}_i$ has in-degree 0. Otherwise, there would be a node α in $\mathcal{D}_i$ such that $(\alpha, \mathcal{N}_i) \in \mathcal{A}_i$. Then, there would be two components α_j and $\alpha_k > 1$, $j < k$, of α that would become some components $\alpha_j + 1$ and $\alpha_k - 1$ of $\mathcal{N}_i$. Then, $\alpha_j + 1$ must be either $\lfloor \frac{D}{d_i} \rfloor$ or $\lceil \frac{D}{d_i} \rceil$. Consequently, α_j must be either $\lfloor \frac{D}{d_i} \rfloor - 1$ or $\lceil \frac{D}{d_i} \rceil - 1$. However, $\alpha_k - 1 \leq \lceil \frac{D}{d_i} \rceil - 2 \leq \lfloor \frac{D}{d_i} \rfloor - 1$, and thus $\alpha_k - 1$ cannot

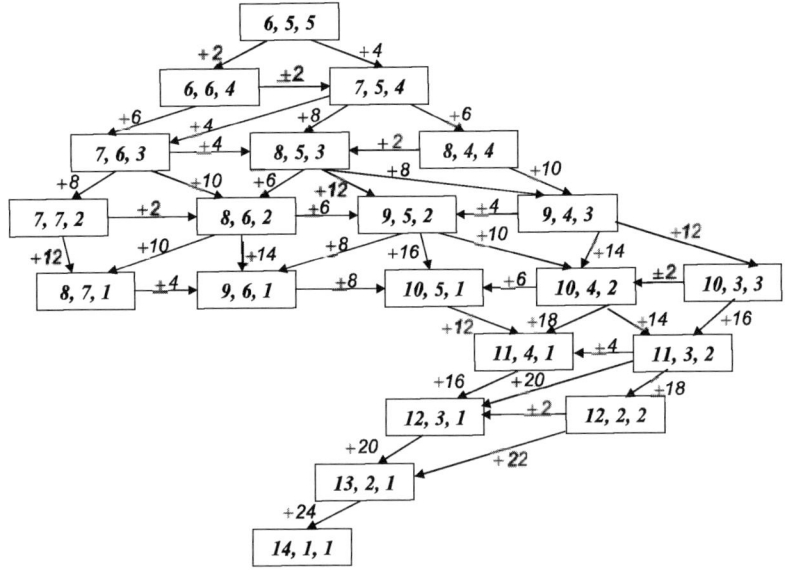

Fig. 7.1 The number decomposition graph for $D = 16$ and $d_i = 3$

be a component of $\mathcal{N}_i$. No other node has in-degree 0 since it can be shown that there is a path from $\mathcal{N}_i$ to any other node in $\mathcal{D}_i$. Finally, by definition the only node with out-degree 0 is the one that has the last d_i components all equal 1. Otherwise a unit exchange from this node would be possible. Moreover, since all components of any node must sum up to d_i, there is only one such a node, namely $\mathcal{M}_i$. □

Figure 7.1 presents the number decomposition graph for $D = 16$ and $d_i = 3$. Consider $\mathcal{N}_i$ and another node α in $\mathcal{D}_i$. The weight of a directed path from $\mathcal{N}_i$ to α is the sum of all the weights along the path. Also, let

$$RTV_i(\alpha) = \sum_{j=1}^{d_i} \left(\alpha_j - \frac{D}{d_i} \right)^2$$

for α. We have the following lemma.

Lemma 7.2. *For any node α in $\mathcal{D}_i$, we have*

$$RTV_i(\alpha) = RTV_i(\mathcal{N}_i) + w_i(p),$$

where p is a path from $\mathcal{N}_i$ to α and $w_i(p)$ is its total weight.

Proof. The proof is by induction on the number of arcs a along a path connecting $\mathcal{N}_i$ and α in $\mathcal{D}_i$. The proof is obvious for $a = 0$ since then $\alpha = \mathcal{N}_i$. Suppose the lemma holds for $a \geq 0$. We show that then it holds for $a + 1$ as well. Let $p = (\alpha^0 = \mathcal{N}_i, \ldots, \alpha^{a+1} = \alpha)$ be a path of $a + 1$ arcs connecting $\mathcal{N}_i$ and α. Then, by the inductive assumption

$$RTV_i(\alpha^a) = RTV_i(\mathcal{N}_i) + w_i(p'),$$

where p' is p without the last node α^{a+1}. By definition of $RTV_i(\alpha^a)$

$$
\begin{aligned}
RTV_i\left(\alpha^{a+1}\right) &= \sum_{l=1}^{d_i}\left(\alpha_l^{a+1} - \frac{D}{d_i}\right)^2 \\
&= \sum_{l=1}^{d_i}\left(\alpha_l^a - \frac{D}{d_i}\right)^2 - \left(\alpha_j^a - \frac{D}{d_i}\right)^2 - \left(\alpha_k^a - \frac{D}{d_i}\right)^2 \\
&\quad + \left(\alpha_j^a + 1 - \frac{D}{d_i}\right)^2 + \left(\alpha_k^a - 1 - \frac{D}{d_i}\right)^2 \\
&= RTV_i(\alpha^a) + 2(\alpha_j^a - \alpha_k^a + 1).
\end{aligned}
$$

Consequently,

$$RTV_i(\alpha^{a+1}) = RTV_i(\mathcal{N}_i) + w_i(p') + 2(\alpha_j^a - \alpha_k^a + 1) = RTV_i(\mathcal{N}_i) + w_i(p),$$

since by definition $2(\alpha_j^a - \alpha_k^a + 1)$ is the weight on the arc (α^a, α^{a+1}). This ends the induction and proves

$$RTV_i(\alpha) = RTV_i(\mathcal{N}_i) + w_i(p).$$

$\square$

Notice that the equation in Lemma 7.2 is independent of the choice of path p for in this equation both $RTV_i(\alpha)$ and $RTV_i(\mathcal{N}_i)$ are constants. Thus, we immediately derive the following result.

Lemma 7.3. *For any node α in $\mathcal{D}_i$, all directed paths from $\mathcal{N}_i$ to α have the same weight.*

By Lemma 7.3 the weight of any $(\mathcal{N}_i, \alpha)$-path in $\mathcal{D}_i$ depends only on α. Therefore, we denote the weight of any such path by $w_i(\alpha)$. The following lemma links the decomposition graphs and the total response time variability.

Lemma 7.4. *Let S be a solution to the Response Time Variability problem with its value equal to RTV. Then there are nodes $\alpha^1, \ldots, \alpha^n$ in the decomposition graphs $\mathcal{D}_1, \ldots, \mathcal{D}_n$ respectively, such that*

$$RTV = \sum_{i=1}^{n} w_i(\alpha^i) + \sum_{i=1}^{n} RTV_i(\mathcal{N}_i),$$

where

$$RTV_i(\mathcal{N}_i) = (D \bmod d_i)\left(\left\lceil \frac{D}{d_i}\right\rceil - \frac{D}{d_i}\right)^2 + (d_i - D \bmod d_i)\left(\left\lfloor \frac{D}{d_i}\right\rfloor - \frac{D}{d_i}\right)^2 \quad (7.2)$$

for $i = 1, \ldots, n$.

Proof. Follows from Lemma 7.2. $\square$

7.2.2 Lower Bounds on Response Time Variability

The following theorem shows that the total response time variability can be 0 only if at least two demands are equal.

Theorem 7.5. *If $d_1 < \ldots < d_n$ for $n > 1$, then $RTV > 0$.*

Proof. If $d_i \nmid D$ for some i, then the theorem holds since the average distance for i equal $\frac{D}{d_i}$ is not integer and all response times are integer. Otherwise, $d_i \mid D$ for each i. By contradiction, if $RTV = 0$, then for each i all response times are equal the average response time $\frac{D}{d_i}$. Consequently, there are non-negative integers $a_1, \ldots, a_n$ such that $(a_1, \frac{D}{d_1}), \ldots, (a_n, \frac{D}{d_n})$ is an exact covering sequence. However, by Lemma 6.12, then $\frac{D}{d_1} = \frac{D}{d_2}$ and consequently $d_1 = d_2$, which leads to a contradiction. This proves the theorem. $\square$

The rates for which it is possible to build a regular sequence for each model that does not overlap with any other model's regular sequence would attain the lower bound of the total response time variability given in (7.2). These sequences would make up an optimal solution to the Response Time Variability problem. We have the following theorem.

Theorem 7.6. *If for the rates $r_i = \frac{d_i}{D}$, $i = 1, \ldots, n$, the problem (6.30) has an affirmative answer then the instance $(d_1, \ldots, d_n)$ of the Response Time Variability problem attains the value*

$$RTV = \sum_{i=1}^{n} RTV_i(\mathcal{N}_i)$$

for the optimal solution which is the exact cover in the problem (6.30).

Proof. The affirmative answer to the problem (6.30) implies that there are phases $\theta_1, \ldots, \theta_n$ such that the sets

$$S_i = \left\{ \left\lceil \frac{j}{r_i} - \frac{\theta_i}{r_i} \right\rceil : j \in \mathbb{N} \right\}$$

for $i = 1, \ldots, n$ make up an exact cover of $\mathbb{N}$, that is each integer is in exactly one set S_i. For each i we have

$$\left\lfloor \frac{1}{r_i} \right\rfloor \leq \left\lceil \frac{j+1}{r_i} - \frac{\theta_i}{r_i} \right\rceil - \left\lceil \frac{j}{r_i} - \frac{\theta_i}{r_i} \right\rceil \leq \left\lceil \frac{1}{r_i} \right\rceil \tag{7.3}$$

see Exercise (6.40) which means that the consecutive copies of i are either at the distance $\left\lfloor \frac{1}{r_i} \right\rfloor$ or $\left\lceil \frac{1}{r_i} \right\rceil$. Moreover, the copy $j = d_i$ is in a position

$$\left\lceil D - \frac{\theta_i}{r_i} \right\rceil \leq D$$

for $i = 1,\ldots,n$. Thus, there are at least d_i copies of i in the first D positions and since $D = \sum_{i=1}^{n} d_i$ there must be exactly d_i copies of i in the first D positions. Therefore, the first D positions make up a just-in-time sequence and by (7.3)

$$RTV_i = RTV_i(\mathcal{N}_i)$$

for $i = 1,\ldots,n$. Thus the theorem holds. $\square$

The opposite claim does not hold, that is there may be just-in-time sequences with their total response time variability equal the lower bound $\sum_{i=1}^{n} RTV_i(\mathcal{N}_i)$ yet not regular for all models. An example is the following optimal just-in-time sequence for $d_1 = 5$ and $d_2 = 12$

$$12221222122122122$$

with $\mathcal{N}_1 = (4,4,3,3,3)$ and $\mathcal{N}_2 = (2,2,2,2,2,1,1,1,1,1,1,1)$. The sequence is not a balanced word, see Sect. 9.8 for definition, since its two factors

$$2221222 \tag{7.4}$$

and

$$1221221 \tag{7.5}$$

of length 7 each differ on model 2 by more than 1. That is model 2 occurs six times in the former and four times in the latter. Thus, the sequence must not be regular on at least one model, see Altman et al. [7].

7.2.3 Two Model Case

This section considers the two model case, $n = 2$. It shows a solution that minimizes both the total response time variability and the bottleneck at the same time, which usually is impossible for more than two models. Prior to giving the details of the solution we point out that the sequence minimizing the total response time variability for two models is quite straightforward to obtain as follows. Let $d_1 < d_2$. We omit the case $d_1 = d_2$ since it is trivial. It follows from Lemma 7.4 that if one can find a solution with

$$RTV = RTV_1(\mathcal{N}_1) + RTV_2(\mathcal{N}_2),$$

where

$$RTV_i(\mathcal{N}_i) = (D \bmod d_i)\left(\lceil \frac{D}{d_i} \rceil - \frac{D}{d_i}\right)^2 + (d_i - D \bmod d_i)\left(\lfloor \frac{D}{d_i} \rfloor - \frac{D}{d_i}\right)^2$$

for $i = 1,2$, then this solution minimizes the total response time variability. Such a solution is always possible, however, since $\lfloor \frac{D}{d_1} \rfloor \geq 2$ and $2 > \frac{D}{d_2} > 1$. Namely, consider a sequence that begins with 1 and that has each next copy of 1 at either a distance $\lceil \frac{D}{d_1} \rceil$ or a distance $\lfloor \frac{D}{d_1} \rfloor$ from the last one. The number of times the distances $\lceil \frac{D}{d_1} \rceil$ and $\lfloor \frac{D}{d_1} \rfloor$ need to be used in this sequence are $(D \bmod d_1)$ and $(d_1 - D \bmod d_1)$ respectively. The sequence separates any two consecutive empty positions by at most one position with a copy of 1. The empty positions can then be filled in with the copies of 2 to ensure the desired distances, either 2 or 1, for this model. The distances 2 and 1 must occur exactly $(D \bmod d_2)$ and $(d_2 - D \bmod d_2)$ times respectively in the sequence, otherwise their total would not be D. The result is a sequence that minimizes RTV which may not, however, minimize TE at the same time. Therefore, we now present another solution that minimizes both RTV and TE simultaneously. We assume that $\gcd\{d_1, d_2\} = 1$. Otherwise, the optimal solution for $\frac{d_1}{g}$ and $\frac{d_2}{g}$ can be repeated $g = \gcd\{d_1, d_2\}$ times resulting into an optimal solution with respect to both metrics for the original instance with demands d_1 and d_2.

We show in Lemmas 7.7–7.9 that the solutions defined in Lemmas 5.14 and 5.16 are optimal for the Response Time Variability problem. We refer to these solutions as the solutions *based on ideal positions*. To that end, we prove that the distance between any two consecutive copies of 1 equals either $\lceil \frac{D}{d_1} \rceil$ or $\lfloor \frac{D}{d_1} \rfloor$ and for 2 equals either $\lceil \frac{D}{d_2} \rceil$ or $\lfloor \frac{D}{d_2} \rfloor$. With definitions of a_j for $j = 1, \ldots, d_1$ and b_k for $k = 1, \ldots, d_2$ as in Lemma 5.13, that is $a_j = \frac{2j-1}{2r_1}$ for $j = 1, \ldots, d_1$ and $b_k = \frac{2k-1}{2r_2}$ for $k = 1, \ldots, d_2$ we have the following results.

Lemma 7.7. *We have* $\lfloor \frac{D}{d_1} \rfloor \leq \lceil a_{j+1} \rceil - \lceil a_j \rceil \leq \lceil \frac{D}{d_1} \rceil$ *for* $j = 1, \ldots, d_1 - 1$ *and* $\lfloor \frac{D}{d_2} \rfloor \leq \lceil b_{j+1} \rceil - \lceil b_j \rceil \leq \lceil \frac{D}{d_2} \rceil$ *for* $j = 1, \ldots, d_2 - 1$.

Proof. We have,

$$\frac{D}{d_1} - 1 = a_{j+1} - a_j - 1 \leq \lceil a_{j+1} \rceil - \lceil a_j \rceil \leq a_{j+1} - a_j + 1 = \frac{D}{d_1} + 1.$$

Since $\frac{D}{d_1}$ is not an integer and $\lceil a_{j+1} \rceil - \lceil a_j \rceil$ is an integer, then

$$\lfloor \frac{D}{d_1} \rfloor \leq \lceil a_{j+1} \rceil - \lceil a_j \rceil \leq \lceil \frac{D}{d_1} \rceil.$$

The proof for model 2 is similar and thus will be omitted. $\square$

Lemma 7.8. *We have* $\lfloor \frac{D}{d_1} \rfloor \leq D - \lceil a_{d_1} \rceil + \lceil a_1 \rceil \leq \lceil \frac{D}{d_1} \rceil$ *and* $\lfloor \frac{D}{d_2} \rfloor \leq D - \lceil b_{d_2} \rceil + \lceil b_1 \rceil \leq \lceil \frac{D}{d_2} \rceil$.

Proof. We have $a_{d_1} = D - \frac{D}{2d_1}$, $a_1 = \frac{D}{2d_1}$ and consequently

$$\frac{D}{d_1} - 1 = D - a_{d_1} + a_1 - 1 \leq D - \lceil a_{d_1} \rceil + \lceil a_1 \rceil \leq D - a_{d_1} + a_1 + 1 = \frac{D}{d_1} + 1.$$

Since $\frac{D}{d_1}$ is not an integer and $D + \lceil a_{d_1} \rceil - \lceil a_1 \rceil$ is an integer, then

$$\lfloor \frac{D}{d_1} \rfloor \le D - \lceil a_{d_1} \rceil + \lceil a_1 \rceil \le \lceil \frac{D}{d_1} \rceil.$$

The proof for model 2 is similar and thus will be omitted. □

Lemma 7.9. *For d_1 and d_2 odd, we have*

$$b_{\frac{d_2+1}{2}} + 1 - \lceil b_{\frac{d_2+1}{2}-1} \rceil = \lceil \frac{D}{d_2} \rceil$$

and

$$\lceil b_{\frac{d_2+1}{2}+1} \rceil - b_{\frac{d_2+1}{2}} - 1 = \lfloor \frac{D}{d_2} \rfloor.$$

Proof. For d_1 and d_2 odd, we have $b_{\frac{d_2+1}{2}} = \frac{D}{2}$, which is an integer. Consequently,

$$\frac{D}{d_2} = b_{\frac{d_2+1}{2}} - b_{\frac{d_2+1}{2}-1} \le b_{\frac{d_2+1}{2}} + 1 - \lceil b_{\frac{d_2+1}{2}-1} \rceil \le b_{\frac{d_2+1}{2}} + 1 - b_{\frac{d_2+1}{2}-1} = \frac{D}{d_2} + 1.$$

Since $\frac{D}{d_2}$ is not an integer and $b_{\frac{d_2+1}{2}} + 1 - \lceil b_{\frac{d_2+1}{2}-1} \rceil$ is an integer, then the first equality holds. Furthermore,

$$\frac{D}{d_2} - 1 = b_{\frac{d_2+1}{2}+1} - b_{\frac{d_2+1}{2}} - 1 \le \lceil b_{\frac{d_2+1}{2}+1} \rceil - b_{\frac{d_2+1}{2}} - 1 \le b_{\frac{d_2+1}{2}+1} - b_{\frac{d_2+1}{2}} = \frac{D}{d_2}.$$

Since $\lceil b_{\frac{d_2+1}{2}+1} \rceil - b_{\frac{d_2+1}{2}} - 1$ is an integer, then the second inequality also holds. This proves the lemma. □

We can now conclude with the following theorem.

Theorem 7.10. *For $n = 2$, the solutions based on ideal positions minimize the total response time variability.*

Proof. Follows immediately from Lemmas 7.4, 7.7, 7.8, and 7.9 □

By Lemmas 5.14 and 5.16, the position of each copy of either model in the optimal just-in-time sequence can be computed in time polynomial in $O(\log(d_1 + 1) + \log(d_2 + 1))$, that is, polynomial in the input size. So it can be the value of *RTV* of the optimal solution by Lemma 7.4.

We close this section by showing that the solution based on ideal positions minimizes bottleneck as well.

Theorem 7.11. *For $n = 2$, the solutions based on ideal positions minimize bottleneck.*

Proof. It suffices to notice that if both demands are odd, then all copies are in their ideal positions, except for copy $\frac{d_2+1}{2}$ of model 2 which is in position $\frac{D}{2} + 1$ whereas its ideal position is $\frac{D}{2}$. This, however, does not change the deviation for the copy, which is $\frac{1}{2}$ in either position. Consequently, the maximum deviation equals $\frac{1}{2}$ which is optimal for both demands being odd. Therefore, the solutions based on ideal positions minimize maximum deviation. □

7.2.4 Fixed Number of Models

This section shows a straightforward dynamic program for the Response Time Variability problem. This program is similar to the algorithm of Anily et al. [77] for the Periodic Scheduling Maintenance problem, see Sect. 7.2.5 for the problem definition, with the difference that in the *RTV* problem one has to maintain also the required number of model copies.

The program proves that the problem can be solved in time polynomial in D for any *fixed* number of models, and thus complements our other results concerning its complexity. However, the program is not intended as a practical alternative for efficiently solving the problem.

The state of the dynamic program is represented by a quadruple $\langle f, \ell, q, d \rangle$. The f is an n dimensional vector $f = (f_1, \ldots, f_n)$, where $f_i = 0, 1, \ldots, D - d_i + 1$ for $i = 1, \ldots, n$ represents the position of the first copy of i. The ℓ is an n dimensional vector $\ell = (\ell_1, \ldots, \ell_n)$, where $\ell_i = 0, d_i + 1, \ldots, D$ for $i = 1, \ldots, n$ represents the position of the last copy of i in the sequence of length d. The q is an n dimensional vector $q = (q_1, \ldots, q_n)$. It represents the number of copies that remain to be sequenced, $q_i = 0, \ldots, d_i$. Finally, d is the length of the current sequence, $d = 0, \ldots, D$. Initially, $f = \ell = 0$, $q = (d_1, \ldots, d_n)$, and $d = 0$. A final state is any state with $q = 0$ and $d = D$. There is a weighted arc from a non-final state $\langle f, \ell, q, d \rangle$ to a state $\langle f', \ell', q', d' \rangle$ if and only if there is an i such that $q'_i = q_i - 1 \geq 0$, $d' = d + 1$, $\ell'_i = d'$. Furthermore, if $f_i = 0$, then $f'_i = d'$. The weight for the arc is calculated as follows for $d_i \geq 2$,

$$
\ell_{\langle f, \ell, q, d \rangle \langle f', \ell', q', d' \rangle} =
\begin{cases}
0, & q_i = d_i; \\
(d' - \ell_i - \frac{D}{d_i})^2, & d_i - 1 \geq q_i > 1; \\
(d' - \ell_i - \frac{D}{d_i})^2 \\
\quad + (D - d' + f_i - \frac{D}{d_i})^2 & q_i = 1
\end{cases}
$$

and it is equal to 0 for $d_i = 1$. Finally, we connect all final states to a dummy state referred to as the destination. All arcs to the destination have weight 0. The shortest path, more precisely the lightest path, between the initial state and the destination obviously defines an optimal solution to the Response Time Variability problem. This path can be found in time which is polynomial in the number of nodes, which is of $O(D^{3n+1})$, therefore the complexity is polynomial in D for a fixed number of products n.

7.2.5 Complexity

We now show that the Response Time Variability problem is NP-hard. The reduction is from the Periodic Maintenance Scheduling problem (PMSP) shown NP-complete by Bar-Noy et al. [79], and motivated by the maintenance problem of Wei and Liu [76], and Anily et al. [77]. The Periodic Maintenance Scheduling problem is defined as follows.

Given m machines and integer service intervals $\ell_1, \ell_2, \ldots, \ell_m$ such that $\sum \frac{1}{\ell_i} < 1$. Does there exist a servicing schedule (or a servicing cycle) $s_1 \ldots s_L$, where $L = \text{lcm}(\ell_1, \ell_2, \ldots, \ell_m)$ is the least common multiple of $\ell_1, \ell_2, \ldots, \ell_m$, of these machines in which the end-to-end distance between any two consecutive servicing of machine i is exactly ℓ_i, no more than one machine is serviced in a single time slot, and it takes a single time slot to perform servicing of any machine? The distance between the last and the first servicing of machine i in the cycle, if any exists, is also ℓ_i. The Periodic Maintenance Scheduling problem has been shown NP-complete by Bar-Noy et al. [79]. Their proof shows that the Periodic Maintenance Scheduling problem is NP-complete in the ordinary sense since the magnitude of distances ℓ_i used in the transformation from the graph coloring problem is not necessarily bounded by any polynomial of the graph size. Thus the transformation of Bar-Noy et al. [79] is not pseudo-polynomial. However, a careful selection of an initial graph coloring problem in their transformation would ensure that the magnitude of the distances remains bounded by a polynomial of the graph size thus ensuring pseudo-polynomial transformation and consequently strong NP-completeness of the Periodic Maintenance Scheduling problem. We shall give details of this new transformation here. We begin with the following lemma.

Lemma 7.12. *The initial positions $v_1, \ldots, v_m$ of a servicing cycle are feasible for the PMSP instance if and only if $v_i - v_j \neq 0 \pmod{\gcd(\ell_i, \ell_j)}$ for any pair $i \neq j$.*

Proof. For any pair $i \neq j$ the numbers

$$\frac{\ell_i}{\gcd(\ell_i, \ell_j)} \quad \text{and} \quad \frac{\ell_j}{\gcd(\ell_i, \ell_j)}$$

are relatively prime. Thus, there are integers x and y that solve the following linear Diophantine equation, see Le Veque [80],

$$\frac{\ell_i}{\gcd(\ell_i, \ell_j)} x + \frac{\ell_j}{\gcd(\ell_i, \ell_j)} y = 1,$$

or equivalently the equation

$$x\ell_i + y\ell_j = \gcd(\ell_i, \ell_j).$$

If $v_i - v_j = 0 \pmod{\gcd(\ell_i, \ell_j)}$ for some pair $i \neq j$, then $v_i - v_j = q \gcd(\ell_i, \ell_j)$ for some integer q. Therefore, there are integers $x' = -xq$ and $y' = yq$ such that

$$y'\ell_j - x'\ell_i = v_i - v_j,$$

and consequently

$$y'\ell_j + v_j = x'\ell_i + v_i$$

and thus the machines i and j are scheduled for maintenance at the same time $y'\ell_j + v_j$. Hence $v_1, \ldots, v_m$ are not feasible.

Now let us assume that $v_1, \ldots, v_m$ are infeasible. Then, there are $i \neq j$ such that machines i and j are scheduled for maintenance at the same time

$$t = x\ell_i + v_i = y\ell_j + v_j,$$

for some integers x and y. Then, however

$$v_i - v_j(\text{mod } \gcd(\ell_i, \ell_j)) = y\ell_j - x\ell_i(\text{mod } \gcd(\ell_i, \ell_j)) = 0$$

which proves the lemma. $\square$

The following restricted graph coloring problem will be used in the transformation.

Theorem 7.13. *The following restricted graph coloring problem remains NP complete in the strong sense: Given n−node graph G with each node degree being at least $n - 7$ and at most $n - 3$. Can the nodes of G be colored with k colors?*

Proof. The Exact Cover by 3-Sets problem remains NP-complete in the strong sense if no element occurs in more than three subsets, see Garey and Johnson [81]. Therefore, the transformation from the Exact Cover by 3-Sets problem to the Partition into Triangles problem given in the Garey and Johnson's book [81] on pages 68 and 69 builds a graph which is K_4-free (K_4-free graph contains no clique of size 4), has a maximum node-degree equal 6, a minimum node-degree equal 2, and the number of nodes being a multiple of 3. Therefore, the Partition into Triangles remains NP-complete for the class of graphs a member of which has just been defined. Let G be any graph of this class with n-nodes. Consider a complement $\overline{G}$ of G. We have a minimum node-degree equal $n - 7$, and a maximum node-degree equal $n - 3$ in $\overline{G}$. We are asking for $k = \frac{n}{3}$ coloring of the complement $\overline{G}$. The transformation just described is polynomial. Moreover, if the sets $T_1, \ldots, T_{\frac{n}{3}}$ make up the partition of G into triangles, then we color all nodes in T_i of $\overline{G}$ with color i. Thus, we get $\frac{n}{3}$−coloring of $\overline{G}$ since the nodes in T_i make up an independent set in $\overline{G}$. On the other hand, let there be a k−coloring, $k \leq \frac{n}{3}$, of $\overline{G}$. Let $C_1, \ldots, C_k$ be the sets of nodes colored with colors $1, \ldots, k$ respectively. Then, no set has more than three nodes, otherwise G would not be K_4-free. Therefore, each set has exactly *three* nodes, and $k = \frac{n}{3}$. Thus, $C_1, \ldots, C_{\frac{n}{3}}$ are triangles in G and they make up the required partition into triangles. This shows that the k-coloring of $\overline{G}$ is NP-complete in the strong sense, which proves the theorem. $\square$

We are now ready to prove.

Theorem 7.14. *The Periodic Maintenance Scheduling problem is NP complete in the strong sense.*

Proof. Lemma 7.12 shows how to verify if the initial positions $v_1, \ldots, v_m$ are feasible for the PMSP instance in polynomial time. Thus, the PMSP is in NP. Now, consider the restricted graph coloring problem defined in Theorem 7.13. We show a pseudopolynomial transformation form this graph coloring problem to the PMSP thus proving that the latter is NP complete in the strong sense. Let graph G with $n-$ nodes and k make up an instance of the restricted graph coloring problem. Let K_n be an $n-$node clique such that its edges in G are painted red and those in $\overline{G}$ are painted white. Since the node-degree in G is at least $n-7$, the number of white-painted edges incident with a given node in K_n is at most 6. Thus, the number of white-painted edges in K_n is at most $3n$. Let us assign a distinct prime number greater than n to each white-painted edge of K_n. Since the number $\pi(x)$ of prime numbers not exceeding $x \geq 2$ satisfies the following inequality, see Le Veque [80]

$$\frac{7}{8} \frac{x}{\log x} < \pi(x),$$

then by taking

$$x = 16n^2$$

we get

$$4n < \pi(16n^2)$$

and thus we can generate the required number of distinct primes greater than n in time which is polynomial in n. The prime numbers do not exceed $16n^2$. Since the node degree in G is at most $n-3$, then the number of white-painted edges incident with a given node is at least 2. Now define a weight of a node in K_n to be the product of all prime numbers assigned to the white-painted edges incident with the node times k.

We are now ready to define the instance of the PMPS problem as follows. There are n machines in this instance, each corresponding to a node of the K_n graph. The ℓ_i for machine i equals the weight of node i. We have

$$\sum_{i=1}^{n} \frac{1}{\ell_i} < 1$$

since $\ell_i > n$. Clearly, the magnitude of ℓ_i is bounded by $k(16n^2)^6$. Consequently, the transformation is pseudopolynomial.

We make the following observation

$$\gcd(\ell_i, \ell_j) = \begin{cases} k & \text{if the edge } (i,j) \text{ of } K_n \text{ is red-painted} \\ pk & \text{if the edge } (i,j) \text{ of } K_n \text{ is white-painted with weight } p. \end{cases} \quad (7.6)$$

We now show that if G is k-colorable, then there exists a feasible maintenance service for the PMPS instance. Let the colors used in the k-coloring be $0, 1, \ldots, k-1$, and let there be n_i nodes colored with the color i. The nodes with color i can then be uniquely numbered $0, 1, \ldots, n_i - 1$. Set the initial position v_i for machine i to $mk + l$, where the l is the color of node i, and the m is the node number in the set with color l. Since $m \leq n-1$ and $l \leq k-1$, we have

$$v_i = mk + l \leq (n-1)k + k - 1 = nk - 1 < nk. \quad (7.7)$$

We show that

$$v_i - v_j \neq 0 (\text{mod } \gcd(\ell_i, \ell_j)) \tag{7.8}$$

for any two machines $i \neq j$. If the edge (i, j) is white-painted, then by (7.6), $\gcd(\ell_i, \ell_j) = pk$ for some prime number $p > n$. However, by (7.7)

$$0 < |v_i - v_j| < nk.$$

Therefore, (7.8) holds for machines i and j along the white-painted edge (i, j). Otherwise, if the edge (i, j) is red-painted, then by (7.6), $\gcd(\ell_i, \ell_j) = k$ and $v_i = mk + l$, $v_j = m'k' + l'$ for some integers m and m'. Furthermore, $l \neq l'$ since $(i, j) \in G$. Hence, $v_i - v_j = l - l' (\text{mod } k)$. However,

$$0 < |l - l'| < k,$$

thus, (7.8) holds again. Therefore, by Lemma 7.12, there is a feasible maintenance service for the PMPS instance.

Now, let us assume that the initial positions $v_1, \ldots, v_n$ are feasible for the PMSP instance. We show that this implies the k-colorability of G. We assign color $v_i \bmod k$ to node i so that the number of colors used does not exceed k. We need to show that

$$v_i \bmod k \neq v_j \bmod k$$

for $(i, j) \in G$. By contradiction, assume that

$$v_i \bmod k = v_j \bmod k$$

and $(i, j) \in G$. Then, by (7.6)

$$v_i - v_j = 0 (\text{mod } \gcd(l_i, l_j)),$$

consequently, by Lemma 7.12, $v_1, \ldots, v_n$ are infeasible for the PMSP instance. Hence, a contradiction and thus G is k-colorable. □

The NP-completeness of PMSP is the point of departure in the main complexity result for the Response Time Variability problem.

Theorem 7.15. *The Response Time Variability problem is NP-hard.*

Proof. For an instance $\ell_1, \ell_2, \ldots, \ell_m$ of the Periodic Maintenance Scheduling problem, define demands $d_i = \frac{L}{\ell_i}$, $i = 1, \ldots, m$, and $d_{m+1} = 2L - \sum d_i$, where $L = \text{lcm}(\ell_1, \ell_2, \ldots, \ell_m)$. Thus, $D = 2L$ and $n = m + 1$. We observe that the average distance $\frac{D}{d_{m+1}} = \frac{2L}{2L - \sum d_i}$ for $m + 1$ is between 1 and 2. Therefore, by Lemma 7.4, the contribution of model $m + 1$ to the total response time variability in any solution is at least

$$V = (D \bmod d_{m+1}) \left(2 - \frac{D}{d_{m+1}} \right)^2 + (d_{m+1} - (D \bmod d_{m+1})) \left(1 - \frac{D}{d_{m+1}} \right)^2.$$

We make V the upper bound on the total response time variability for all models $1,\ldots,m,m+1$, which is intended to force the models from 1 to m to keep their response time variabilities equal 0 in any solution that respects the upper bound V. That is, all their consecutive copies must be at (end-to-end) distances $\frac{D}{d_i} = 2\ell_i$, respectively, in such a solution. We now prove that the instance of the Periodic Maintenance Scheduling problem has a solution if and only if its corresponding instance of the Response Time Variability problem has variability not exceeding V.

(If) Assume that there is a solution

$$s_1 s_2 \ldots s_L \tag{7.9}$$

for the Periodic Maintenance Scheduling problem. Then, consider the sequence

$$S = s_1' n s_2' n \ldots s_L' n$$

with every other time slot occupied by n, and $s_j' = s_j$ if s_j is a machine or $s_j' = n$ if s_j is empty in the solution (7.9). In such a sequence, consecutive copies of i are exactly $2\ell_i$ time slots apart in S for $i = 1,\ldots,m$. Therefore, the response time variability for each of these is is 0. Furthermore, any two consecutive copies of n are either next to each other, thus at a distance 1, or separated by a single copy of $i \neq n$, thus at a distance 2. Consequently, the total response time variability of S equals V.

(Only if) Now, let us assume that there is a solution Q to the Response Time Variability problem with variability $RTV \leq V$. Firstly, we observe from Lemma 7.4 that $RTV_n \geq V$ in any solution to the Response Time Variability problem, and consequently $RTV = RTV_n$ and $RTV_i = 0$ for $i = 1,\ldots,m$ in Q. Secondly, we observe that if any two copies of n were three or more slots apart in Q, that is, if they were separated by two or more copies of $i \neq n$, then $RTV_n > RTV$ for Q, which would lead to a contradiction. Therefore, no $i \leq m$ and $j \leq m$ are next to each other in Q. The first observation implies that all distances between consecutive copies of model i, $i = 1,\ldots,m$, are equal in Q. Furthermore, since there are d_i copies of i in the sequence of length D, then the distance between the consecutive copies of i is $\frac{D}{d_i} = 2\ell_i$ in Q. Now, let $c_i, c_i + 2\ell_i, \ldots, c_i + 2(d_i - 1)\ell_i$ be the ends of time slots in Q occupied by i, $i = 1,\ldots,m$. Of course all these numbers are different as no time slot is occupied by more than one i. Consequently, all numbers, $\frac{c_i}{2}, \frac{c_i}{2} + \ell_i, \ldots, \frac{c_i}{2} + (d_i - 1)\ell_i$, $i = 1,\ldots,m$, are different. Furthermore, by the second observation, if for some i, j, h and k, $\frac{c_k}{2} + h\ell_k > \frac{c_i}{2} + j\ell_i$, then $\frac{c_k}{2} + h\ell_k - (\frac{c_i}{2} + j\ell_i) > \frac{1}{2}$. Consequently, all numbers $\lceil \frac{c_i}{2} \rceil, \lceil \frac{c_i}{2} \rceil + \ell_i, \ldots, \lceil \frac{c_i}{2} \rceil + (d_i - 1)\ell_i$, $i = 1,\ldots,m$, are different. Therefore, by servicing machine i in time slots $\lceil \frac{c_i}{2} \rceil, \lceil \frac{c_i}{2} \rceil + \ell_i, \ldots, \lceil \frac{c_i}{2} \rceil + (d_i - 1)\ell_i, i = 1,\ldots,m$ we obtain a solution to the Periodic Maintenance Scheduling problem. This proves the theorem. $\square$

The transformation presented in the proof of Theorem 7.15 is not pseudo-polynomial since the largest number $L = \text{lcm}(\ell_1, \ell_2, \ldots, \ell_m)$ in the instance of the Response Time Variability problem may not be bounded by any polynomial of $\max\{\ell_i\}$ and $m \log \max\{\ell_i\}$, the latter being the input size of the Periodic Maintenance Scheduling problem. Consequently, this transformation does not ensure that the Response Time Variability problem is NP-hard in the strong sense even if the Periodic Maintenance Scheduling problem is NP-complete in the strong sense. Therefore, it remains an open question whether the Response Time Variability problem is NP-hard in the strong sense or whether there is a pseudo-polynomial, that is polynomial in D, time algorithm that solves it to optimality. It also remains open if the decision counterpart of the Response Time Variability problem is in NP.

7.3 Heuristics

This section describes heuristics for the total response time variability problem. Each of these heuristics uses the Exchange Heuristic, described in Sect. 7.3.1, to exchange the neighboring copies of a sequence in order to reduce its response time variability. The exchanges can be applied to any feasible sequence and their outcome is sensitive to the selection of the initial sequence. Section 7.3.2 describes a number of different ways the initial sequence can obtained.

7.3.1 The Exchange Heuristic

Consider a sequence $s = s_1 \ldots s_D$ and a model i with $d_i \geq 2$. Let i be in position p of s, that is $s_p = i$. We define the closest to position p *clockwise i* as the first i in s encountered when moving clockwise away from p, similarly, we define the closest to p *counter-clockwise i* as the first i encountered when moving counter-clockwise away from p. Notice that if $d_i = 1$ or 2, then the clockwise i is the same as the counter-clockwise i. The Exchange Heuristic does the exchange of two neighboring copies whenever it reduces the total response time variability. More precisely, consider copies of i and j, $i \neq j$, that are next to each other in a given sequence s, assume the i is in position p and the j in position $p + 1$. Let R_i and L_i be distances to the closest to p clockwise and counter-clockwise, respectively, i in s. Similarly, let R_j and L_j be distances to the closest to $p + 1$ clockwise and counter-clockwise, respectively, j in s. Then, for $d_i, d_j \geq 2$,

$$B = L_i^2 + R_i^2 + L_j^2 + R_j^2$$

before the exchange and

$$A = (L_i + 1)^2 + (R_i - 1)^2 + (L_j - 1)^2 + (R_j + 1)^2$$

after the exchange. Since all the other distances remain unchanged, we have the following net change in the value of the total response time variability,

$$\Delta = B - A = 2(R_i - L_i - 1) + 2(L_j - R_j - 1).$$

If $d_j = 1$ and $d_i \geq 2$, then $\Delta = 2(R_i - L_i - 1)$. If $d_i = 1$ and $d_j \geq 2$, then $\Delta = 2(L_j - R_j - 1)$. Finally, if $d_i = d_j = 1$, then $\Delta = 0$. The exchange takes place only if Δ is positive. The Exchange Heuristic starts with position 1 and passes clockwise through the sequence checking each couple $s_p s_{p+1}$. If the exchange within the couple reduces the response time variability, then it is made and the algorithm proceeds to position $p + 1$. Position D is immediately followed by position 1. If position 1 is reached (again) without any reduction in the total response time variability, then the algorithm stops. Otherwise, the next pass through the sequence begins. We observe that the heuristic eventually stops since each pass either reduces the total response time variability or proves that no exchange improves the total response time variability. In fact, the Exchange Heuristic goes a bit further, namely, even if an exchange attempt results in $\Delta = 0$ it is done after all as long as the maximum distance over all distances for both models being exchanged does not increase and at least one of the two maxima decreases. This last condition ensures that the heuristic actually terminates. A single pass through the sequence can be done in $O(D^2)$ time. The upper bound on the value of RTV that can be easily derived from the number decomposition graph is $O(nD^2)$. Therefore the exchange heuristic runs in $O(nD^4)$. Fortunately, computational experiments, Corominas et al. [82], show that the heuristic never took longer than 20 s for D as large as 1,500 to do the exchanges.

The Exchange Heuristic is applied to an initial sequence which is generated in a number of different ways. These will be detailed in the subsequent section.

7.3.2 The Initial Sequences

Bottleneck (Minimum TE) Sequences

The bottleneck sequences have been obtained by solving the bottleneck problem (5.2) from Chap. 5 to optimality with the algorithm implementation of Moreno [83]. However, other algorithms could be used to obtain, possibly different, bottleneck sequences (see Steiner and Yeomans [41], Kubiak [23], and Bautista et al. [29]).

Random Sequences

The bottleneck sequence S has been randomized as follows. For each position x in $[1, D]$, get a random integer number ran in the range $[1, D]$, then swap S_x with S_{ran}.

Webster's Sequences

These sequences have been obtained by applying the parametric method of apportionment, described in Sect. 2.4, with parameter $\delta = \frac{1}{2}$. The Webster's method . The sequence is generated as follows. Consider x_{it}, the number of model i copies in the sequence of length t, $t = 0, 1, \ldots$. Assume $x_{i0} = 0$, $i = 1, \ldots, n$. The model to be sequenced in position $t + 1$ can be computed as follows $i^* = \arg\max_i\{\frac{d_i}{(x_{it}+\delta)}\}$.

Jefferson's Sequences

These sequences have been generated by applying the parametric method of apportionment with $\delta = 1$. The Jefferson's parametric method. The stride scheduling technique produces the same sequences, see Chap. 10.

Insertion Sequences

Let again $d_1 \leq \cdots \leq d_n$. Consider $n - 1$, two-product instances $I_2 = (d_2, d_1)$, $I_3 = (d_3, \sum_{j=1}^{2} d_j), \ldots, I_n = (d_n, \sum_{j=1}^{n-1} d_j)$. In each of the instances $I_n, I_{n-1}, \ldots, I_3$, the first model comes from the original instance, that is $n, n - 1, \ldots$, and 3 respectively, and the second model is the same fictitious product for all problems, denoted by $*$. Let sequences $S_n, S_{n-1}, \ldots, S_2$ be the optimal solution for the instances $I_n, I_{n-1}, \ldots, I_2$ respectively. They can be obtained by the algorithm for two models described in Sect. 7.2.3. Notice that the sequence S_j, $j = n, \ldots, 3$, is made up of the model j and $*$. Next, the sequence for the original problem is built recursively by first replacing $*$ in S_n by S_{n-1} to obtain S'_n. Notice that the latter is made up of models n, $n - 1$, and $*$. Next, $*$ are replaced by S_{n-2} in S'_n to obtain a sequence S''_n made up of models n, $n - 1$, $n - 2$ and $*$. Finally, sequence S_2 replaces all the remaining $*$ and thus we obtain a sequence, referred to as the *insertion* sequence, where model i occurs exactly d_i times.

All these initial sequences, except the bottleneck, can be generated in $O(D)$ steps. The bottleneck needs $O(D \log D)$ steps.

7.3.3 Computational Experiment

The Webster's and Jefferson's sequences have resulted in the final RTV comparable to the bottleneck and insertion sequences only for very small values of n. For larger n, on average, their final RTV was higher than that of the bottleneck, random and insertion sequences. The same has been observed about the change in the bottleneck TE. For small n, their TE has been in the same range as the TE for the bottleneck and insertion sequences, while their TE significantly grew when n increased. Therefore, we do not include the computational results with either the Webster's or the Jefferson's sequences in Fig. 7.2. For each of the three initial sequences in Fig. 7.2 , the first column includes the averages of the initial RTV. The second column includes the averages of the final RTV. The averages are rounded to integers for convenience. The third column contains the averages of the initial TE and the fourth column the averages of the final TE.

n:	Bottleneck				Random				Insertion			
	Init RTV	Final RTV	Init TE	Final TE	Init RTV	Final RTV	Init TE	Final TE	Init RTV	Final RTV	Init TE	Final TE
10	17904	601	0.837	1.503	253685	5435	13.327	10.985	1546	625	1.461	1.406
50	180503	3208	0.936	2.004	2706679	7407	9.704	6.219	49101	4242	1.761	1.570
100	636545	4946	0.964	2.289	7130615	6375	8.812	5.564	253319	5866	2.207	1.825
200	1020847	8412	0.978	2.931	13476818	3517	7.662	4.596	873579	5032	2.039	1.977
300	1418033	6120	0.985	3.555	16271934	2266	8.605	6.400	1376361	2717	2.166	2.153
400	3574364	3913	0.991	4.754	35047637	1195	6.339	4.203	3584645	3395	3.512	3.409
500	6232671	1162	0.992	2.327	35083775	862	5.883	4.006	3290617	1313	1.856	1.574
600	7681994	496	0.992	1.648	21386442	962	7.549	5.924	1321738	375	1.049	1.014
700	12206931	179	0.993	1.407	20496095	952	6.716	5.093	706786	130	1.007	1.001
800	18083157	334	0.994	1.466	15952258	697	5.431	3.676	184422	84	1.163	1.071
900	15549203	82	0.996	1.181	8299817	472	4.421	2.685	14832	18	1.004	1.001

Fig. 7.2 The exchange improvements on the bottleneck, insertion and random initial sequences for $D = 1,000$

7.4 Mathematical Programming Formulation

Optimal solutions to the Response Time Variability problem can be obtained by means of the dynamic program described in Sect. 7.2.4 but this program is too time and space consuming to be practical. Another approach to get optimal solutions uses mathematical programming approach described in this section and off the shelf optimization software CPLEX. We assume for simplicity that $d_1 \geq 2$ in this and next sections. We begin by defining some input parameters that would reduce the size of the program.

7.4.1 Input Parameters

Given

$M = \{1, \ldots, n\}$ – The set of models.

X – The best solution obtained by applying the Exchange Heuristic to five greedy initial sequences (see Sect. 7.3 for details).

Z – The total response time variability of X.

Z_i – An upper bound on the value of RTV_i defined as follows

$$Z_i = Z - \sum_{j \in M, j \neq i} RTV_j(\mathcal{N}_j).$$

Let $\lambda \geq 1$ be an integer such that

$$\left\lfloor \frac{D - \lambda}{d_i - 1} \right\rfloor \geq \lambda.$$

Let $\mathcal{N}'_i$ be the top node in the number decomposition graph for $D' = D - \lambda$ and $d'_i = d_i - 1$. We set LB_i, the lower bound on the distance between the copies of i, equal to the smallest λ such that

$$RTV_i(\mathcal{N}'_i, \lambda) \leq Z_i.$$

For instance, let $D = 52$, $d_i = 5$, $\overline{\alpha}_i = \frac{52}{5} = 10.4$ and $Z_i = 33.2$. For $\lambda = 5$, the best decomposition vector is $N = (\mathcal{N}'_i, \lambda) = (12, 12, 12, 11, 5)$ which results in $RTV_i(N) = 37.2 > Z_i$. For, $\lambda = 6$ the best decomposition vector is $L = (\mathcal{N}'_i, \lambda) = (12, 12, 11, 11, 6)$, which gives $RTV_i(L) = 25.2 \leq Z_i$ and thus $LB_i = \lambda = 6$.

For a given LB_i, the upper bound on the distance between the copies of i, UB_i, can be set as follows: $UB_i = D - ((d_i - 1) \cdot LB_i)$. However, this bound can be possibly improved as follows. Let $\omega \geq 1$ be such that

$$\left\lceil \frac{D - \omega}{d_i - 1} \right\rceil \leq \omega.$$

Let $\mathcal{N}'_i$ be the top node in the number decomposition graph for $D' = D - \omega$ and $d'_i = d_i - 1$. We find the largest ω such that

$$RTV_i(\omega, \mathcal{N}'_i) \leq Z_i$$

and set $UB_i = \min(D - ((d_i - 1) \cdot LB_i); \omega)$. For instance, $D = 52$, $d_i = 5$, $\overline{\alpha}_i = \frac{52}{5} = 10.4$ and $Z_i = 7.6$. For, $\omega = 13$, the best decomposition vector is $N = (13, 10, 10, 10, 9)$, which gives $RTV_i(N) = 9.2 > Z_i$. Then, for $\omega = 12$ the best decomposition vector is $L = (12, 10, 10, 10, 10)$, which gives $RTV_i(L) = 3.2 \leq Z_i$ and thus $UB_i = \min(D - ((d_i - 1) \cdot LB_i) = 28; 12) = 12$.

Therefore, the earliest, E_{ik}, and the latest, L_{ik}, positions to be occupied by copy k of i can be set as follows

$$E_{ik} = \max(1 + LB_i \cdot (k - 1); D - UB_i \cdot (d_i - k + 1) + 1)$$

$$L_{ik} = \min(D - LB_i \cdot (d_i - k); UB_i \cdot k)$$

for $i = 1, \ldots, n; k = 1, \ldots, d_i$, and consequently the set of positions available for copy k of i can be set as follows

$$H_{ik} = \{h : L_{ik} \leq h \leq E_{ik}\}.$$

7.4.2 The Objective

The Response Time Variability problem can be considered as a special case of the quadratic assignment problem and therefore formulated as a quadratic integer program. The well known separable convex programming technique (see, e.g., Wagner [84]) can be applied with variables $\gamma^j_{ik} \in \{0, 1\}$, where $\gamma^j_{ik} = 1$ if and only if the distance between copies k and $k + 1$ of i is greater than or equal to $LB_i + j$, where $i \in M = \{1, \ldots, n\}; k = 1, \ldots, d_i; j = 1, \ldots, UB_i - LB_i$, to recast the response time variability objective function as a linear one as follows:

$$RTV = \sum_{i \in M, k, j} \left((LB_i + j)^2 - (LB_i + j - 1)^2 \right) \cdot \gamma_{ik}^j$$

$$+ \sum_{i \in M} d_i \cdot LB_i^2 - \sum_{i \in M} d_i \cdot \overline{\alpha}_i^2. \tag{7.10}$$

The consistency of values γ_{ik}^j is imposed, though already guaranteed for the convexity of the objective function, by the following constraint:

$$\gamma_{ik}^j \geq \gamma_{ik}^{j+1} \quad (i \in M; k = 1, \ldots, d_i; j = 1, \ldots, UB_i - LB_i - 1) \tag{7.11}$$

Moreover, for each $i \in M$, it is required from variables γ_{ik}^j, that the sum of the distances between its units is equal to D:

$$\sum_{j=1}^{UB_i - LB_i} \sum_{k=1}^{d_i} \gamma_{ik}^j = D - d_i \cdot LB_i \tag{7.12}$$

$$\sum_{j=1}^{UB_i - LB_i} \sum_{k=1}^{d_i} \left(1 - \gamma_{ik}^j \right) = d_i \cdot UB_i - D. \tag{7.13}$$

Finally, using (7.12) we can rewrite the RTV in (7.10) so that the minimization of RTV is equivalent to the minimization of

$$rtv = \sum_{i \in M, k, j} j \cdot \gamma_{ik}^j.$$

7.4.3 Eliminating Symmetries

Let us introduce additional integer *position* variables sl_{ik}, $i \in M; k = 1, \ldots, d_i$, where sl_{ik} equals the position index of copy k of i. The first copy of n can be fixed in the first position of the sequence, thus

$$sl_{n,1} = 1.$$

Distance Mirror Reflection Elimination

Consider sequence S, then

$$S = nS_1 nS_2 \cdots nS_{d_n},$$

where S_j $j = 1, \ldots, d_n$ are sequences, possibly empty, of copies of $i = 1, \ldots, n - 1$. Then, the mirror reflection

$$S_{d_n} n \cdots S_2 n S_1 n$$

of S does not change the total response time variability, neither does the cyclic rotation by a single position

$$S' = nS_{d_n}n\cdots S_2 nS_1.$$

To eliminate this mirror reflection symmetry we introduce the following constraint

$$\sum_{k=1}^{d_n-1} k^2 \cdot \left(sl_{n,k+1} - sl_{n,k}\right) + d_n^2 \cdot \left(D - sl_{n,d_n} + sl_{n,1}\right)$$

$$\leq \sum_{k=1}^{d_n-1} (d_n - k + 1)^2 \cdot \left(sl_{n,k+1} - sl_{n,k}\right)$$

$$+ \left(D - sl_{n,d_n} + sl_{n,1}\right). \tag{7.14}$$

Cyclic Rotation Elimination

Consider S again

$$S = nS_1 nS_2 \cdots nS_{d_n},$$

and its $d_n - 1$ cyclic rotations

$$nS_2 \cdots nS_{d_n} nS_1$$

$$\vdots$$

$$nS_{d_i} nS_1 \cdots nS_{d_n-1}.$$

All these rotations have the same value of the total response time variability. We introduce the following constraints to eliminate these cyclic rotations.

$$\sum_{k=1}^{d_n-1} k^2 \cdot \left(sl_{n,k+1} - sl_{n,k}\right) + d_n^2 \cdot \left(D - sl_{n,d_n} + sl_{n,1}\right)$$

$$\leq \sum_{k=1}^{d_n-1} (1 + ((k+j) \bmod d_n))^2 \cdot \left(sl_{n,k+1} - sl_{n,k}\right)$$

$$+ (1 + j)^2 \cdot \left(D - sl_{n,d_n} + sl_{n,1}\right) (j = 0, \ldots, d_n - 2) \tag{7.15}$$

Consistency with γ_{ik}^j

The consistency of the sl_{ik} and γ_{ik}^j variables is ensured by the following constraints (7.16) and (7.17):

$$sl_{i,k+1} - sl_{i,k} = LB_i + \gamma_{ik}^1 + \cdots + \gamma_{ik}^j + \cdots + \gamma_{ik}^{UB_i-LB_i}$$

$$(\forall i \in M; k = 1, \ldots, d_i - 1) \tag{7.16}$$

$$D - sl_{i,d_i} + sl_{i,1} = LB_i + \gamma_{i,d_i}^1 + \cdots + \gamma_{i,d_i}^j + \cdots + \gamma_{i,d_i}^{UB_i-LB_i} \quad (i \in M) \tag{7.17}$$

For each $i \in M$, it is required that the sum of the distances between its units is equal D:

$$\sum_{k=1}^{d_i-1} \left(sl_{i,k+1} - sl_{i,k}\right) + \left(D - sl_{i,d_i} + sl_{i,1}\right) = D \quad (i \in M) \tag{7.18}$$

7.4.4 Feasibility of Position Variables

Finally, we introduce the variables $y_{ikh} \in \{0,1\}$, where $y_{ikh} = 1$ if and only if copy k of i is in position h, where $i \in M; k = 1, \ldots, d_i; h \in H_{ik}$, and we link them with the position variables sl_{ik} by the constraint (7.19). These y_{ikh} variables are necessary to ensure that all position variables sl_{ik} have distinct values in a feasible solution. This requirement is ensured by adding the following assignment type constraints (7.20) and (7.21):

$$\sum_{h \in H_{ik}} h \cdot y_{ikh} = sl_{ik} \quad (i \in M; k = 1, \ldots, d_i) \tag{7.19}$$

$$\sum_{(i,k)|i \in M \wedge h \in H_{ik}} y_{ikh} = 1 \quad (h = 1, \ldots, D) \tag{7.20}$$

$$\sum_{h \in H_{ik}} y_{ikh} = 1 \quad (i \in M; k = 1, \ldots, d_i). \tag{7.21}$$

7.5 The Algorithm

We now put all constraints and the objective together.

$$\min \ rtv = \sum_{i \in M, k, j} j \cdot \gamma_{ik}^j \tag{7.22}$$

Subject to

$$\sum_{h \in H_{ik}} h \cdot y_{ikh} = sl_{ik} \quad (i \in M; k = 1, \ldots, d_i) \tag{7.23}$$

$$sl_{i,k+1} - sl_{i,k} = LB_i + \gamma_{ik}^1 + \cdots + \gamma_{ik}^j + \cdots + \gamma_{ik}^{UB_i - LB_i}$$
$$(i \in M; k = 1, \ldots, d_i - 1) \tag{7.24}$$

$$D - sl_{i,d_i} + sl_{i,1} = LB_i + \gamma_{i,d_i}^1 + \cdots + \gamma_{i,d_i}^j + \cdots + \gamma_{i,d_i}^{UB_i - LB_i} \quad (i \in M) \tag{7.25}$$

$$\gamma_{ik}^j \geq \gamma_{ik}^{j+1} \quad (i \in M; k = 1, \ldots, d_i; j = 1, \ldots, UB_i - LB_i - 1) \tag{7.26}$$

$$\sum_{(i,k)|i \in M \wedge h \in H_{ik}} y_{ikh} = 1 \quad (h = 1, \ldots, D) \tag{7.27}$$

$$\sum_{h \in H_{ik}} y_{ikh} = 1 \quad (i \in M; k = 1, \dots, d_i) \tag{7.28}$$

$$\sum_{k=1}^{d_n-1} k^2 \cdot \left(sl_{n,k+1} - sl_{n,k}\right) + d_n^2 \cdot \left(D - sl_{n,d_n} + sl_{n,1}\right)$$

$$\leq \sum_{k=1}^{d_n-1} (d_n - k + 1)^2 \cdot \left(sl_{n,k+1} - sl_{n,k}\right)$$

$$+ \left(D - sl_{n,d_n} + sl_{n,1}\right) \tag{7.29}$$

$$\sum_{k=1}^{d_n-1} k^2 \cdot \left(sl_{n,k+1} - sl_{n,k}\right) + d_n^2 \cdot \left(D - sl_{n,d_n} + sl_{n,1}\right)$$

$$\leq \sum_{k=1}^{d_n-1} \left(1 + ((k+j) \bmod d_n)\right)^2 \cdot \left(sl_{n,k+1} - sl_{n,k}\right)$$

$$+ (1+j)^2 \cdot \left(D - sl_{n,d_n} + sl_{n,1}\right) \quad (j = 0, \dots, d_n - 2) \tag{7.30}$$

$$\sum_{j=1}^{UB_i - LB_i} \sum_{k=1}^{d_i} \gamma_{ik}^j = D - d_i \cdot LB_i \quad (i \in M) \tag{7.31}$$

$$\sum_{j=1}^{UB_i - LB_i} \sum_{k=1}^{d_i} \left(1 - \gamma_{ik}^j\right) = d_i \cdot UB_i - D \quad (i \in M). \tag{7.32}$$

Corominas et al. [85] use off the shelf software, CPLEX, to show that the practical limit for obtaining optimal solutions is $D = 40$ units to be scheduled in 2,000 s or less. The exact algorithms capable of solving the Response Time Variability problem for lager problem sizes in comparable amount of time remain to be found.

7.6 Exercises

Exercise 7.16. Develop a branch and bound algorithm for the Response Time Variability problem.

Exercise 7.17. Let α_i be a given node in a number decomposition graph $\mathcal{D}_i$ for d_i, $i = 1, \dots, n$. Formulate an integer program to test the feasibility of the nodes $\alpha_1, \dots, \alpha_n$. The feasibility means that there exists a just-in-time sequence for $(d_1, \dots, d_n)$ with distance of copies of i as in the node α_i. What is the complexity of this test? Hint: The complexity appears an open problem.

Exercise 7.18. What is the total response time variability for the power-of-two instances?

7.7 Comments and References

This chapter is based to a large extend on Corominas et al. [82], where more results of computational experiments with the Exchange Heuristic can be found. The mathematical programming formulation and algorithm come from Corominas et al. [82,85]. The total response time variability is suggested as a main metric of the stride schedule by Waldspurger and Weihl [13], see also Chap. 10. The idea of Theorem 7.13 comes from [86]. Lemma 7.12 is from Bar-Noy et al. [79]. The proof of Theorem 7.14 was given in Bar-Noy et al. [79].

Chapter 8
Applications to the Liu–Layland Problem and Pinwheel Scheduling

8.1 Introduction

This chapter presents a fresh approach to two hard real-time classical sequencing problems: the Liu–Layland periodic sequencing problem and the (generalized) pinwheel scheduling problem. We show that a number of important results obtained in the literature on either of these two problems follow quite easily from the properties of just-in-time sequences with small bottleneck deviations. We also present new solutions to the Liu–Layland problem based on the quota-divisor methods of apportionment. This fresh approach sheds a new light on the connections between the just-in-time optimization and the apportionment problem on one side and the hard two real-time scheduling problems on the other.

The Liu-Layland periodic sequencing problem [87] is one of the most fundamental problems studied in hard real-time computing systems. A system is said to be real-time if the correctness of its operation depends not only on the logical correctness of the tasks making it but also on the time at which the tasks are performed. In a hard real-time system, the completion of a task after its deadline is considered useless and may ultimately lead to a critical failure of the whole system. For instance, a task may calculate a current position of an aircraft in C seconds but the position must be updated every T seconds. Missing a deadline may prove fatal for the aircraft and generally for the system, thus deadlines must not be missed. More details about hard real-time systems can be found in Cheng [88], and Butazzo [89].

The second problem discussed in this chapter is the pinwheel scheduling problem. Its motivation comes from the real-time satellite communication with a ground station without data loss. This problem was introduced by Holte et al. [90] in 1989 and since then it is referred to as the *pinwheel* scheduling problem. In this problem, the ground station receives data from a number of satellites. In each time slot the station can only receive a single data packet from a single satellite. Each satellite, due to its orbital characteristics, has a possibly different window of time of length b, measured in the number of time slots, to repeatedly broadcast the same data packet to the ground station. The station then has to allocate time slots to satellites so as

W. Kubiak, *Proportional Optimization and Fairness,* International Series in Operations Research & Management Science 127, DOI 10.1007/978-0-387-87719-8_8,
© Springer Science+Business Media LLC 2009

to ensure that no data packet is lost, that is a satellite with time window b must be allocated *at least* one time slot out of any consecutive b time slots. A generalized pinwheel scheduling problem introduced by Baruah and Bestavros [91] requires that a sliding time window of size b always includes *at least a* ≥ 1 time slots allocated to a satellite, which can speed up error recovery in case of possible broadcast problems.

The plan for this chapter is as follows. Section 8.2 introduces the Liu–Layland problem in detail and briefly reviews well-known results on this problem. Section 8.3 shows that the Liu–Layland problem can be solved by any optimization algorithm for the bottleneck problem. Actually, even stronger results holds, namely, any algorithm that finds a just-in-time sequence with bottleneck deviation *less* than 1 solves the Liu–Layland problem. Such algorithms may be, for instance, the algorithm of Tijdeman presented in Chap. 5 or the quota-divisor methods of apportionment discussed in Sect. 8.4. We also show that the quota satisfaction is the necessary condition for a devisor method to solve the Liu–Layland problem. Therefore, no divisor method solves the Liu–Layland problem, and the culprit is population monotonicity that characterizes any divisor method. Section 8.5 introduces the pinwheel and the generalized pinwheel scheduling problems and reviews well-known results on these problems. Section 8.6 presents additional properties of just-in-time sequences with bottleneck deviation *less* than 1. The properties lead to elegant and simple proofs of a number of results on these problems known in the literature. For instance the problems with the total density not exceeding $\frac{1}{2}$ or the generalized pinwheel problem with $n = 2$.

8.2 The Liu–Layland Problem

Following Liu and Layland [87], we define the periodic sequencing problem as follows. Consider n independent, preemptive and periodic tasks $1, \ldots, n$ with their request periods being $T_1, \ldots, T_n$ and their run-times being $C_1, \ldots, C_n$ respectively. The execution of the kth request of task i, which occurs at moment $(k-1)T_i$, must finish by moment kT_i when the next request for the task begins, $k = 1, 2, \ldots$. Missing a deadline is fatal to the system, therefore the deadlines $T_i, 2T_i, \ldots$ are considered hard for task i. All numbers are positive integers and $C_i \leq T_i$ for $i = 1, \ldots, n$. We need to find an infinite word $S = s_1 s_2 \ldots$ on the alphabet $\{1, 2, \ldots, n\}$ such that i occurs exactly C_i times in each subsequence $s_{(k-1)T_i+1} \cdots s_{kT_i}$ for $k = 1, \ldots$ and $i = 1, \ldots, n$. We call S a periodic schedule for tasks $1, \ldots, n$. For any periodic schedule S, its L-prefix $s = s_1 \ldots s_L$, where $L = \text{lcm}(T_1, \ldots, T_n)$, defines a periodic schedule s^∞ which is an infinite concatenation of s. Consequently, we shall assume that s^∞ represents all periodic schedules with their L-prefixes being s, and our discussion assumes for simplicity that the term periodic schedule S refers to the L-prefix s of S rather than to S itself. We begin with an example of three tasks and their periodic schedule.

Example 8.1. Consider a three task instance with request periods $T_1 = 3, T_2 = 4$ and $T_3 = 5$, and run-times $C_1 = C_2 = 1$ and $C_3 = 2$ respectively. The periodic schedule for this instance is shown in Fig. 8.1 where $L = \text{lcm}(T_1, T_2, T_3) = 60$. Notice that the

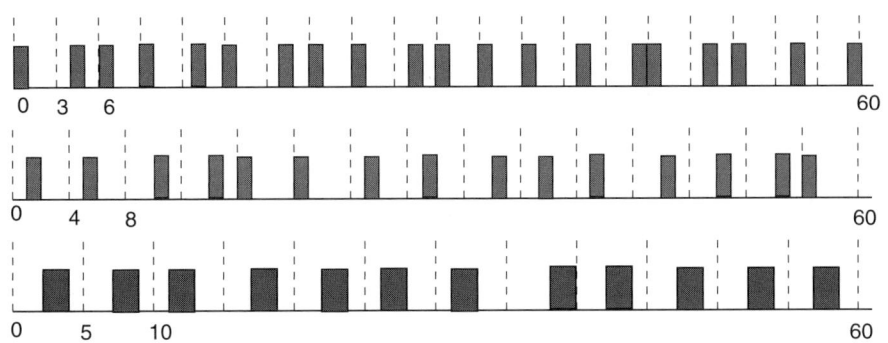

Fig. 8.1 The periodic schedule for tasks with periods $T_1 = 3$, $T_2 = 4$, and $T_3 = 5$, and run-times $C_1 = C_2 = 1$ and $C_3 = 2$ respectively

total load between 0 and L is $1 \times 20 + 1 \times 15 + 2 \times 12 = 59$ which leaves a single time slot, between 35 and 36, empty. $\square$

The solution in Fig. 8.1 can be obtained by the deadline driven algorithm of Liu and Layland [87], see also Dertouzos [92]. The algorithm assigns priorities to tasks according to the deadlines of their current requests. Therefore, a task with the highest priority at a unit time slot k will be the one with the deadline of its current request being the nearest to k, and a task will be assigned the lowest priority if the deadline of its current request is the furthest from k. In any unit time slot k a task with the highest priority and still incomplete current request will be executed. This deadline driven algorithm always produces a periodic schedule as long as

$$\sum_{i=1}^{n} \frac{C_i}{T_i} \leq 1. \tag{8.1}$$

Clearly, no periodic schedule exists with (8.1) being violated.

We view the Liu–Layland periodic sequencing problem from a rather different angle in this chapter. Our solution is based on the solution to the bottleneck problem and it can be explained as follows. Assume that a task i can be allocated processor so that it can progress at a rate as close to $r_i = \frac{C_i}{T_i}$ as possible. More precisely, consider the straight line

$$\frac{C_i}{T_i} k$$

in the interval $[0, L]$ where variable k takes on integer values $0, 1, \ldots, L$. Let x_{ik} be the total time allocated by a given schedule to task i in $[0, k]$, notice that the schedule may allocate time to multiple requests for i in this interval. Then, Theorem 5.8 ensures that there is a schedule that keeps x_{ik} within distance less than 1 from the line $\frac{C_i}{T_i} k$ *simultaneously* for each task $i = 1, \ldots, n$. We show that being this close to the lines $\frac{C_i}{T_i} k$ automatically ensures that the schedule is periodic, that is each request for task i falls inside its request period.

Our main goal is to show that any just-in-time sequence with $r_i = \frac{C_i}{T_i}$ and its bottleneck deviation being less than one (but not necessarily optimal) solves the Liu–Layland problem for tasks $1, \ldots, n$ with their request periods being $T_1, \ldots, T_n$ and their run-times being $C_1, \ldots, C_n$ respectively. Formally, the ratios $r_i = \frac{C_i}{T_i}$ translate into total demand $D = L$ and the individual demands $d_i = \frac{LC_i}{T_i}$. Moreover, if $\sum_{1 \leq i \leq n} \frac{C_i}{T_i} < 1$, then a dummy model $n + 1$ with its demand

$$d_{n+1} = L \left(1 - \sum_{1 \leq i \leq n} \frac{C_i}{T_i} \right)$$

can be added to complete the instance of the bottleneck problem. It is worth noticing that our proofs are formulated in terms of the ratios $r_i = \frac{C_i}{T_i}$ rather than the actual demands d_i which frees our approach from working with potentially large demands d_i and L. For the proof, it is sufficient to show that any interval $[(k-1)T_i + 1, kT_i]$ of the bottleneck just-in-time sequence includes at least C_i copies of i. We begin with the following two results shown in Lemmas 8.2 and 8.3, and Lemmas 8.4 and 8.5 respectively. The first shows that the inequalities

$$E(i, j) \leq L(i, j) \leq E(i, j+1)$$

hold for all $i = 1, 2, \ldots, n$ and $j = 1, \ldots, d_j - 1$ as long as $B < 1$. Hence, copies $(k-1)C_i + 1, \ldots, kC_i$ of i occupy C_i positions in the interval

$$[E(i, (k-1)C_i + 1), L(i, kC_i)].$$

The second proves that

$$(k-1)T_i < E(i, (k-1)C_i + 1) \text{ and } L(i, kC_i) \leq kT_i \tag{8.2}$$

for $B < 1$. Notice that the first inequality in (8.2) must hold as a strict inequality since by definition the $E(i, (k-1)C_i + 1)$ is the earliest *position* that copy $(k-1)C_i + 1$ can occupy. Thus the starting moment of that copy can be as early as $E(i, (k-1)C_i + 1) - 1$ which must not be sooner than the release of the kth request that is the moment $(k-1)T_i$.

The above definition of periodic schedule tacitly implies preemptions at integer points only. This is justified, however, by the fact that the existence of preemptive periodic schedule implies the existence of preemptive periodic schedule with preemptions at integer points only. This claim holds since the Liu–Layland problem can be reduced to the $1|pmtn, r_i, \overline{d}_i|-$ scheduling problem, we refer the reader to Błażewicz et al. [93] for the definition of the three field scheduling notation, by taking the release dates $0, T_i, \ldots, (\frac{L}{T_i} - 1)T_i$ and the deadlines $T_i, 2T_i, \ldots, L$ for the $\frac{L}{T_i}$ requests of task i with run-time C_i in the L-prefix. The latter problem can in turn be reduced to the maximal network flow problem, Bratley et al. [94], which by the integrality of all input data and the Integral Flow Theorem of network flows, see Lawler [95], is solved by an integral flow. The flow can then be readily turned into a periodic schedule with preemptions at integer points only. These observations also imply that the L-prefix, if it exists, can be found in time polynomial in L.

8.3 Just-In-Time Solution of the Liu–Layland Problem

We now give the details of the solution of the Liu–Layland periodic sequencing problem based on the solution to the bottleneck problem studied in Chap. 5. The following is the summary of the correspondences between the two problems alluded in the previous section:

$$\text{number tasks } n \longleftrightarrow \text{ number of models } n$$
$$\text{task } i \longleftrightarrow \text{ model } i$$
$$\text{total run-time of task } i \text{ in } [0, L] \longleftrightarrow \text{ demand } d_i \text{ for model } i$$
$$\text{cycle time } L \longleftrightarrow \text{ total demand } D = \sum_{i=1}^{n} d_i$$
$$\text{the ratio } \frac{C_i}{T_i} \longleftrightarrow r_i = \frac{d_i}{D} = \frac{C_i}{T_i}.$$

We begin with some implications of Theorem 5.1 for the bottleneck problem solutions.

Lemma 8.2. *For any feasible bottleneck deviation* B, $i = 1, 2, \ldots, n$ *and* $j = 1, \ldots, d_j$, *we have*
$$E(i, j) \le L(i, j).$$

Proof. Follows immediately from Theorem 5.1. $\square$

We now show that for $B < 1$ the intervals for consecutive copies of model i can overlap at the ends only.

Lemma 8.3. *For* $B < 1$ *and* $k = 1, 2, \ldots j = 1, \ldots$, *we have*
$$L(i, j) \le E(i, j+1).$$

Proof. By Theorem 5.1
$$L(i, j) = \lfloor \frac{j - 1 + B}{r_i} + 1 \rfloor,$$

and
$$E(i, j+1) = \lceil \frac{j + 1 - B}{r_i} \rceil = \lceil \frac{j - 1 + B}{r_i} + \frac{2(1 - B)}{r_i} \rceil.$$

If $\frac{j-1+B}{r_i}$ is an integer, then $L(i, j) = \frac{j-1+B}{r_i} + 1 \le \frac{j-1+B}{r_i} + \lceil \frac{2(1-B)}{r_i} \rceil = E(i, j+1)$ since $\frac{2(1-B)}{r_i} > 0$ for $B < 1$. If $\frac{j-1+B}{r_i}$ is not an integer, then $L(i, j) = \lceil \frac{j-1+B}{r_i} \rceil \le \lceil \frac{j-1+B}{r_i} + \frac{2(1-B)}{r_i} \rceil = E(i, j+1)$ since again $\frac{2(1-B)}{r_i} > 0$ for $B < 1$. Thus, the lemma holds. $\square$

Moreover, for $B < 1$, the copy kC_i of model i must be in a position not later than kT_i. That is the completion of the k−th request of task i is not later than the deadline of this request.

Lemma 8.4. *For any bottleneck $B < 1$ and $i = 1, \ldots, n$, we have*

$$kT_i \geq L(i, kC_i).$$

Proof. By Theorem 5.1

$$L(i, kC_i) = \lfloor \frac{kC_i - 1 + B}{r_i} + 1 \rfloor.$$

Since $r_i = \frac{C_i}{T_i}$, we have

$$L(i, kC_i) = \lfloor kT_i + 1 + \frac{B - 1}{r_i} \rfloor.$$

Finally, since $B < 1$, then
$$L(i, kC_i) \leq kT_i$$

which ends the proof.   □

Finally, for $B < 1$, the copy $(k - 1)C_i + 1$ of model i must be in a position later than $(k - 1)T_i$. That is the earliest start time of the $k - 1$-st request of task i is not earlier than its release date.

Lemma 8.5. *For any $B < 1$ and $i = 1, \ldots, n$, we have*

$$(k - 1)T_i < E(i, (k - 1)C_i + 1).$$

Proof. By Theorem 5.1

$$E(i, (k - 1)C_i + 1) = \lceil \frac{(k - 1)C_i + 1 - B}{r_i} \rceil.$$

Since $r_i = \frac{C_i}{T_i}$, we have

$$E(i, (k - 1)C_i + 1) = \lceil (k - 1)T_i + \frac{1 - B}{r_i} \rceil.$$

Finally, since $B < 1$, then

$$E(i, (k - 1)C_i + 1) > (k - 1)T_i$$

which ends the proof.   □

We are now ready to prove the main result.

Theorem 8.6. *Any solution to the bottleneck problem with rates $r_i = \frac{C_i}{T_i}$, $i = 1, \ldots, n$ and the bottleneck deviation less than 1 is a periodic schedule.*

Proof. Consider any solution to the bottleneck problem with its bottleneck deviation B being less than 1. By Theorem 5.6, such a solution always exists. Then, copy j of model i occupies a position in the interval $[E(i,j), L(i,j)]$, where $E(i,j)$ and $L(i,j)$ are defined as in Theorem 5.1. Therefore, by Lemmas 8.2 and 8.3, C_i copies $(k-1)C_i+1, \ldots, kC_i$ of model i occupy C_i positions in the interval $[E(i,(k-1)C_i+1), L(i,kC_i)]$. However, by Lemmas 8.4 and 8.5, we have $(k-1)T_i < E(i,(k-1)C_i+1) \leq L(i,kC_i) \leq kT_i$ and thus at least C_i copies of i occupy positions in $[(k-1)T_i+1, kT_i]$. Consequently, the schedule is periodic which proves the theorem. □

The following example shows the just-in-time solution to the instance of Example 8.1.

Example 8.7. The instance of the Liu–Layland problem from Example 8.1 translates into an instance of the bottleneck problem with $n=4$ models, demands $d_1 = 20, d_2 = 15, d_3 = 24$ and $d_4 = 1$, and rates $r_1 = \frac{C_1}{T_1} = \frac{1}{3}, r_2 = \frac{C_2}{T_2} = \frac{1}{4}, r_3 = \frac{C_3}{T_3} = \frac{2}{5}$, and $r_4 = \frac{1}{60}$ respectively. Notice that

$$\sum_{i=1}^{3} \frac{C_i}{T_i} = \frac{59}{60} < 1$$

thus a dummy task with the request period $T_4 = 60$ and the run-time $C_4 = 1$ is added. Table 8.1 shows the time windows in which to run a unit of tasks 1, 2, and 3

Table 8.1 The position windows for tasks 1, 2, and 3

j	$E(1,j)$	$L(1,j)$	$E(2,j)$	$L(2,j)$	$E(3,j)$	$L(3,j)$
1	1	3	1	4	1	2
2	4	6	5	8	4	5
3	7	9	9	12	6	7
4	10	12	13	16	9	10
5	13	15	17	20	11	12
6	16	18	21	24	14	15
7	19	21	25	28	16	17
8	22	24	29	32	19	20
9	25	27	33	36	21	22
10	28	30	37	40	24	25
11	31	33	41	44	26	27
12	34	36	45	48	29	30
13	37	39	49	52	31	32
14	40	42	53	56	34	35
15	43	45	57	60	36	37
16	46	48			39	40
17	49	51			41	42
18	52	54			44	45
19	55	57			46	47
20	58	60			49	50
21					51	52
22					54	55
23					56	57
24					59	60

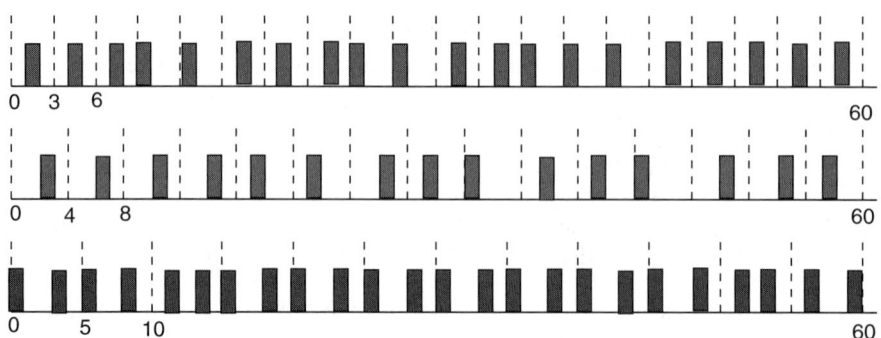

Fig. 8.2 The periodic schedule obtained by the just-in-time sequencing

obtained for $B = 1 - \frac{1}{2(n-2)} = \frac{3}{4}$. The sequence is shown in Fig. 8.2. The sequence ignores a dummy task 4. $\square$

8.4 Divisor Methods for the Liu–Layland Problem

8.4.1 The Necessary Conditions

Theorem 8.6 proves that in fact staying within the quota is a sufficient condition for a house monotone method to solve the Liu–Layland periodic sequencing problem. This section goes further and studies the apportionment divisor methods as a possible approach to solving this problem. These methods are chosen because of their extreme computational efficiency, which makes them especially attractive for real-time scheduling, their well know properties, which make them a cornerstone of the apportionment theory, see Chap. 2, and because of their practical significance as shown by their applications in operating systems and networking, see Chap. 10. Our main result shows that staying within the quota is a necessary condition for a divisor method to solve the Liu–Layland problem. This in fact proves that the divisor methods can not solve the Liu–Layland problem since no divisor method stays within the quota by Theorem 2.4. Though this result can be deemed negative it provides interesting insights into the solutions of the Liu–Layland problem, for instance, they are not population monotone. Consequently, no population monotone method solves the Liu–Layland problem and therefore any method that solves this problem must suffer from the population paradox. That is sometimes even if task is run-time increases and task js run-time decreases, the task i may get fewer allocations and task j may get more allocations with the new run-times than they do with the original run-times in the same time interval $[0, t]$. Finally, we show that slight modifications of the Adams's and Jefferson's divisor methods solve the Liu–Layland problem. These modifications result into the quota-Adams and the quota-Jefferson divisor methods introduced by Balinski and Young [2], and Still [96]. These particular quota methods are chosen in this chapter because they are conceptually the simplest and computationally the most efficient ones among all quota divisor methods. The following summarizes the correspondence between the Liu–Layland problem and the apportionment problem discussed in Chap. 2:

$$\text{number of tasks } n \longleftrightarrow \text{number of states } n$$
$$\text{task } i \longleftrightarrow \text{state } i$$
$$\text{total run-time of task } i \text{ in } [0,L] \longleftrightarrow \text{population } p_i(=d_i) \text{ of state } i$$
$$\text{position in sequence } h \longleftrightarrow \text{size of house } h$$
$$\text{total time } x_{ih} \text{ allocated to task } i \text{ in } [0,h] \longleftrightarrow a_i \text{ for a house of size } h$$
$$\text{cycle time } L \longleftrightarrow \text{total population } P = \Sigma_{i=1}^{n} p_i.$$

We now show two necessary conditions for divisor methods to solve the Liu–Layland periodic scheduling problem.

Lemma 8.8. *Staying above lower quota is a necessary condition for any divisor method to solve the Liu–Layland problem.*

Proof. By contradiction. Let us consider any divisor method **M** that does not stay above lower quota and solves the Liu–Layland problem at the same time. Since **M** does not stay above lower quota, then there exist a vector of populations $\mathbf{p} = (p_1, p_2, \ldots, p_n)$, $n \geq 3$ since all divisor methods stay within the quota for all $2-$state problems, see Theorem 2.5, and the house size h for which the lower quota (2.3) is not satisfied for some state k, that is

$$y_{kh} < \lfloor q_k \rfloor = \left\lfloor \frac{p_k h}{P} \right\rfloor. \tag{8.3}$$

By Lemma 2.7, we can assume that q_k is fractional.

Define an instance of the Liu–Layland problem as follows. For the task k corresponding to the state k, let $T_k = h$. Let the run-time of task k be

$$C_k = \lfloor q_k \rfloor = \left\lfloor \frac{p_k h}{P} \right\rfloor \leq h. \tag{8.4}$$

At the moment $t = h \times P$, the total run time of all P requests for task k equals

$$\alpha_k = P \left\lfloor \frac{p_k h}{P} \right\rfloor < P \frac{p_k h}{P} = p_k h. \tag{8.5}$$

For all other tasks $i \neq k$ the run time is defined as follows

$$C_i = p_i h, \tag{8.6}$$

and their request periods $T_i = h \times P$. Thus all other tasks, except k, can be scheduled at any moment up to $t = h \times P$. Notice that this Liu–Layland instance has a solution by (8.1), since

$$\sum_{i=1}^{n} \frac{C_i}{T_i} = \sum_{i=1,i\neq k}^{n} \frac{hp_i}{h\times P} + \frac{\left\lfloor \frac{p_k h}{P} \right\rfloor}{h} < \sum_{i=1}^{n} \frac{p_i}{P} = 1.$$

Since **M**, by assumption, solves the Liu–Layland problem, then it obtains a periodic sequence S_{LL} with $x_{kh}=C_k$ positions assigned to the task k in the first request period $T_k = h$ of this task. By (8.3) and (8.4), $y_{kh} < C_k$. Moreover, the instance of the apportionment problem **r** with $r_i = hp_i$ for $i \neq k$ and $r_k = P \left\lfloor \frac{p_k h}{P} \right\rfloor < hp_k$ changes the populations according to the following proportions:

$$\frac{r_i}{r_k} > \frac{hp_i}{hp_k} = \frac{p_i}{p_k} \quad \text{for all } i \neq k.$$

However, since $x_{kh} > y_{kh}$, and $h = \sum_{i=1}^{n} x_{ik} = \sum_{i=1}^{n} y_{ik}$, then there is $i \neq k$ such that $x_{ih} < y_{ih}$, which contradicts the population monotonicity of method **M**. This ends the proof. □

Though staying above the lower quota is a necessary condition for a population monotone method to solve the Liu–Layland problem, which we just proved, it is not a necessary condition for a *house monotone* method. To show this let us consider the following instance of the Liu–Layland problem:

$$C_1 = 3, T_1 = 5, C_2 = 2, T_2 = 5 \tag{8.7}$$

and a *house monotone* (however not a divisor) method, for instance, the original Liu–Layland algorithm [87]. Let us have this algorithm to break ties whenever there is more then one task with the same request period T_i by assigning the current time slot to a task with the longest run-time. The sequence produced by this algorithm is as follows:

$$1 \rightarrow 1 \rightarrow 1 \rightarrow 2 \rightarrow 2 \tag{8.8}$$

where 1 and 2 denote the unit time allocations of tasks 1 and 2 respectively. This sequence solves the Liu–Layland problem, however, it violates the lower quota

$$\lfloor q_2 \rfloor = \left\lfloor \frac{2 \times 3}{5} \right\rfloor = \left\lfloor \frac{6}{5} \right\rfloor = 1 \tag{8.9}$$

for task 2 at position 3.

The following lemma presents the second necessary condition.

Lemma 8.9. *Staying below upper quota is a necessary condition for any divisor method to solve the Liu–Layland problem.*

Proof. By contradiction. Let us consider any divisor method **M** that does not stay below upper quota and solves the Liu–Layland problem at the same time. Since **M** does not stay below upper quota, then there exists a vector of populations $\mathbf{p} = (p_1, p_2, \ldots, p_n)$, $n \geq 3$, and the house size h for which the upper quota (2.4) is not satisfied for some state k. That is

$$h \geq y_{kh} > \lceil q_k \rceil = \left\lceil \frac{p_k h}{P} \right\rceil. \tag{8.10}$$

By Lemma 2.8, we can assume that q_k is fractional.

Define an instance of the Liu–Layland problem as follows. For the task k corresponding to the state k, let $T_k = h$. Let the run-time of task k be

$$C_k = \left\lceil \frac{p_k h}{P} \right\rceil = \lceil q_k \rceil < h. \tag{8.11}$$

At the moment $t = h \times P$, the total run time of all P requests of task k equals

$$\alpha_k = P \left\lceil \frac{p_k h}{P} \right\rceil < P \left(\frac{p_k h}{P} + 1 \right) = p_k h + P. \tag{8.12}$$

For all other tasks $i \neq k$ the *initial* run-time is defined as follows

$$C_i = p_i h, \tag{8.13}$$

and their request periods $T_i = h \times P$. The run time C_i of task i is now (possibly) reduced by an integer δ_i, $\delta_i \geq 0$, to $C_i' = C_i - \delta_i$ so that $C_i' \geq 1$ and

$$\sum_{i \neq k} \delta_i = P \left\lceil \frac{p_k h}{P} \right\rceil - p_k h > 0. \tag{8.14}$$

This reduction is feasible, though it can be done in a number of different ways, since by (8.11)

$$\sum_{i \neq k} C_i' = h \sum_{i \neq k} p_i - \left(P \left\lceil \frac{p_k h}{P} \right\rceil - p_k h \right) = hP - P \left\lceil \frac{p_k h}{P} \right\rceil$$
$$= P \left(h - \left\lceil \frac{p_k h}{P} \right\rceil \right) \geq P \geq n. \tag{8.15}$$

Notice that this reduced Liu–Layland instance has a feasible solution by (8.1), since

$$\sum_{i=1}^{n} \frac{C_i'}{T_i} = \sum_{i=1,i \neq k}^{n} \frac{h p_i - \left(P \left\lceil \frac{p_k h}{P} \right\rceil - p_k h \right)}{h \times P} + \frac{\left\lceil \frac{p_k h}{P} \right\rceil}{h} = \sum_{i=1}^{n} \frac{p_i}{P} = 1. \tag{8.16}$$

Finally, since **M**, by assumption, solves the Liu–Layland problem, then it obtains a periodic sequence S_{LL} with $x_{kh} = C_k$ positions assigned to the task k in the first request period $T_k = h$. However, by (8.10) and (8.11) $y_{kh} > x_{kh}$, and the instance **r** where $r_i = h p_i - \delta_i$ for $i \neq k$ and $r_k = P \left\lceil \frac{p_k h}{P} \right\rceil > h p_k$ satisfies the following condition:

$$\frac{r_i}{r_k} = \frac{h p_i - \delta_i}{P \left\lceil \frac{p_k h}{P} \right\rceil} < \frac{p_i}{p_k} \quad \text{for all } i \neq k.$$

However, since $x_{kh} < y_{kh}$, and $h = \sum_{i=1}^{n} x_{ik} = \sum_{i=1}^{n} y_{ik}$, then there is $i \neq k$ such that $x_{ih} > y_{ih}$, which contradicts the population monotonicity of the method **M**. This ends the proof. □

Though staying below upper quota is a necessary condition for a population monotone method to solve the Liu–Layland problem, it is not a necessary condition for a *house monotone* method. Consider again the instance (8.7) and a *house monotone* (however not divisor) method that is the original Liu–Layland algorithm [87]. Let us now have this algorithm to break ties whenever there is more then one task with the same request period T_i by assigning the current time slot to a task with the shortest run-time. The sequence produced by this algorithm is as follows:

$$2 \rightarrow 2 \rightarrow 1 \rightarrow 1 \rightarrow 1 \tag{8.17}$$

where $1, 2$ denote the unit time allocations of tasks 1 and 2 respectively. This sequence solves the Liu–Layland problem, however, it violates the upper quota

$$\lceil q_2 \rceil = \left\lceil \frac{2 \times 2}{5} \right\rceil = \left\lceil \frac{4}{5} \right\rceil = 1 \tag{8.18}$$

for task 2 at position 2.

The following theorem summarizes our results.

Theorem 8.10. *Satisfying quota is a necessary condition for any divisor method to solve the Liu–Layland problem.*

Proof. Follows immediately from Lemmas 8.8 and 8.9. □

Thus, by Theorem 2.4 we have an immediate conclusion formulated as Corollary 8.11.

Corollary 8.11. *No divisor method solves the Liu–Layland problem.*

There are, however, divisor methods that stay above the lower quota, or below the upper quota as stated in the following lemmas, see Balinski and Young [2].

Lemma 8.12. *The Jefferson's method is a unique divisor method that stays above lower quota.*

Lemma 8.13. *The Adams's method is a unique divisor method that stays below upper quota.*

The following examples show that indeed neither Jefferson's nor Adams's method solve the Liu–Layland problem.

Example 8.14. The Jefferson's method of apportionment does not solve the Liu–Layland problem.

Proof. Consider an instance of four tasks.

$$T_1 = 5, C_1 = 3, T_2 = 5, C_2 = 1, T_3 = 10, C_3 = 1, T_4 = 10, C_4 = 1. \tag{8.19}$$

The rate monotonic scheduler proposed by Liu–Layland [87] gives the following sequence for the tasks:

$$1 \rightarrow 1 \rightarrow 1 \rightarrow 2 \rightarrow 3 \rightarrow 1 \rightarrow 1 \rightarrow 1 \rightarrow 2 \rightarrow 4. \tag{8.20}$$

The transformation of the instance to the apportionment problem results into four state instance with populations

$$p_1 = \frac{LC_1}{T_1} = 6, p_2 = \frac{LC_2}{T_2} = 2, p_3 = \frac{LC_3}{T_3} = 1, p_4 = \frac{LC_4}{T_4} = 1, \tag{8.21}$$

where $L = \mathrm{lcm}(5, 10) = 10$. The Jefferson's method results in the following sequence.

$$1 \rightarrow 1 \rightarrow 1 \longleftrightarrow 2 \rightarrow 1 \rightarrow 1 \rightarrow 1 \longleftrightarrow 2 \longleftrightarrow 3 \longleftrightarrow 4, \tag{8.22}$$

where $x \longleftrightarrow y$ means that the x and y can be interchanged in the sequence. In (8.22) only two (instead of required three) positions between positions 6 and 10 are occupied by task 1. Thus, (8.22) is not periodic. □

Example 8.15. The Adams's method of apportionment does not solve the Liu–Layland problem.

Proof. Consider again the instance (8.19). The Adams's method results in the following sequence.

$$1 \rightarrow 2 \rightarrow 3 \longleftrightarrow 4 \rightarrow 1 \rightarrow 1 \rightarrow 2 \longleftrightarrow 1 \rightarrow 1 \rightarrow 1 \tag{8.23}$$

for the apportionment problem (8.21). In (8.23) only two (instead of required three) positions between positions 1 and 5 are occupied by task 1. Thus, (8.23) is not periodic. □

8.4.2 Adjusting the Jefferson's Method to Solve the Liu–Layland Problem

We now show that it is possible to adjust the Jefferson's method to solve the Liu–Layland problem. We begin with the following observation.

Lemma 8.16. *The Jefferson's method assures that for every task i the cumulative run-time a_i of this task at the end of its kth request period kT_i fulfils the following condition:*

$$a_i \geq kC_i. \tag{8.24}$$

Proof. The Jefferson's method is the only divisor method that satisfies the lower quota q_i of any task i. By Lemma 8.12, we have

$$a_i \geq \lfloor q_i \rfloor = \left\lfloor \frac{h \times p_i}{\sum_{i=1}^{n} p_i} \right\rfloor, \tag{8.25}$$

for any house size h. By setting $p_i = \frac{LC_i}{T_i}$, $\sum_{i=1}^{n} p_i = L$ and finally $h = kT_i$ for some $k = 1, 2, \ldots$ and by substituting those values in (8.25) we obtain

$$a_i \geq \left\lfloor \frac{(kT_i) \times (\frac{LC_i}{T_i})}{L} \right\rfloor = kC_i. \tag{8.26}$$

Since kC_i is an integer, this completes the proof. $\square$

Lemma 8.16 guarantees that the task i runs for at least kC_i time units by the end of the kth request period kT_i which however does not guarantee that i runs for C_i time units in each interval $[(\ell - 1)T_i, \ell T_i]$, $\ell = 1, \ldots, k$. We illustrated this problem in Example 8.14. To rectify the problem we can modify the Jefferson's method so that before allocating the next time unit to task i according to the standard Jefferson's method the upper quota of that task is verified and if it is not violated the task is allocated the unit. Otherwise the task must be put on hold. This rule is nothing but the quota method of Balinski and Young [2]. More formally, their method for any apportionment $\mathbf{a}$, vector of population $\mathbf{p}$ and house size h defines the set of states that can receive the next seat at house size $h + 1$ without exceeding their upper quota, $U(\mathbf{p}, \mathbf{a})$, as follow:

$$U(\mathbf{p}, \mathbf{a}) = \left\{ i : a_i < \frac{p_i(h + 1)}{P} \right\}. \tag{8.27}$$

We have $U(\mathbf{p}, \mathbf{a}) \neq \emptyset$ for every $\mathbf{p}$ and $\mathbf{a}$ because otherwise at the house of size h the sum of apportioned seats would exceed h. Thus the quota satisfying solution can be defined by the following recursive procedure:

1. $M(\mathbf{p}, 0) = \mathbf{0}$
2. If $M(\mathbf{p}, h) = \mathbf{a}$ and k maximizes $\frac{p_k}{a_k + 1}$ over all $k \in U(\mathbf{p}, \mathbf{a})$, then $\mathbf{b} \in M(\mathbf{p}, h + 1)$ with $b_k = a_k + 1$ and $b_j = a_j$ for all $j \neq k$

Let us now consider the instance from Example 8.14. Using the procedure of Balinski and Young [2] just presented we obtain the following possible sequences:

$$1 \rightarrow 1 \rightarrow 2 \rightarrow 1 \rightarrow 3 \rightarrow 1 \rightarrow 1 \rightarrow 2 \rightarrow 1 \rightarrow 4$$
$$1 \rightarrow 1 \rightarrow 2 \rightarrow 1 \rightarrow 3 \rightarrow 1 \rightarrow 1 \rightarrow 4 \rightarrow 1 \rightarrow 2$$
$$1 \rightarrow 1 \rightarrow 2 \rightarrow 1 \rightarrow 3 \rightarrow 1 \rightarrow 1 \rightarrow 2 \rightarrow 4 \rightarrow 1$$
$$1 \rightarrow 1 \rightarrow 2 \rightarrow 1 \rightarrow 3 \rightarrow 1 \rightarrow 1 \rightarrow 4 \rightarrow 2 \rightarrow 1.$$

Notice that each is a periodic sequence required by the Liu–Layland problem.

8.4.3 Adjusting the Adams's Method to Solve the Liu–Layland Problem

For the Adams's method following lemma can be proved.

Lemma 8.17. *The Adams's method assures that for every task i the cumulative run time a_i of this task at the end of its kth request period kT_i fulfils the following condition:*

$$a_i \leq kC_i. \tag{8.28}$$

Proof. The proof is similar to the proof of Lemma 8.16 and thus will be omitted. □

Lemma 8.17 guarantees that the task i runs for at least $(\frac{L}{T_i} - k)C_i$ time units from the end of the kth request period kT_i until L which however does not guarantee that i runs for C_i time units in each interval $[(\ell - 1)T_i, \ell T_i]$, $\ell = k + 1, \ldots, \frac{L}{T_i}$. We illustrated this problem in Example 8.15. To correct the problem we can modify the Adams's method to solve the Liu–Layland in a similar fashion as was shown for the Jefferson's method, Balinski and Young [2], we omit details of this modification here.

We finally remark that in general any divisor method can be modified to satisfy quota and thus to solve the Liu–Layland problem using the general Still's procedure [96].

8.5 The Pinwheel Scheduling

The pinwheel schedule is defined as follows:

Definition 8.18 (Pinwheel Schedule). A *pinwheel schedule* on alphabet $\{1, 2, \ldots, n\}$ is an infinite sequence $S = s_1 s_2 \ldots$ such that

1. $s_j \in \{1, 2, \ldots, n\}$ for all $j \in \mathbb{N}$ and
2. Each $i \in \{1, 2, \ldots, n\}$ occurs at least once in any subsequence σ consisting of b_i consecutive elements of S

The pinwheel scheduling problem is then to find a pinwheel schedule for given $b_i, i = 1, \ldots, n$, or show that the schedule does not exist. Although the definition of pinwheel schedule requires it to be infinite, it is well-known that if the schedule exists, then there exists a periodic schedule whose period does not exceed the product $\prod b_i$. A pinwheel schedule is given in the following example.

Example 8.19. The pinwheel schedule for $b_1 = 3, b_2 = 4, b_3 = 7$, and $b_4 = 10$ is shown below,

$$(1 \to 3 \to 2 \to 1 \to 4 \to 2 \to 1 \to 3 \to 2 \to 1 \to 4 \to 2)^\infty. \qquad □$$

However, the pinwheel schedule may not always exist. We then say that the instance for which this happens is not schedulable. The following example gives one such instance.

Example 8.20. Consider the $b_1 = 3, b_2 = 4, b_3 = 5, b_4 = 5$ instance of the pinwheel problem. This instance is not schedulable. To show this let us assume that the instance is schedulable. Then, there is a periodic schedule s^∞ whose period s is not longer than $3 \times 4 \times 5 \times 5$. The s^∞ must have some two adjacent copies of $i = 1$ separated by exactly two other symbols different than 1. Otherwise, the s^∞ would solve the instance $(2, 4, 5, 5)$ which leads to a contradiction since $\frac{1}{2} + \frac{1}{4} + \frac{1}{5} + \frac{1}{5} > 1$. Therefore, without loss of generality, the following subsequence must occur in s^∞

$$** \rightarrow * \rightarrow 4 \rightarrow 1 \rightarrow 2 \rightarrow 3 \rightarrow 1 \rightarrow 4$$

which can not be extended to the left since the *two* positions $**$ and $*$ are contested by 1, 2, and 3. This proves that there is no sequence s^∞ such that its any subsequence of three includes at least one 1, any subsequence of four includes at least one 2, and any subsequence of five includes at least one 3 and at least one 4. However, there is a sequence s such that there is a 1 in any subsequence $s_{3i-2}s_{3i-1}s_{3i}$, a 2 in any subsequence $s_{4i-3}s_{4i-2}s_{4i-1}s_{4i}$, and a 3 and a 4 in any subsequence $s_{5i-4}s_{5i-3}s_{5i-2}s_{5i-1}s_{5i}$, $i = 1, 2, \ldots$This sequence is simply the periodic schedule of the Liu–Layland problem for $C_1 = 1, T_1 = 3, C_2 = 1, T_2 = 4, T_3 = 2, C_3 = 5$ which is shown in Example 8.1 in Chap. 8. Therefore, synchronization of the satellite orbital characteristics simplifies the scheduling of satellite communication with a ground station without data loss and guarantee a solution as long as

$$\Sigma \frac{1}{b_i} \leq 1. \qquad \square$$

In Example 8.20 we have

$$\frac{1}{b_1} + \frac{1}{b_2} + \frac{1}{b_3} + \frac{1}{b_4} = \frac{1}{3} + \frac{1}{4} + \frac{1}{5} + \frac{1}{5} = \frac{59}{60}.$$

Generally, the higher the instance density

$$\Sigma \frac{1}{b_i}$$

the more difficult, if possible at all, is to find a pinwheel schedule. Fishburn and Lagarias [97] prove the following theorem.

Theorem 8.21. *There is a pinwheel schedule for any instance with*

$$\Sigma \frac{1}{b_i} \leq \frac{3}{4}.$$

Example 8.20 shows that there are instances with

$$\Sigma \frac{1}{b_i} > \frac{5}{6}$$

for which no pinwheel schedule exists, actually the simplest such an instance is $b_1 = 2$, $b_2 = 3$, and b_3 any positive integer grater than 5. Chan and Chin [98] conjecture that

Conjecture 8.22. There is a pinwheel schedule for any instance with

$$\Sigma \frac{1}{b_i} \le \frac{5}{6}.$$

The conjecture is supported by the three-value theorem of Lin and Lin [99].

Theorem 8.23. *There is a pinwheel schedule for any instance with at most three different values in the multiset $\{b_1, \ldots, b_n\}$ and*

$$\Sigma \frac{1}{b_i} \le \frac{5}{6}.$$

As well as by the following theorem of Fishburn and Lagarias [97].

Theorem 8.24. *There is a pinwheel schedule for any instance with value 2 in the multiset $\{b_1, \ldots, b_n\}$ and*

$$\Sigma \frac{1}{b_i} \le \frac{5}{6}.$$

The pinwheel scheduling problem is still not shown to be NP-hard, either ordinary or strong, see Chen and Mok in the Handbook of Scheduling [100]. However, it is conjectured that the problem is NP-hard. The proof of Theorem 9.3 is insufficient to prove the NP-hardness of the pinwheel scheduling problem since the latter specifies each satellite individually and therefore model $m+1$ with demand $(K+2)L(1 - \Sigma \frac{1}{b_i})$ in this proof would have to be turned into $(K+2)L(1 - \Sigma \frac{1}{b_i})$ satellites each with time window L. This, however, may result into an exponential number of satellites, since we can not generally limit the value $(K+2)L(1 - \Sigma \frac{1}{b_i})$ by the polynomial of the input size being $O(\log(b_i + 1))$, and thus the transformation of Theorem 9.3 would not be polynomial.

The security and fault tolerance in computer networks lead to a more general pinwheel scheduling problem introduced by Baruah and Bestavros [91] where a message is split into b equal size blocks by using the Information Dispersal Algorithm developed by Rabin [101], so that any $a \le b$ out of these b suffices to reconstruct a message. We illustrate the problem by the following example with three messages $i = 1, 2, 3$ where message 1 has 5 blocks, message 2 has 3 blocks, and message 3 has 2 blocks. Without dispersion the 10 blocks of the 3 messages can be transmitted according to the following sequence

$$1 \to 2 \to 3 \to 1 \to 2 \to 1 \to 1 \to 3 \to 2 \to 1$$

for instance. However, then if any block is lost or gets corrupted during transmission, then the delay in retrieving the block is the time needed to transmit 10 blocks. By dispersing message 1 in 15 blocks of which *any* 5 would suffice to reconstruct it, message 2 in 9 blocks of which *any* 3 would suffice to reconstruct it, and message 3 in 6 blocks of which *any* 2 would suffice to reconstruct it we get a sequence

$$(1 \rightarrow 2 \rightarrow 3 \rightarrow 1 \rightarrow 2 \rightarrow 1 \rightarrow 1 \rightarrow 3 \rightarrow 2 \rightarrow 1)^3.$$

The delay to recover any single block lost during transmission does not exceed 3, 4, and 5 for messages 1, 2, and 3 respectively in the sequence with the dispersion. Furthermore, there are *at least* 5 copies of 1 in any subsequence of 10 consecutive blocks, *at least* 3 copies of 2 in any subsequence of 10 consecutive blocks, and *at least* 2 copies of 3 in any subsequence of 10 consecutive blocks. The delay can be controlled by setting a size b_i of a time window independently for each message and requesting that a sliding time window of that size always includes enough, that is a_i, blocks of that message to recover it completely. This leads the generalized pinwheel scheduling problem defined as follows.

Definition 8.25 (Generalized Pinwheel Schedule). A *generalized pinwheel schedule* on alphabet $\{1, 2, \ldots, n\}$ is an infinite sequence $S = s_1 s_2 \ldots$ such that

1. $s_j \in \{1, 2, \ldots, n\}$ for all $j \in \mathbb{N}$ and
2. Each $i \in \{1, 2, \ldots, n\}$ occurs at least a_i times in any subsequence σ consisting of b_i consecutive elements of S

The problem is NP-hard.

Theorem 8.26. *The generalized pinwheel scheduling problem is NP-hard.*

Proof. Follows immediately from the proof of Theorem 9.3. □

8.6 Applications to Pinwheel Scheduling

We now discuss some additional properties of bottleneck optimal sequences and then apply them to the pinwheel scheduling problem later in this section.

8.6.1 Additional Properties

Let $\mathbf{d} = d_1, \ldots, d_n$ be an instance of the bottleneck problem. In this and following sections we shall use the term letter (or symbol) instead of model. Let $r_i = \frac{d_i}{D} = \frac{a_i}{b_i}$, where a_i and b_i are relatively prime, be the rate for letter i. We shall consider an infinite periodic sequence $S = s^\infty$ with cycle s obtained by the algorithm, for instance, described in Chap. 5 with a given bound B on the bottleneck deviation.

First, we show that any subsequence w of S that starts with an i, ends with an i, and includes exactly ka_i copies of i is not *longer* than $kb_i + 1$, $k = 1, 2, \ldots$. Let the first occurrence of i in the subsequence w be the copy $j + 1$ of i, and the last occurrence of i in the subsequence w be the copy $j + ka_i$ of i. In S, copy $j + 1$ can not be in any position prior to $E(i, j+1)$ and copy $j + ka_i$ can not be in any position later than $L(i, j + ka_i)$ for a given feasible B. We have the following upper bound on the difference between the latter and the former in S with $B < 1$.

Lemma 8.27. *For $B < 1$ and $k = 1, 2, \ldots j = 0, 1, \ldots$, we have*

$$L(i, j + ka_i) - E(i, j+1) \le kb_i,$$

and

$$L(i, ka_i) \le kb_i,$$

where $i = 1, \ldots, n$.

Proof. By Theorem 5.1

$$L(i, j + ka_i) = \left\lfloor \frac{j + ka_i - 1 + B}{r_i} + 1 \right\rfloor,$$

and

$$E(i, j+1) = \left\lceil \frac{j + 1 - B}{r_i} \right\rceil.$$

Thus,

$$L(i, j + ka_i) - E(i, j+1) \le kb_i + 1 + \frac{2B - 2}{r_i}. \tag{8.29}$$

Since $B < 1$, then $1 + \frac{2B-2}{r_i} < 1$. Thus, since the left hand side of the inequality (8.29) is integral, then, we have

$$L(i, j + ka_i) - E(i, j+1) \le kb_i.$$

Finally,

$$L(i, ka_i) = \left\lfloor \frac{ka_i - 1 + B}{r_i} + 1 \right\rfloor = kb_i + 1 - \left\lceil \frac{1 - B}{r_i} \right\rceil \le kb_i,$$

for $B < 1$. $\quad\square$

The bound can be slightly reduced for the is with small ratios $\frac{a_i}{b_i}$. More precisely, we have.

Lemma 8.28. *If $\frac{a_i}{b_i} \le \frac{1}{n-1}$, then there is a sequence S with*

$$L(i, j + ka_i) - E(i, j+1) \le kb_i - 1,$$

for $k = 1, 2, \ldots j = 0, 1, \ldots$.

Proof. Consider the inequality (8.29) for $B = 1 - \frac{1}{2(n-1)}$. By Theorem 5.8 a sequence S with its bottleneck not exceeding this B always exists. For the S, we have

$$1 + \frac{2B - 2}{r_i} < 0$$

as long as $r_i = \frac{a_i}{b_i} < \frac{1}{n-1}$. Then, however,

$$L(i, j + ka_i) - E(i, j + 1) < kb_i, \tag{8.30}$$

The inequality (8.30) holds for $\frac{a_i}{b_i} = \frac{1}{n-1}$ as well. We then have

$$L(i, j + ka_i) = kb_i + j(n - 1)$$

and

$$E(i, j + 1) = j(n - 1) + 1,$$

thus

$$L(i, j + ka_i) - E(i, j + 1) = kb_i - 1.$$

This proves the lemma.   □

By Lemmas 8.27 and 8.28 we prove.

Theorem 8.29. *Let S be a sequence with bottleneck deviation $B < 1$. Then any its subsequence w that starts with an i, ends with an i, and has ka_i copies of i is not longer than $kb_i + 1$, $k = 1, 2, \ldots$. Moreover, for any i with $\frac{a_i}{b_i} \leq \frac{1}{n-1}$ the subsequence w is not longer than kb_i, $k = 1, 2, \ldots$ for $B = 1 - \frac{1}{2(n-1)}$.*

Proof. Let the first occurrence of i in the subsequence w be the copy $j + 1$ of i, and the last occurrence of i in the subsequence w be copy $j + ka_i$ of i. In S, copy $j + 1$ can not be in any position prior to $E(i, j + 1)$ and copy $j + ka_i$ can not be in any position after $L(i, j + ka_i)$ if the bottleneck of S is at most B. By Lemma 8.27, the difference between the latter and the former is at most kb_i but since a copy of i can occupy the position $E(i, j + 1)$, the sequence can not be longer than $kb_i + 1$ which proves the first part of the theorem. The second part follows from Lemma 8.28.   □

Second, we show that any subsequence w of s with at least $ka_i + 2$ copies of i is not *shorter* than $kb_i + 1$, $k = 1, 2, \ldots$. Let the first copy of i and the last copy of i in subsequence w be copies j and l respectively. Obviously, $l \geq j + ka_i + 1$, otherwise there would be less than $ka_i + 2$ copies of i in subsequence w. In S, copy j can not be in any position higher than $L(i, j)$ and copy $j + ka_i + 1$ can not be in any position prior to $E(i, j + ka_i + 1)$ for a given feasible B. We have the following lower bound on the difference between the latter and the former in S with $B < 1$.

Lemma 8.30. *For $B < 1$ and $k = 1, 2, \ldots$ $j = 1, \ldots$, we have*

$$E(i, j + ka_i + 1) - L(i, j) \geq kb_i,$$

where i=1,…,n.

Proof. By Theorem 5.1

$$E(i,j+ka_i+1) = \lceil \frac{j+ka_i+1-B}{r_i} \rceil,$$

and

$$L(i,j) = \lfloor \frac{j-1+B}{r_i} + 1 \rfloor.$$

Thus,

$$E(i,j+ka_i+1) - L(i,j) \geq kb_i - 1 + \frac{2-2B}{r_i}. \tag{8.31}$$

Since $B < 1$, then $\frac{2-2B}{r_i} - 1 > -1$. Thus, since the left hand side of the inequality (8.31) is integral, thus, we have

$$E(i,j+ka_i+1) - L(i,j) \geq kb_i.$$

$\square$

The bound can be slightly increased for the is with small ratios $\frac{a_i}{b_i}$. More precisely, we have.

Lemma 8.31. *If* $\frac{a_i}{b_i} \leq \frac{1}{n-1}$, *then there is a sequence S with*

$$E(i,j+ka_i+1) - L(i,j) \geq kb_i + 1,$$

for $k = 1,2,\dots$ $j = 0,1,\dots$ *for* $B = 1 - \frac{1}{2(n-1)}$.

Proof. Consider the inequality (8.31) for $B = 1 - \frac{1}{2(n-1)}$. By Theorem 5.8 a sequence s with its bottleneck not exceeding this B always exists. For the S, we have

$$-1 + \frac{2-2B}{r_i} > 0$$

as long as $r_i = \frac{a_i}{b_i} < \frac{1}{n-1}$. Then, however,

$$E(i,j+ka_i+1) - L(i,j) > kb_i. \tag{8.32}$$

The inequality (8.32) holds for $\frac{a_i}{b_i} = \frac{1}{n-1}$ as well. We then have

$$E(i,j+ka_i+1) = kb_i + j(n-1) + 1$$

and

$$L(i,j) = j(n-1),$$

thus

$$E(i,j+ka_i+1) - L(i,j) = kb_i + 1$$

which proves the lemma. $\square$

Lemmas 8.30 and 8.31 imply the following theorem.

Theorem 8.32. *Let S be a sequence with bottleneck deviation $B < 1$. Then, any its subsequence w with at least $ka_i + 2$ copies of i is not shorter than $kb_i + 1$, $k = 1, 2, \ldots$. Moreover, for any i with $\frac{a_i}{b_i} \leq \frac{1}{n-1}$ the subsequence w is not shorter than $kb_i + 2$, $k = 1, 2, \ldots$.*

Proof. Let the first occurrence of i in the subsequence w be the copy j of i, and the last occurrence of i in the subsequence w be copy l of i. We have, $l \geq j + ka_i + 1$ since there are at least $ka_i + 2$ occurrences of i in w. In s, copy j can not be in any position after $L(i, j)$ and copy $j + ka_i + 1$ can not be in any position prior to $E(i, j + ka_i + 1)$ if the bottleneck of s is at most B. By Lemma 8.30, the difference between the latter and the former is at least kb_i but since a copy of i can occupy the position $L(i, j)$, the sequence can not be shorter than $kb_i + 1$ which proves the first part of the theorem. The second part follows from Lemma 8.31. □

We use the two lemmas to characterize the distribution of letter i in S with $B < 1$.

Theorem 8.33. *Let $s_{j+1} \ldots s_{j+b_i}$ be any subsequence of b_i consecutive letters of $S = s^\infty$ with $B < 1$. Then, letter i occurs either $a_i - 1$, or a_i, or $a_i + 1$ times in the subsequence. Furthermore, if the closest subsequences $s_{j+1} \ldots s_{j+b_i}$ and $s_{j+1+kb_i} \ldots s_{j+(k+1)b_i}$, for some $k \geq 1$ have $a_i - 1$ copies of letter i each, then $k \geq 2$ and there are exactly $(k-1)a_i + 1$ copies of i in the sequence $s_{j+1+b_i} \ldots s_{j+kb_i}$.*

Proof. Consider a subsequence $w = s_{j+1} \ldots s_{j+b_i}$, $j \geq 0$, of S. We first show that there are at least $a_i - 1$ copies of i in w. This claim obviously holds for $a_i = 1$. Thus, let $a_i \geq 2$. Assume that the number of copies of i in w is less than $a_i - 1$. Then, if there is no i in $s_1 \ldots s_j$, then there is $l > j + b_i + 1$ such that $s_l = i$ and the sequence $s_1 \ldots s_l$ has exactly a_i copies of i but this contradicts Lemma 8.27 since $L(i, a_i) \leq b_i$ for $B < 1$. Now, if i occurs in $s_1 \ldots s_j$, then there are $k \leq j$ and $l > j + b_i$ such that $s_k = s_l = i$ and the sequence $s_k \ldots s_l$ has exactly a_i copies of i. However, $l - k + 1 > b_i + 1$ which contradicts Theorem 8.29. Therefore, there is at least $a_i - 1$ copies of i in $s_{j+1} \ldots s_{j+b_i}$.

Let us now assume that there are at least $a_i + 2$ copies of i in w. This, however, contradicts Theorem 8.32. Therefore, the only possible numbers of copies of i in $s_{j+1} \ldots s_{j+b_i}$ are $a_i - 1$, a_i, and $a_i + 1$, which completes the proof of the first part of the theorem.

Now, let $s_{j+1} \ldots s_{j+b_i}$ and $s_{j+kb_i+1} \ldots s_{j+(k+1)b_i}$ for $k \geq 1$ be two sequences with $a_i - 1$ copies of i. By contradiction. Assume that each sequence $s_{j+lb_i+1} \ldots s_{j+(l+1)b_i}$, $0 < l < k$, if any, in between the two has exactly a_i copies of i. Then, if there is no i in $s_1 \ldots s_j$, then sequence $s_1 \ldots s_{j+(k+1)b_i}$ has no more than $(k+1)a_i - 2$ copies of i and thus there is $l > j + (k+1)b_i + 1$ such that $s_l = i$ and the sequence $s_1 \ldots s_l$ has exactly $(k+1)a_i$ copies of i but this contradicts Lemma 8.27 which implies $L(i, (k+1)a_i) \leq (k+1)b_i$ for $B < 1$. If there is an i in $s_1 \ldots s_j$, then there are $h \leq j$ and $l > j + (k+1)b_i + 1$ such that $s_h = s_l = i$ and the sequence $s_h \ldots s_l$ has exactly $(k+1)a_i$ copies of i. However, $l - h + 1 > (k+1)b_i + 1$ which contradicts Theorem 8.29. Therefore, in both cases $k \geq 2$, and there must be a sequence $s_{j+lb_i+1} \ldots s_{j+(l+1)b_i}$ $0 < l < k$ with $a_i + 1$ copies of i.

To complete the proof we observe that there is a positive integer α_i such that $\alpha_i b_i = D$ and $\alpha_i a_i = d_i$ consequently there are $\alpha_i a_i$ copies of i in $s = s_1 \ldots s_D$. For any j, $j = 0, \ldots, b_i - 1$, consider sequences $w_k = s_{j+kb_i+1} \ldots s_{j+(k+1)b_i}$ for $k = 0, \ldots, \alpha_i - 2$, and

$$w_{\alpha_i-1} = s_{j+(\alpha_i-1)b_i+1} \ldots s_D s_1 \ldots s_j.$$

Let $w_{j_0}, \ldots, w_{j_m}$ for $j_0 < \ldots < j_m$ be all the sequences with $a_i - 1$ copies of i. Then there must be *exactly* one sequence w_j with $j_{k \bmod (m+1)} < j < j_{k \bmod (m+1)+1}$ with $a_i + 1$ copies of i. This claim follows from the fact that there must be at *least* one such a sequence, which we have already shown, and the fact that there cannot be more than one since the number of is is $\alpha_i a_i$. $\square$

8.6.2 The Applications

We now apply the properties of just-in-time sequences with small bottleneck deviations to the pinwheel scheduling problems. We begin with the following sufficient condition for the existence of the generalized pinwheel schedule.

Theorem 8.34. *If*

$$\sum_{1 \le i \le n} \frac{a_i}{b_i} + \frac{1}{b_i} \le 1,$$

then there is a generalized pinwheel schedule for pairs $(a_1, b_1), \ldots, (a_n, b_n)$. The schedule can be found any optimization algorithm for the bottleneck problem.

Proof. Let $(a_1, b_1), \ldots, (a_n, b_n)$ be an instance of the generalized pinwheel scheduling problem such that $\sum_{1 \le i \le n} \frac{a_i}{b_i} + \frac{1}{b_i} \le 1$. Define $d_i = \frac{L(a_i+1)}{b_i}$, for $i = 1, \ldots, n$, where $L = \mathrm{lcm}\,(b_1, \ldots, b_n)$. Then, $\sum_{i=1}^{n} d_i \le L$, and if $\sum_{i=1}^{n} d_i < L$, then add a dummy with $d_{n+1} = L - \sum_{i=1}^{n} d_i$. Theorem 8.33 ensures that any optimization algorithm for the bottleneck problem when applied to this instance with the ratios $\frac{d_i}{L} = \frac{a_i+1}{b_i}$, for $i = 1, \ldots, n$, delivers a sequence with at least $(a_i + 1) - 1 = a_i$ copies of i in any subsequence of b_i consecutive letters, and therefore a generalized pinwheel schedule for $(a_1, b_1), \ldots, (a_n, b_n)$. $\square$

A similar result was independently obtained by Baruah and Lin [102]. Notice that as a corollary from Theorem 8.34 we have that, see also Holte et al. [90], the following holds.

Corollary 8.35. *There exists a pinwheel schedule for $(1, b_1), \ldots, (1, b_n)$ if only* $\sum_{1 \le i \le n} \frac{1}{b_i} \le \frac{1}{2}$.

Theorem 6.7 characterizes all instances of the bottleneck problem for $n = 2$ with the bottleneck deviation $B^* < \frac{1}{2}$. This small B^* is key in the proof that the generalized pinwheel schedule always exists for $n = 2$.

Theorem 8.36. *The generalized pinwheel schedule always exists for $n = 2$.*

Proof. Consider a generalized pinwheel problem instance $(a_1, b_1), (a_2, b_2)$ for $n = 2$. Without loss of generality assume that $\gcd(a_1, b_1) = \gcd(a_2, b_2) = 1$. Observe that a pinwheel schedule for $(\frac{a_1}{\gcd(a_1,b_1)}, \frac{b_1}{\gcd(a_1,b_1)}), (\frac{a_2}{\gcd(a_2,b_2)}, \frac{b_2}{\gcd(a_2,b_2)})$ is a pinwheel schedule for $(a_1, b_1), (a_2, b_2)$. If

$$\frac{a_1}{b_1} + \frac{a_2}{b_2} = 1,$$

then $b_1 = b_2$ and any sequence with a_1 copies of 1 and a_2 copies of 2 solves the pinwheel problem.

If

$$\frac{a_1}{b_1} + \frac{a_2}{b_2} < 1,$$

then if the product $b_1 b_2$ is odd, then there are *nonnegative* α_1 and α_2 such that $\alpha_1 + \alpha_2 = b_1 b_2 - (a_1 b_2 + a_2 b_1) \geq 1$

$$\frac{a_1}{b_1} \leq \frac{a_1 b_2 + \alpha_1}{b_1 b_2} \quad \text{and} \quad \frac{a_2}{b_2} \leq \frac{a_2 b_1 + \alpha_2}{b_1 b_2}.$$

Let $\beta = \gcd(a_1 b_2 + \alpha_1, a_2 b_1 + \alpha_2, b_1 b_2)$. Define

$$d_1 = \frac{a_1 b_2 + \alpha_1}{\beta},$$

$$d_2 = \frac{a_2 b_1 + \alpha_2}{\beta},$$

$$D = \frac{b_1 b_2}{\beta}.$$

We have $D = d_1 + d_2$ and odd since $b_1 b_2$ is odd. Thus, exactly one of d_1 or d_2 is odd and the other even.

If the product $b_1 b_2$ is even, then there are *positive* α_1 and α_2 such that $\alpha_1 + \alpha_2 = 1 + b_1 b_2 - (a_1 b_2 + a_2 b_1) \geq 2$

$$\frac{a_1}{b_1} \leq \frac{a_1 b_2 + \alpha_1}{b_1 b_2 + 1} \quad \text{and} \quad \frac{a_2}{b_2} \leq \frac{a_2 b_1 + \alpha_2}{b_1 b_2 + 1}.$$

Let $\beta = \gcd(a_1 b_2 + \alpha_1, a_2 b_1 + \alpha_2, b_1 b_2 + 1)$. Define

$$d_1 = \frac{a_1 b_2 + \alpha_1}{\beta},$$

$$d_2 = \frac{a_2 b_1 + \alpha_2}{\beta},$$

$$D = \frac{b_1 b_2 + 1}{\beta}.$$

We have $D = d_1 + d_2$ and odd since $b_1 b_2 + 1$ is odd. Thus, exactly one of d_1 or d_2 is odd and the other even.

By Theorem 6.7, $B^* < \frac{1}{2}$ for (d_1, d_2) if and only if one of d_1 or d_2 is odd and the other even. However, for $B^* < \frac{1}{2}$ copy j of $i = 1, 2$ is in position

$$\left\lceil \frac{2j-1}{2r_i} \right\rceil$$

and copy $j + a_i$ in position

$$\left\lceil \frac{2(j+a_i)-1}{2r_i} \right\rceil \le b_i + \left\lceil \frac{2j-1}{2r_i} \right\rceil .$$

Therefore, there are at least a_i copies of i in any subsequence of length b_i. This ends the proof. $\square$

We now illustrate Theorem 8.36 with an example.

Example 8.37. Consider an instance $(a_1 = 2, b_1 = 5), (a_2 = 4, b_2 = 7)$ of the generalized pinwheel scheduling problem. We have

$$\frac{a_1}{b_1} = \frac{2}{5} \quad \text{and} \quad \frac{a_2}{b_2} = \frac{4}{7},$$

consequently $b_1 b_2 = 35$ is odd and $\alpha = b_1 b_2 - (a_1 b_2 + a_2 b_1) = 35 - (14 + 20) = 1$. Thus, we can take

$$d_1 = \frac{15}{5} = 3 \ d_2 = \frac{20}{5} = 4 \ \text{and} \ D = \frac{35}{5} = 7,$$

where $\beta = \gcd(a_1 b_2 + \alpha_1, a_2 b_1 + \alpha_2, b_1 b_2) = 5$. Consider the instance $(d_1 = 3, d_2 = 4)$ of the bottleneck problem. By Theorem 8.36, the three copies of 1 will be in positions

$$\left\lceil \frac{(2j-1)}{2r_1} \right\rceil = \left\lceil \frac{7(2j-1)}{6} \right\rceil$$

for $j = 1, 2, 3$ and the four copies of 2 will be in positions

$$\left\lceil \frac{(2j-1)}{2r_2} \right\rceil = \left\lceil \frac{7(2j-1)}{8} \right\rceil$$

for $j = 1, 2, 3, 4$ in an optimal solution for this instance. Therefore, 1 occupies all even-indexed positions, and 2 all odd-indexed positions and the sequence is

$$2 \to 1 \to 2 \to 1 \to 2 \to 1 \to 2. \quad \square$$

Theorem 8.33 applies to $n = 3$ provided the following condition is met

$$\min\{\frac{a_1 + a_2}{b_1} + \frac{a_3}{b_3}, \frac{a_1}{b_1} + \frac{a_2 + a_3}{b_2}\} \le 1, \tag{8.33}$$

where $(a_1, b_1), (a_2, b_2)$ and (a_3, b_3) is an instance of the generalized pinwheel scheduling problem with $n = 3$. Without loss of generality we assume $b_1 < b_2 < b_3$. The condition allows to reduce the $n = 3$ case to the $n = 2$ case with either $(a_1 + a_2, b_1)$ and (a_3, b_3) or (a_1, b_1) and $(a_2 + a_3, b_2)$ depending on whichever of the two instances is feasible.

Example 8.38. Consider the instance $(2, 6), (4, 15)$, and $(3, 19)$ given in Lin and Lin [99]. We have

$$\frac{2}{6} + \frac{4+3}{15} \le 1,$$

thus the instance $(2, 6), (7, 15)$ is feasible. Since $\gcd(2, 6) = 2$, consider $(1, 3), (7, 15)$ instead. We have $3 \times 15 = 45$ odd, and $\alpha = 45 - (15 + 21) = 9$. Thus, we can take

$$d_1 = \frac{15}{15} = 1 \quad d_2 = \frac{30}{15} = 2 \text{ and } D = \frac{45}{15} = 3,$$

where $\beta = \gcd(a_1 b_2 + \alpha_1, a_2 b_1 + \alpha_2, b_1 b_2) = \gcd(15 + 0, 21 + 9) = 15$. By Theorem 8.36, the single copy of 1 will be in position $\left\lceil \frac{(2j-1)}{2r_1} \right\rceil = 2$ and the two copies of 2 will be in positions

$$\left\lceil \frac{(2j-1)}{2r_2} \right\rceil = \left\lceil \frac{3(2j-1)}{4} \right\rceil$$

for $j = 1, 2$ in an optimal solution for this instance. Therefore, the sequence is

$$2 \to 1 \to 2.$$

Turning one of the $2's$ into a 3 we get the following pinwheel schedule

$$(2 \to 1 \to 3)^\infty,$$

which actually meets a condition "1 out of 3" for each of the three. Observe, the a different than 0 and 9 choice of α_1 and α_2 respectively could lead to a different sequence. □

Recall from Sect. 8.5 that we can not guarantee the existence of a generalized pinwheel schedule for density higher than $\frac{5}{6}$ for $n = 3$. More applications can be found in the exercises.

8.7 Exercises

Exercise 8.39. Adjust the Adams's method to solve the Liu–Layland problem. Hint: See Balinski and Young [2].

Exercise 8.40. Show that any divisor method can be adjusted to solve the Liu–Layland problem. Hint: See Still [96]

Exercise 8.41. Prove Lemmas 8.12 and 8.13.

Exercise 8.42. Show that there is a pinwheel schedule for any instance with at most two different values in the multiset $\{b_1, \ldots, b_n\}$ and

$$\sum \frac{1}{b_i} \leq 1.$$

Hint: See Holte et al. [103] and Theorem 8.36.

Exercise 8.43. Prove Theorem 8.23 for $\min\{b_1, b_2, b_3\} \geq 18$.

Exercise 8.44. Show that

$$L(i, j) + 1 \geq E(i, j+1) \geq L(i, j),$$

for $B = 1 - \frac{1}{D}$, $i = 1, \ldots, n$ and $j = 1, \ldots, d_i (\geq 2)$.

Exercise 8.45. Let (a_i, b_i), $i = 1, \ldots, n$ be an instance of the generalized pinwheel scheduling problem. Consider any just-in-time sequence s with the rates $r_i = \frac{a_i}{b_i}$ and bottleneck deviation not exceeding $B = 1 - \frac{1}{2(n-1)}$. show that any sequence of length at least

$$b_i + 1 + \left\lfloor \frac{n-2}{n-1} \times \frac{b_i}{a_i} \right\rfloor$$

must have at least a_i copies of $i = 1, \ldots, n$.

Exercise 8.46. Find solutions to the generalized pinwheel scheduling problem for each of the following two instances: $(1, 3), (7, 15), (1, 16)$ and $(1, 4)$, $(1, 5), (2, 6)$, see Lin and Lin [99].

8.8 Comments and References

The Liu–Layland periodic scheduling problem was introduced by Liu and Layland [87], see also Devillers and Goossens [104]. Its solution via just-in-time sequencing presented in Sect. 8.3 was given by Kubiak [105]. The divisor methods for the Liu–Layland problem were studied by Józefowska et al. [106] on which Sect. 8.4 is based. Some of the properties of the bottleneck sequences with $B < 1$ have been shown by Kubiak [107]. The pinwheel scheduling was introduced by Holte et al. [90] in 1989, and the generalized pinwheel scheduling problem by Baruah and Bestavros [91], see also Baruah and Lin [102]. Theorem 8.34 was given by Kubiak [107], its application to the pinwheel scheduling problem gives a short proof of Corollary 8.35 which was originally proven by Holte et al. [90], see also Baruah and Lin [102]. The pinwheel scheduling problem for two distinct numbers was studied by Holte et al. [103]. Theorem 8.36 gives a short proof that the problem is always feasible, that is a pinwheel schedule with two distinct numbers always exists.

Chapter 9
Temporal Capacity Constraints and Supply Chain Balancing

9.1 Introduction

This chapter discusses leveling off (smoothing out) demand in mixed-model, pull supply chains. These chains respond to customer demand by setting forecast-based demands for each model produced, and pulling supplies required for model production whenever they are needed. To level off and synchronize these supply chains, it is crucial to design model delivery sequence for given demands for models. Two main goals shape this sequence. The external, that is meeting the model demands, and, the internal, that is satisfying the chain temporal capacity constraints. These constraints may render a model delivery sequence difficult to implement through the pull mechanism of the chain just-in-time material flow for the sequence may *temporarily* impose too much strain on supplier's resources by setting too high a temporary delivery rate for their supplies.

The chapter addresses the temporal supplier capacity constraints. It presents in Sect. 9.7 a framework for a mixed-model, pull supply chain, and proposes using the model of temporal capacity constraints based on the car sequencing problem to level off demand in the chain. It claims that this model of temporal capacity constraints though typically used to better balance the workload of mixed-model assembly lines may serve a more important and general goal of leveling off demand in the entire supply chain. We assume in particular that supplier s is a subject to a *capacity* constraint in the form $p_s : q_s$, which means that *at most* p_s models of the model delivery sequence S in each consecutive sequence of q_s models of S may need options supplied by s. A digression: it seems natural here and later in the chapter to use the small letter s to denote a *supplier*, we hope this notation will not be confused with the notation that uses the letter to denote sequences in other chapters of the book.

The leveling off problem then consists in finding a just-in-time sequence S of length D over models $\{1, \ldots, n\}$, where i occurs exactly d_i times, and which respects the $p_s : q_s$ capacity constraints for each supplier s. This type of constraint is an Artificial Intelligence model of temporal capacity constraints intended originally

W. Kubiak, *Proportional Optimization and Fairness,* International Series in Operations Research & Management Science 127, DOI 10.1007/978-0-387-87719-8_9,
© Springer Science+Business Media LLC 2009

for sequencing models on mixed-model assembly lines. The problem has been a *cause célèbre* for the constraint programming methods for about 20 years now, see ILOG [108].

Section 9.2 introduces the car sequencing problem, and Sect. 9.3 studies its computational complexity. Section 9.3 also points out a similarity between the car sequencing problem and the generalized pinwheel problem discussed in Chap. 8. While the former seeks sequences with models being sparsely spread throughout the sequence so that not too many copies of the same model are close to one another – thus the "at most q out of p" constraint, the latter seeks the opposite, that is a sequence with models being densely packed throughout the sequence – thus the "at least q out of p" constraint. Though similar to the car sequencing problem, the pinwheel scheduling problem has lead its independent live in Computer Science for its applications come from the satellite scheduling and the information dispersal algorithms. To the author's knowledge the literature does not report solutions to the pinwheel problem by constrained programming methods. Section 9.4 presents a dynamic programming algorithm for the car sequencing problem. Section 9.6 presents an integer programming formulation of the car sequencing problem, and Sect. 9.5 gives some characteristics of the instances for which a feasible car sequence exists.

Section 9.8 presents other method of leveling off the chains that has its roots in the theory of regular words. Recall from Chap. 6 that regular words balance workloads in events graphs which can also naturally model flows in mixed-model, pull supply chains. The section investigates certain properties of the model and option delivery sequences. The properties are based on the combinatorics on words, especially on the concept of balanced words which is a more general concept than regular words, Tijdeman [5]. The chapter introduces and explores a link between the model delivery sequences and balanced words, and shows that though balanced words result in optimal workload balancing, see Chap. 6, they can not be obtained for all possible sets of demand rates. The obtainable model delivery sequences are either 2-balanced or 3-balanced at best, that is if they disregard temporal capacity constraints, and thus they are more complex than balanced sequences. Section 9.9 observes that moving downstream of the supply chain the model delivery sequences translate in the option delivery sequences, and thus the variability of the sequences as measured by the degree of their balance increases, a similar effect of increased variability down the supply chain has been well know in the theory of supply chain, see Daganzo [109]. We show bounds on the degree of balance based on minimum bottleneck model delivery sequences studied in Chap. 5.

Finally, the chapter discusses two other techniques to level off the suppliers demand for parts. Section 9.10 discusses the periodic synchronized delivery, and Sect. 9.11 the synchronized delivery models for constructing the model delivery sequence to minimize safety stocks of parts in the mixed-model, pull supply chains. They also show bounds for the safety stocks based on the model delivery sequences with minimum bottleneck.

9.2 The Car Sequencing Problem: A Model of Temporal Capacity Constraints

The car sequencing problem was first introduced by Parello et al. [110] in 1986 to sequence a mix of models to be produced on a single assembly line. In their illustrative example a car assembly line assembles cars with slightly different sets of options, for instance transmissions or air conditioning. Assume, for instance, that the demand for the car models with air conditioning is estimated to be 60% of the total demand. If the cars are moving through the line with a given cycle time so that exactly five of them will pass the air-conditioning workstation in the time it takes to install air-conditioning on a single car, then three teams are the minimum number required at the air-conditioning workstation in order to meet the demand for models with the air-conditioning option. With the three teams whenever a model with air conditioning enters the air-conditioning workstation, an available team starts working on it walking along with the car until it reaches the end of the workstation. At this moment, four more cars will have passed the workstation entrance. If no more than two of them require the air conditioning installation, then the other two teams will be able to handle the workload. If, however, at least one more car requires air conditioning installation, the workstation will be unable to respond fast enough and either the workstation operator will have to push a stop button to slow down the line or a utility team will be called for to finish the work. Thus, having three teams may prove insufficient and a proper sequencing of models must be done to deal with the temporal capacity shortage. That is to avoid the disruption and related costs, the sequence of models that enters the line must meet the 3 out of 5 *capacity constraint,* also denoted by 3:5, for models with air conditioning. That is no more than three models may require air conditioning out of any consecutive five that enter the assembly line.

The car sequencing problem is defined by a set $M = \{1,\ldots,n\}$ of n models and a set $O = \{1,\ldots,m\}$ of m options. Each model i differs from all other models in M by its subset $O_i \subseteq O$ of options, that is $O_i \neq O_k$ for $i \neq k$. Since there are 2^m different subsets of O, then the number of models n does not exceed 2^m, or $\log n \leq m$. The demand vector $\mathbf{d} = (d_1,\ldots,d_n)$ for models in M is also given, where d_i is the number of copies of model i to be produced. The problem then is to find a sequence S, if any exists or return NO otherwise, of length $D = \sum_{i=1}^{n} d_i$ where model i occurs exactly d_i times that meets the option capacity constraints for each option. The constraints are defined for each option $j \in O$ by giving a pair $p_j : q_j$, $p_j < q_j$, that stipulates that no more than p_j models out of any consecutive q_j models in S may include option j. The option content of the models will be also represented by an $m \times n$ binary matrix C, where $C_{ji} = 1$ if and only if model i includes option j, that is $j \in O_i$.

Example 9.1. Table 9.1 shows an instance of the car sequencing problem with $n = 6$ models and $m = 5$ options. Table 9.2 shows a feasible sequence for the instance in Table 9.1. $\square$

Table 9.1 An instance of the car sequencing problem

Option	Capacity	Models					
		1	2	3	4	5	6
1	2:3	1	0	0	0	1	1
2	2:3	0	0	1	1	0	1
3	1:2	1	0	0	0	1	0
4	3:5	1	1	0	1	0	0
5	2:5	0	0	1	0	0	0
Demands		2	3	1	1	2	2

Table 9.2 A feasible sequence of models

Option	Sequence										
	2	2	1	3	5	2	1	6	4	5	6
1	0	0	1	0	1	0	1	1	0	1	1
2	0	0	0	1	0	0	0	1	1	0	1
3	0	0	1	0	1	0	1	0	0	1	0
4	1	1	1	0	0	1	1	0	1	0	0
5	0	0	0	1	0	0	0	0	0	0	0

An interesting feature of the car sequencing problem is that it is relatively easy to prove that the problem is NP-hard even if one significantly constraints the problem instances. This issue will be explored further in the next section.

9.3 The Complexity of the Car Sequencing Problem

Interestingly, the complexity of the car sequencing problem was rightly suspected to be in the class of the NP-hard in the strong sense problems from the time the problem was first formulated, however, it was only in 1998 when Gent [111] showed the problem NP-hard in the strong sense by an elegant transformation from the Hamiltonian path problem, though his transformation requires different capacity constraints for different car options. We now show that the car sequencing problem remains NP-hard in the strong sense even if all demands are unit, that is $d_i = 1$, and all options have the same capacity constraints $1 : \alpha$ for some positive integer α, that is for each option j at most 1 in each consecutive α models of the sequence may require the option j. We have the following theorem.

Theorem 9.2. *The car sequencing problem is NP-hard in the strong sense even if the demand for each model is unit and each option capacity constraint is the same* $1 : \alpha$ *for some* $\alpha > 1$.

Proof. The transformation is from the graph coloring problem, see Garey and Johnson [81]. Let graph $G = (V, E)$ and $k \geq 2$ make up an instance of the graph coloring problem. Let $|V| = n$ and $|E| = m$. Take k *disjoint* isomorphic copies of

G, $G^1 = (V^1, E^1), \ldots, G^k = (V^k, E^k)$. Let $\mathcal{G} = (\mathcal{V} = \bigcup_{i=1}^{k} V^i, \mathcal{E} = \bigcup_{i=1}^{k} E^i)$ be the union of the k copies. Now, consider an independent set S on n nodes, that is the graph $S = (N = \{1, \ldots, n\}, \emptyset)$. Take $k+1$ *disjoint* isomorphic copies of S, $S^1 = (N^1, \emptyset), \ldots, S^{k+1} = (N^{k+1}, \emptyset)$. Add an edge between any two nodes of $\mathcal{N} = \bigcup_{i=1}^{k+1} N^i$ being in different copies of S to make a graph $\mathcal{S} = (\mathcal{N}, \mathcal{X} = \bigcup_{i \neq j} N_i \times N_j)$. Notice that $N^1, \ldots, N^{k+1}$ are independent sets of $\mathcal{N}$ each with cardinality n. No independent set of $\mathcal{N}$ with cardinality $n+1$ exists. Finally, consider a *disjoint* union of $\mathcal{G}$ and $\mathcal{N}$, that is $\mathcal{H} = \mathcal{G} \cup \mathcal{N} = (\mathcal{V} \cup \mathcal{N}, \mathcal{E} \cup \mathcal{X})$. Clearly, the union has $nk + n(k+1)$ nodes and $mk + \frac{k(k+1)}{2} n^2$ edges, and thus its size is polynomially bounded in n, m and k and consequently polynomial in the size of the input instance of the graph coloring problem. Consider the *arc-node* incidence matrix I of graph $\mathcal{H}$. The columns of I correspond to the nodes of $\mathcal{H}$ and they, in turn, correspond to the *models*. The rows of I correspond to the edges of $\mathcal{H}$ and they, in turn, correspond to the *options*. The demand for each model equals one. The capacity constraint for each option in $\mathcal{E}$ is $1 : (n+1)$, and so is the capacity constraint for each option in $\mathcal{X}$ is $1 : (n+1)$. We shall refer to any option in $\mathcal{E}$ as the $\mathcal{E}$-option, and to any option in $\mathcal{X}$ as $\mathcal{X}$-option.

(*if*) Assume there is a coloring of G using no more than k colors. Then, obviously, there is a coloring of G using *exactly* k colors. The coloring defines a partition of V into k independent sets $W_1, \ldots, W_k$. Let $W_j^i \subseteq V^i$ be a copy of the independent set W_j inside of the copy G^i of G. Define the sets

$$A_1 = W_1^1 \cup W_2^2 \cup \ldots \cup W_k^k,$$
$$A_2 = W_2^1 \cup W_3^2 \cup \ldots \cup W_1^k,$$
$$\vdots$$
$$A_k = W_k^1 \cup W_1^2 \cup \ldots \cup W_{k-1}^k.$$

These sets partition set $\mathcal{V}$, moreover, each of them is an independent set of $\mathcal{G}$ of cardinality n. Let us sequence these sets as follows

$$N^1 A_1 N^2 A_2 \ldots A_k N^{k+1}, \tag{9.1}$$

then to obtain a sequence of models we sequence models in each set arbitrarily. Next, we observe that each set N^j is independent thus no $\mathcal{X}$-option is used twice by models in N^j. Furthermore, there are n models with no $\mathcal{X}$-option between N^j and N^{j+1}, $j = 1, \ldots, k$. Consequently, any two models with an $\mathcal{X}$-option are separated by at least n models without this $\mathcal{X}$-option, and therefore the sequence (9.1) respects the $1 : (n+1)$ capacity constraint for each $\mathcal{X}$-option. Finally, we observe that each set A_j, $j = 1, \ldots, n$ is independent, thus no $\mathcal{E}$-option is used twice by models in A_j. Moreover, there are n models with no $\mathcal{X}$-option between A^j and A^{j+1}, $j = 1, \ldots, k-1$. Thus, any two models with an $\mathcal{E}$-option are separated by at least n models without this $\mathcal{E}$-option, and therefore the sequence (9.1) respects the $1 : (n+1)$ capacity constraint for each $\mathcal{E}$-option. Therefore, sequence (9.1) is a feasible model sequence in the car sequencing problem.

(*only if*) Let S be a feasible sequence of models. Let us assume for the time being that S is of the following form

$$S = N^1 M_1 N^2 M_2 \dots M_k N^{k+1} \tag{9.2}$$

where $\bigcup_{j=1}^k M_j = \mathcal{V}$ and $|M_j| = n$ for $j = 1, \dots, k$. Consider models in V^1 and the intersections

$$V_i = M_i \cap V^1, i = 1, \dots, k.$$

Obviously, $\bigcup_{i=1}^k V_i = V^1$ and each set V_i is an independent set. Otherwise, there would be an edge (a,b) between some models a and b of some V_i. Then, however, the $\mathcal{E}$-option (a,b) would be used by both a and b models in M_i of length n which would make S infeasible by violating the $1 : (n+1)$ capacity constraint for the $\mathcal{E}$-option (a,b). Consequently, coloring each V_i with a distinct color would provide a coloring of G^1 using k colors. Since G^1 is an isomorphic copy of G, then the coloring would be a required coloring of G itself. It remains to show that a feasible sequence of the form (9.2) always exists. To this end, let us consider the following decomposition of S into $2k+1$ subsequences of equal length n,

$$S = \gamma_1 \gamma_2 \dots \gamma_{2k+1},$$

where

$$\gamma_i = S_{(i-1)n+1} \dots S_{in}, i = 1, \dots, 2k+1. \tag{9.3}$$

For each γ_i there is at most one N^j whose models are in γ_i. Otherwise, the $1 : (n+1)$ constraint for some $\mathcal{X}$-option would be violated. Consequently, no N^j can share γ_i, $i = 1, \dots, 2k+1$ with any other N^l, $j \neq l$. However, since there are only $2k+1$ subsequences γ_i, then there must be N^{j^*} which models *completely* fill in one of the subsequences γ_i. Let us denote this sequence by γ. Neither the subsequence γ_i immediately to the left of γ, if any, nor to the right of γ, if any, may include models from $\bigcup_{j \neq j^*, j=1}^{k+1} N^j$. Otherwise, the $1 : (n+1)$ constraint for some $\mathcal{X}$-option would be again violated. Consequently, there are at most $2k-1$ subsequences with models from $\bigcup_{j \neq j^*, j=1}^{k+1} N^j$ in S, but this again implies the existence of $N^{j^{**}}$, $j^* \neq j^{**}$, which models *completely* fill in one of the subsequences γ_i, say γ^*. Furthermore, neither the subsequence γ_i immediately to the left of γ^*, if any, nor to the right of γ^*, if any, may include models from $\bigcup_{j \neq j^{**}, j=1}^{k+1} N^j$. By continuing this argument we reach a conclusion that for any feasible S there is an injection f of $\{N^1, \dots, N^{k+1}\}$ into $\{\gamma_1, \dots, \gamma_{2k+1}\}$ such that the sequence $f(N^i)$ is made up of models from N^i only, $i = 1, \dots, k+1$. Also, if γ_i and γ_j are mapped into then $|i - j| \geq 2$. This injection f is only possibly if s is of the form (9.1), which we needed to prove. $\quad\square$

The car sequencing remains NP-hard even if different models do not share any option, that is $O_i \cap O_j = \varnothing$ for $i \neq j$ and $|O_i| = 1$ for $i = 1, \dots, n$. We will refer to this subproblem as the *singular* car sequencing problem. This subproblem is NP-hard as the following theorem shows.

Theorem 9.3. *The car sequencing problem is NP-hard even if each model requires a single option unique for the model.*

Proof. The reduction is from the Periodic Maintenance Scheduling Problem. We recall from Chap. 7 that the Periodic Maintenance Scheduling Problem is defined as follows. Given m machines and integer service intervals $\ell_1, \ell_2, \ldots, \ell_m$ such that $\sum \frac{1}{\ell_i} < 1$. Does there exist a servicing schedule (or a servicing cycle) $s_1 \ldots s_L$, where $L = \text{lcm}(\ell_1, \ell_2, \ldots, \ell_m)$ is the least common multiple of $\ell_1, \ell_2, \ldots, \ell_m$, of these machines in which the end-to-end distance between any two consecutive servicing of machine i is exactly ℓ_i, no more than one machine is serviced in a single time slot, and it takes a single time slot to perform servicing of any machine? The distance between the last and the first servicing of machine i in the cycle, if any exists, is also ℓ_i. The Periodic Maintenance Scheduling Problem is NP-complete in the strong sense, see Theorem 7.14. Without loss of generality we assume $L > 2\ell_i$ for $i = 1, \ldots, m$. This problem naturally transforms to the car sequencing problem as follows. We have a model for each machine i, the demand for the model is $d_i = (K+2)\frac{L}{\ell_i}$, where

$$K = \prod_{i=1}^{m} (\ell_i + 1).$$

There is a unique option for model i, also denoted by i, with the capacity constraint $1{:}\ell_i$. Moreover, there is one more model $m+1$ with demand $(K+2)L(1 - \sum \frac{1}{\ell_i})$ and its unique option $m+1$ having capacity constraint $L(1 - \sum \frac{1}{\ell_i}) : L$. This transformation is polynomial but not pseudo-polynomial since $(K+2)L$ is not bounded by a polynomial of the input size which is of $O(\sum \log(\ell_i + 1))$. Now, any servicing cycle can be easily turned in into a car sequence that respects the capacity constraints of each option $i = 1, \ldots, m$ since there is exactly one servicing of machine i in each ℓ_i time slots. The insertion of $L(1 - \sum \frac{1}{\ell_i})$ copies of model $m+1$ into this sequence and repeating it $K+2$ times, then results into a feasible solution to the car sequencing problem. On the other hand, consider any feasible solution

$$S = S_1 \cdots S_{K+2} = s_1 \cdots s_{(K+2)L}$$

to the cars sequencing problem, where

$$S_j = s_{(i-1)L+1} \cdots s_{iL}$$

for $j = 1, \ldots, (K+2)$. We observe that the distance between any two consecutive copies of model $i = 1, \ldots, m$ may not exceed $2\ell_i$ for any i in S, otherwise the demand for i would not be met, and it may not be less than ℓ_i, otherwise the capacity constraint $1{:}\ell_i$ would be violated. Therefore, there are no more than $2\ell_i - \ell_i + 1 = \ell_i + 1$ possible values of the distance between any two consecutive copies of i in S. Consequently, the number of possible consecutive distance configurations between the last copy of model i in S_j and its first copy in S_{j+1} for all m models is K and the length $(K+2)L$ of the sequence S ensures the repetition of some of them in S. Thus there must be a subsequence

$$S' = S_j \cdots S_{j+k}$$

of S for some j and $k \geq 0$ such that the distance between the last copy of i in S_{j+k} and the first copy of i in S_j is at least ℓ_i. Since there are at most $L(1 - \sum \frac{1}{\ell_i})$ copies of model $m+1$ in each subsequence $S_j, \ldots, S_{j+k}$, due to the capacity constraint $L(1 - \sum \frac{1}{\ell_i}) : L$ for $m+1$, and since for each i the distance between any consecutive copies of i in S', including the first and the last, is at least ℓ_i, then they must be at the distance exactly ℓ_i, otherwise there would be two of them at a distance shorter than ℓ_i to satisfy the demand for i which would violate the capacity constraints for i. Thus, any two consecutive copies of i must be at a distance ℓ_i in S' which results in a feasible servicing cycle S_i and proves that the car sequencing problem is NP-hard. □

The singular car sequencing problem requires that any two consecutive copies of the same model be sufficiently distant from one another in a feasible sequence in order not to violate temporal capacity limitations. On the other hand the generalized pinwheel scheduling problem requires that the copies of the same model be not too far away from one another, see Chap. 8. The proof of Theorem 9.3 works for both problems since the capacity constraints in the proof actually require that there is *exactly* one copy of a model with a given option out of any consecutive ℓ_i for $i = 1, \ldots, m$ and exactly $L(1 - \sum \frac{1}{\ell_i})$ out of L for model $m+1$. Thus, minor modifications to the algorithms for the car sequencing problem presented later in the chapter lead to algorithms for the generalized pinwheel problem. However, it is worth pointing out an important difference between the two problems. The car sequencing problem is formulated as a finite acyclic problem whereas the pinwheel problem as an infinite one, that is cyclic.

9.4 Dynamic Programming for the Car sequencing Problem

We sketch a straightforward dynamic program to test the existence of a just-in-time sequence of length D that meets all option capacity constraints. Let $q = \max_j(q_j)$, without loss of generality we assume $q \leq D$. Moreover, to simplify the exposition, we assume for the time being that $D = k(q-1)$. The sequence $v = v_1 \cdots v_{q-1}$ of length $q - 1 > 0$ of models from M that meets all option capacity constraints will be referred to as a feasible $(q-1)$-*pattern*. For a feasible $(q-1)$-pattern v, let $i(v)$ be an n-dimensional vector with the coordinate j equal to the number of copies of model $j \in M$ in v. Define a node of a directed graph $G = (N, A)$ as being either a pair (Δ, v), where Δ is an n-dimensional vector with its coordinate j, $0 \leq \Delta_j \leq d_j$, keeping track of the number of model j copies already in the current prefix of the sequence, and v being a feasible $(q-1)$-pattern, or a sink node $(\mathbf{d}, \times)$ with an empty sequence $\times$. An arc $((\Delta, v), (\Delta', v')) \in A$ if and only if

$$\text{the concatenation } vv' \text{ is feasible and } \Delta' = \Delta + i(v).$$

The concatenation vv' is feasible if it meets all option capacity constraints. The inequality of arithmetic and geometric means implies the following upper bounds on the numbers of nodes

$$|N| \leq \prod_{i=1}^{n} (d_i + 1) \times n^{q-1} \leq \left(\frac{D+n}{n} \right)^n \times n^{q-1}$$

and the number of arcs

$$|A| \leq \left(\frac{D+n}{n} \right)^{2n} \times n^{2(q-1)}$$

of $G = (N, A)$. We observe that there is a feasible sequence if and only if there is a directed path from a node $(0, v)$ for some v to $(\mathbf{d}, \times)$ in the graph $G = (N, A)$. The existence of this directed path can be checked in time $O(|A|)$ thus the car sequencing problem can be solved in time

$$O\left(\left(\frac{D + 2^m}{2^m} \right)^{2^{m+1}} 2^{2m(q-1)} \right).$$

The algorithm is thus polynomial in D as long as the number of options m is fixed, that is it is not part of the problem's input, and 2^q is bounded from above by a polynomial of D. This observation as well as the proof of Theorem 9.2 strongly suggest that the computational complexity of the car sequencing problem often hinges only on the option contents of models represented by the binary matrix C that encodes the option subsets O_i of models $i = 1, \ldots, n$. Therefore, a natural first step in the quest for efficient algorithms for the car sequencing problem is to understand the logic used in the model design and reflected in the model option content. The logic is shaped both by the market research determination of the attributes most valued by customers and by the car design team that translates these attributes at minimum cost into possible options and models, see Lockledge et al. [112]. However, Theorem 9.3 shows that the car sequencing problem is a much more complex problem than that since even if the option content of models is made trivial by having a single unique option per model, the car sequencing problem still exhibits its NP-hardness through the number theoretic features embedded in the option capacity constraints. This relative ease with which even very special cases of the car sequencing problem can be shown NP-hard explains why there are no, to the author's knowledge, special cases of the car sequencing problem with polynomial time solutions published in the literature. Consequently, the problem computational intractability provides a very fertile ground for computational competitions, for instance the 2005 ROADEF (Société Française de Recherche Opérationnelle et d'Aide à la Décision) challenge sponsored by the car maker Renault, and the study of formulations, search methods and especially the constraint programming techniques. These formulations will be briefly discussed in Sect. 9.6.

We note that if $D = k(q-1) + r$, $0 < r < q - 1$, then the sink node must be immediately preceded by the nodes $(\mathbf{d} - i(w), w)$, where w is a sequence of length r that meets all option capacity constraints. Each of these nodes is connected to a sink node $(\mathbf{d}, \times)$. Again, there is a feasible sequence if and only if there is a directed path from a node $(0, v)$ for some v to $(\mathbf{d}, \times)$.

For $q = 2$, all option capacity constraints are of type $1 : 2$. Hence, the car sequencing problem reduces to an undirected Hamiltonian path problem. since the $(q-1)$-pattern boils down to a single model. Therefore, there is not need for having the vector Δ as a part of the node. Instead, the number of nodes corresponding to model i can be made equal to the demand d_i for the model. Thus, $N = \bigcup_{i=1}^{n} D_i$, where D_i the set of d_i nodes being copies of model i. The edge between nodes v and u belongs to E if and only if the two model sequence vu meets all option capacity constraints. A Hamiltonian path in $G = (N, E)$ exists if and only if there is a feasible sequence. The reduction does not help to solve this case of the car sequencing problem in polynomial time generally since the Hamiltonian path problem is well-known to be NP-hard in the strong sense. However, it can be used to prove that the subproblem with all option constraints being $1 : 2$ still renders the cars sequencing problem NP-hard, see Exercise 9.20.

9.5 Simple Necessary Conditions

We give two necessary conditions for a feasible model sequence to exist. The first condition states that the total demand for models with a given option may not be too high. Otherwise the demand for the option would certainly violate its capacity constraint.

Lemma 9.4. *In order for a feasible model sequence S to exist the capacity constraints must satisfy the condition $\left\lceil \frac{D}{q_j} \right\rceil p_j \geq \sum_{i \in \{k : C_{jk}=1\}} d_i$ for all j.*

Proof. For option j, the sequence can be rewritten as follows

$$S = \alpha_1 \cdots \alpha_{\left\lceil \frac{D}{q_j} \right\rceil},$$

where all sequences α_i except perhaps the last one $\alpha_{\left\lceil \frac{D}{q_j} \right\rceil}$, which can be shorter, include exactly q_j models. By the option capacity constraint, $p_j : q_j$, each α_i may include at most p_j copies of models that require option j. Therefore, the total demand for all models that require option j, that is

$$\sum_{i \in \{k : C_{jk}=1\}} d_i,$$

must not exceed

$$\left\lceil \frac{D}{q_j} \right\rceil p_j,$$

which proves the lemma.   □

For instance, in the example from Table 9.1 demand for option 1 equals 6 which is less than the $\left\lceil \frac{11}{3} \right\rceil \times 2$ with the $2 : 3$ capacity constraint for option 1.

The second category of conditions is based on the observation that for a given set of options ω if there is too much demand for the models that require *all* options in ω and too little demand for the models that require *none* of the options in ω, then the capacity constraints for the options in ω may again be violated. An example of the necessary condition that falls into this category is given in the following lemma.

Lemma 9.5. *Let $\lambda \subseteq M$ with $d_\lambda = \sum_{i \in \lambda} d_i \geq 2$ be a set of models and ω a set of options such that each model in λ requires all options from ω. Let $\overline{\lambda}$ be a set of all models that require no option from ω. If $|\omega| < q - 1$ and the capacity requirement for each $j \in \omega$ falls in the set $\{1 : q, 2 : q\}$ and at least one of them is $1 : q$, then we must have $d_{\overline{\lambda}} \geq d_\lambda - 1$ for a feasible sequence S to exist.*

Proof. Let S be a feasible sequence with the copies of models from λ in positions $j_1, \ldots, j_{d_\lambda}$. Consider a subsequence α_i of S between positions j_i and j_{i+1}. By the $1 : q$ constraint, $|\alpha_i| \geq q - 1$. If none of the models in α_i belongs to $\overline{\lambda}$, then by the pigeon-hole argument for some $k \in \omega$ we would have at least three models in $S_{j_i} \alpha_i$ that require k as long as $|\omega| < q - 1$. Therefore, at least one model in α_i must come from $\overline{\lambda}$. Consequently, $d_{\overline{\lambda}} \geq d_\lambda - 1$, which proves the lemma. $\square$

The lemma holds if the capacity constraints $2 : q$ are replaced by stronger constraint $2 : q_j$ where $q_j \leq q$ and possibly different for different options.

9.6 IP Formulation and Heuristics for the Car Sequencing Problem

The car sequencing problem is often solved by constraint satisfaction programming, see Dincbas et al. [113], and ILOG [108]. However, as Gravel et al. [114], and Drexl and Kimms [115] point out this method performance quickly deteriorates as the problem size grows. Furthermore, the former observes that the constraint satisfaction programming is rather sensitive to the existence of a feasible solution. Thus confirming once again a common experimental observation for any NP-hard problem that the absence of a feasible solution is usually a much harder nut to crack, see Exercise 9.19 as an example, than showing one if any exists.

A similar observation has been made by Gravel et al. [114] on solvers for a linear integer programming model of the car sequencing model.

The following assignment-based formulation comes from Drexl and Kimms [115] and Gravel et al. [114], it uses binary variables

$$y_{i,k} = \begin{cases} 1 & \text{if model } i \text{ is in position } k \text{ of the sequence,} \\ 0 & \text{otherwise} \end{cases}$$

for $i = 1, \ldots, n$ and $k = 1, \ldots, D$. The formulation is as follows

$$\sum_{i=1}^{n} y_{i,k} = 1 \qquad k = 1, \ldots, D \tag{9.4}$$

$$\sum_{k=1}^{D} y_{i,k} = d_i \qquad i = 1, \ldots, n \tag{9.5}$$

$$\sum_{\ell=0}^{q_j-1} \sum_{i=1}^{n} C_{ji} y_{i,k+\ell} \le p_j \qquad j = 1, \ldots, m \quad k = 1, \ldots, D - q_j + 1 \tag{9.6}$$

The constraints (9.4) and (9.5) are standard assignment constraints. The constraint (9.6) is a sliding window constraint that ensures that a window of size q_j sliding along the sequence of length D never includes more than p_j copies with option j, $j = 1, \ldots, m$.

This formulation is to test a feasibility of an instance of the car sequencing problem – usually a harder task. However, an objective function can be added and the constraints (9.6) modified by the well-known big M technique to turn the integer program into the minimization of the number of capacity constraints violations, see Gravel et al. [114] who also point out that the number of violations may itself be difficult to calculate as the following example illustrate.

Example 9.6. Consider a sequence

$$111111$$

for an option with 2:3 capacity constraint. This constraints is violated four times if we use the sliding window method of calculating the violations – we simply have four factors 111 when sliding the window of size three from the left to the right along the sequence. However, replacing the second and the and the fourth models in the sequence by the ones that do not require the option, and do not cause any violations for other options, would reduce the number of violations to 0. Thus, it may be argued that the number of violation in this example should be two rather than four since in practice changing the sequence two positions fixes all violations for the option. □

We conjecture that this problem may be NP-hard itself, see Exercise 9.22 for details .

Drexl et al. [116] propose an integer programming model to minimize maximum deviation of model copies from their optimal positions, which is different from though somehow related to the bottleneck problem discussed in Sect. 5, over all sequences satisfying capacity constraints. The LP-relaxation of their model is then solved by *column generation* technique, see Desrochers and Soumis [117] for details of this technique, to provide lower bound which is reported tight in their computational experiments.

The ant colony optimization has been often proposed heuristic approach to the car sequencing model, see Solnon [118], and Gravel et al. [114]. The latter computational study of this heuristic concludes that it is able to more efficiently find optimal solutions than the integer programming solver for those instances the solver was able to find optimal solutions for. Moreover, the quality of solutions found by the heuristic for the benchmark instances equalled the quality of the best solutions known for these instances.

We close this section by again observing that straightforward modifications of the formulations and heuristics discussed here would readily apply to both the pinwheel scheduling and the generalized pinwheel scheduling problems, see Exercise 9.21. The research in this direction seems unexplored yet promising.

9.7 Mixed-Model, Pull Supply Chains

The goal of this section is to present a framework for a mixed-model, pull supply chains which is a point of departure for our discussion in the following sections. A mixed-model supply chain has a set $\{0, 1, \ldots, \mathcal{S}\}$ of suppliers. The supplier s offers supplies from its list $\mathcal{S}_s = \{(s, 1), \ldots, (s, n_s)\}$ of n_s supplies. The supplies of different suppliers are connected by directed arcs as follows. There is an arc from (s_i, x) to (s_j, y) if and only if supplier s_i requires supplier s_j to supply y for its x. The arc $((s_i, x), (s_j, y))$ is weighted by the number (or amount) of y needed for a *unit* of x. The set of supplies $\bigcup_{s=0}^{\mathcal{S}} \mathcal{S}_s$ and the set of arcs $\mathcal{A}$ between supplies make up a weighted, acyclic digraph – called the model-supplier graph. Without loss of generality we shall assume that $s_i < s_j$ for any arc $((s_i, x), (s_j, y))$ in this graph. The supplies $S_0 = \{(0, 1), \ldots, (0, n_0)\}$ at *Level 1* will be called *models*. For simplicity, we denote model $(0, j)$ by j and the number of models n_0 by n. To avoid duplicates in the supply chain, we assume that any two nodes of the digraph have different out-sets and no node has out-degree 1. In fact we assume that the digraphs are *multistage* digraphs, as virtually all supply chains appear to have this structure simplifying feature, see Shapiro [119], and Bowersox et al. [120]. For an example of the mixed-model supply chain please see Fig. 9.1 where all arcs are directed from the top down.

Each path π from model m to (s, j) represents a demand for (s, j) originating from a single copy of model m. The size of this demand equals the *product* of all weights along the path π. Therefore, the total demand for the supply (s, j) originating from a single copy of model m is the sum of path demands over all paths from m to (s, i) in the model-supplier graph and it is denoted by $b_{m,(s,j)}$. The matrix **b** of the size $n \times (\sum_{s \in S} n_s)$ contains all model-supplier contents.

The model-supplier graph shown in Fig. 9.1 has two paths from model 1 to $(4, 1)$ both with weight 1, therefore the total demand for $(4, 1)$ originating from 1 equals 2. Each supplier s aggregates its demand over all supplies on its list $\mathcal{S}_s$. For supplier 4 the demand originating from model 1 is $[112]$, from model 2, $[12233]$, and from model 3, $[233]$. In our notation supply i for a given model is listed the number of times equal to the unit demand for i originating from the model. Each of these lists will be referred to as a *kit* to emphasize the fact that suppliers may not deliver an individual part or a subassembly required by models but rather a complete collection required by the model, a common practice in manufacturing Bowersox et al. [120]. Thus, model 1 needs the entire kit $[112]$ delivered in a single delivery from supplier 4 rather than two 1s and one 2 delivered separately. We shall also refer to a kit as

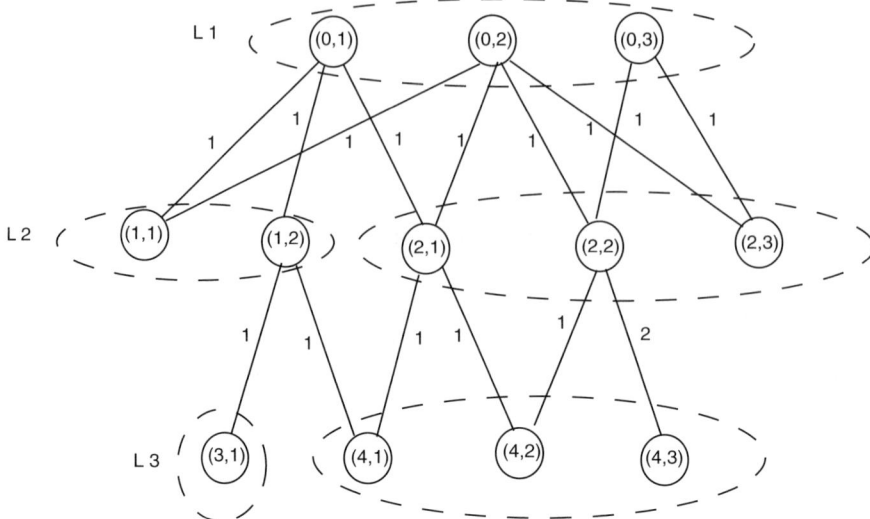

Fig. 9.1 Mixed-model supply chain with three levels and five suppliers (or chain nodes): one at level 1 supplying three models, two at level 2, and two at level 3

Table 9.3 Matrix **b** for the supply chain in Fig. 9.1

$m/(s,j)$	$(1,1)$	$(1,2)$	$(2,1)$	$(2,2)$	$(2,3)$	$(3,1)$	$(4,1)$	$(4,2)$	$(4,3)$
1	1	1	1	0	0	1	2	1	0
2	1	0	1	1	1	0	1	2	2
3	0	0	0	1	1	0	0	1	2

an *option*. Notice that a model may require at most one kit from a supplier. The supplier *content* of models is defined by an n by $S+1$ matrix C, where

$$C_{si} = \begin{cases} 1 & \text{if model } i \text{ requires a kit (option) from supplier } s, \\ 0 & \text{otherwise.} \end{cases} \tag{9.7}$$

We assume that the supply chain operates in a *pull* mode. That is any supply at a *higher* level is drawn as needed by a *lower* level. Therefore, it is a sequence of models at *Level 1* (L1), called the model delivery sequence, that determines the *delivery* sequence of each supplier, called the option delivery sequence, at every level higher than 1 (downstream) and the supplier must exactly follow this delivery sequence when delivering its options. For instance, the model delivery sequence

$$1 \rightarrow 2 \rightarrow 3 \rightarrow 1 \rightarrow 1 \rightarrow 2 \rightarrow 1 \rightarrow 3 \rightarrow 2 \rightarrow 1 \tag{9.8}$$

at L1 results in the option delivery sequence

$$[1] \rightarrow [123] \rightarrow [23] \rightarrow [1] \rightarrow [1] \rightarrow [123] \rightarrow [1] \rightarrow [23] \rightarrow [123] \rightarrow [1] \tag{9.9}$$

for supplier 2 at the level L2 and the option delivery sequence

$$[12] \to [1] \to * \to [12] \to [12] \to [1] \to [12] \to * \to [1] \to [12] \qquad (9.10)$$

for supplier 1 at the level L2, where the "$*$" in a given position of the option delivery sequence denotes an empty kit, the option delivery sequence

$$[112] \to [12233] \to [233] \to [112] \to [112]$$
$$\to [12233] \to [112] \to [233] \to [12233] \to [112] \qquad (9.11)$$

for supplier 4 at L3, and finally the option delivery sequence

$$[1] \to * \to * \to [1] \to [1] \to * \to [1] \to * \to * \to [1] \qquad (9.12)$$

for supplier 3 at L3.

Observe that the kit $[112]$ to be delivered first by supplier 4 according to the option delivery sequence (9.11) is in fact made up of kits $[1]$ and $[12]$ the former to be delivered to supplier 1 at L2 as it is needed for its supply $(1,2)$ and the latter to be delivered to supplier 2 at L2 as it is needed for its $(2,1)$ though both eventually end up in a copy of model 1 which is to be produced first. Therefore, the option delivery sequence has in fact the following nested structure

$$[[1][12]] \to [[12][233]] \to [233] \to [[1][12]] \to [[1][12]]$$
$$\to [[12][233]] \to [[1][12]] \to [233] \to [[12][233]] \to [[1][12]]$$

for supplier 4 at L3. This nested structure corresponds naturally to the levels of the multistage model-supplier graph.

The demand for model j is denoted by d_j and it is assumed given. The demand for any other supply can easily be *derived* from the demand vector for models $\mathbf{d} = (d_1, \ldots, d_n)$ and the kit content of each model given in the matrix $\mathbf{b}$.

The supplier content matrix $\mathcal{C}$ defined in (9.7) corresponds to the option content matrix of the car sequencing problem. Moreover, supplier s is a subject to a capacity constraint in the form $p_s : q_s$, which requires that at most p_s models of the model delivery sequence S in each consecutive sequence of q_s models of S may need options supplied by s. Thus, the leveling of demand problem can be recast as a car sequencing problem.

9.8 Balanced Words and Model Delivery Sequences

We show in Chap. 6 that regular words minimize multimodular functions and thus balance workloads in event graphs. We further explore this line of research in this section by having a closer look at balanced words, which are a slightly more general concept than regular words, to explore insights they can provide into the complexity of model delivery sequences. We use the terminology and the notation borrowed from the combinatorics on words which is briefly recall from the preliminary Chap. 1.

The models $\{1,\ldots,n\}$ will be viewed as the letters of a finite alphabet $\mathcal{A} = \{1,\ldots,n\}$. We consider both finite and infinite words over $\mathcal{A}$. A model delivery sequence will then be viewed a finite word of length D on $\mathcal{A}$, where the letter i occurs exactly d_i times. This word can be concatenated *ad infinitum* to obtain a periodic, infinite word on $\mathcal{A}$.

We recall from Chaps. 5 and 3 that sequencing copy j of model i in its ideal position $\lceil \frac{2j-1}{2r_i} \rceil$ minimizes both the total deviation and the maximum deviation, however, it leads to an infeasible solution whenever more than one copy *competes* for the same ideal position in the sequence. The algorithms discussed in these chapters show how to efficiently resolve the conflicts so that the outcome is an optimal sequence, minimizing either total or maximum deviations.

Let us now consider an infinite, periodic sequence of the ideal vertices

$$\frac{2j-1}{2r_i} = \frac{jD}{d_i} - \frac{D}{2d_i} = \frac{(j-1)D}{d_i} + \frac{D}{2d_i}.$$

We build an infinite word on $\mathcal{A}$ using these numbers as follows. Label the points $\{\frac{(j-1)D}{d_i} + \frac{D}{2d_i}, j \in \mathbb{N}\}$ by the letter i. Consider

$$\bigcup_{i=1}^{n} \left\{ \frac{(j-1)D}{d_i} + \frac{D}{2d_i}, j \in \mathbb{N} \right\}$$

and the corresponding sequence of labels. Each time there is a tie we chose i over j whenever $i < j$. Notice that here higher priority is always given to a lower index whenever a conflict needs to be settled. This way we obtain what Vuillon [46] refers to as an *hypercubic billiard* word with the angle vector $\alpha = (\frac{D}{d_1}, \frac{D}{d_2}, \ldots, \frac{D}{d_n})$ and the starting point $\beta = (\frac{D}{2d_1}, \frac{D}{2d_2}, \ldots, \frac{D}{2d_n})$. Vuillon [46] proves the following theorem.

Theorem 9.7. *Let x be an infinite hypercubic billiard word in dimension n of angle α and starting point β. Then x is $(n-1)$-balanced word.*

The c-balanced words, $c > 0$ is referred to as the *degree of balance*, are defined as follows.

Definition 9.8 (c-Balanced Word). A *c-balanced word* on alphabet $\{1, 2, \ldots, n\}$ is an infinite sequence $S = s_1 s_2 \ldots$ such that

1. $s_j \in \{1, 2, \ldots, n\}$ for all $j \in \mathbb{N}$, and
2. If x and y are two factors of S of the same size, then $||x|_i - |y|_i| \leq c$, for all $i = 1, 2, \ldots, n$.

Theorem 9.7 shows that the *priority* based conflict resolution applied whenever there is a competition for an ideal position results in c being almost of the size of the alphabet, in fact 1 less than this size. However, Jost [121] proves that the conflict resolution provided by any optimization algorithm for the bottleneck problem leads to the c being a small constant. He proves the following theorem.

Theorem 9.9. *For a word S^∞ obtained be infinitely repeating a sequence S with bottleneck deviation B for n models with demands $d_1,\ldots,d_n$. We have:*

If $B < \frac{1}{2}$, then S is 1-balanced.
If $B < \frac{3}{4}$, then S is 2-balanced.
If $B < 1$, then S is 3-balanced.

Proof. Consider a solution x_{ik}, $i = 1,\ldots,n$ and $k = 1,\ldots,D$ to the bottleneck problem, (5.2) subject to (5.3). Let B be the bottleneck deviation of this solution. Define,

$$y_{ik} = \left\lfloor \frac{k}{D} \right\rfloor d_i + x_{i,k \bmod D}$$

$i = 1,\ldots,n$ and $k = 1,\ldots$ for sequence S. Thus,

$$\left| y_{i,k} - k r_i \right| = \left| \left\lfloor \frac{k}{D} \right\rfloor d_i + x_{i,k \bmod D} - \left(\left\lfloor \frac{k}{D} \right\rfloor D + (k \bmod D) \right) r_i \right|$$
$$= \left| x_{i,k \bmod D} - (k \bmod D) r_i \right| \le B$$

and

$$\left| y_{i,(k+h)} - (k+h) r_i \right| = \left| \left\lfloor \frac{k+h}{D} \right\rfloor d_i + x_{i,(k+h) \bmod D} \right.$$
$$\left. - \left(\left\lfloor \frac{k+h}{D} \right\rfloor D + (k+h) \bmod D \right) r_i \right|$$
$$= \left| x_{i,(k+h) \bmod D} - ((k+h) \bmod D) r_i \right| \le B.$$

Therefore,

$$-B + k r_i \le y_{i,k} \le B + k r_i$$

and

$$-B + (k+h) r_i \le y_{i,(k+h)} \le B + (k+h) r_i.$$

Consequently,

$$-2B + h r_i \le y_{i,(k+h)} - y_{i,k} \le 2B + h r_i.$$

Since k is arbitrary, then any factor of S of length $h \ge 1$ includes no less than $\lceil -2B + h r_i \rceil$ and no more that $\lfloor 2B + h r_i \rfloor$ occurrences of i. Therefore, the difference between the number of i occurrences in any two factors of S of length $h \ge 1$ does not exceed $4B$. Thus, it is *less* than 2 for $B < \frac{1}{2}$, *less* than 3 for $B < \frac{3}{4}$, and *less* than 4 for $B < 1$. Therefore, S is 1-balanced for $B < \frac{1}{2}$, 2-balanced for $B < \frac{3}{4}$, and 3-balanced for $B < 1$. This proves the theorem. $\square$

The opposite claim does not hold, for instance, any sequence for n models with their demands all equal 1 is a 1-balanced word though its maximum deviation equals $1 - \frac{1}{n}$, and thus it is greater than 0.5 for $n \ge 3$. Theorem 5.8 shows that there always is an optimal solution with $B < 1$, and Theorem 9.9 shows that such solutions are 3-balanced. These two ensure that 3-balanced words can be obtained for any set of demands $d_1, \ldots, d_n$. It remains an open question to show whether or not there always is a 2-balanced word for *any* given set of demands $d_1, \ldots, d_n$. However, the infinite word

$$(1 \rightarrow 2 \rightarrow 3 \rightarrow 1 \rightarrow 2 \rightarrow 1 \rightarrow 1 \rightarrow 3 \rightarrow 2 \rightarrow 1)^{\infty}$$

is 2-balanced as its maximum deviation equals $\frac{1}{2}$ but not 1-balanced, factors 23 and 11 differ by 2 on the latter 1. It is a challenging open problem to find a characterization of the instances with bottleneck $B < \frac{3}{4}$. The characterization of instances with $B < \frac{1}{2}$ is given in Chap. 6 and Theorem 5.8 gives the characterization of instances with $B < 1$.

In the hierarchy of balanced words, the 1-balanced words, or just balanced words, have attracted most attention thus far, see Vuillon [46], Altman et al. [7], and Tijdeman [60] for reviews of recent results on balanced words. Berthe and Tijdeman [122] observe that the number of balanced words of length m is bounded by a *polynomial* of m, which makes the balanced words very rare. They furthermore observe that the number of c-balanced words of length m is *exponential* in m for any $c > 1$. The polynomial complexity of balanced words reduces the number of possible factors of delivery sequences, and thus the variability of demand through the supply chain which could have obvious advantages for their management. As well balanced words would optimally balance suppliers workload according to the results of Altman et al. [7], see also Chap. 6. However, the balanced sequences may turn out to be out of reach in practice. Indeed, according to the famous Frankel's Conjecture, Altman et al. [7] and Tijdeman [5] and Chap. 6, there is only *one* such word on n letter alphabet with *distinct* demands. The demands must be powers-of-two. Though this conjecture remains open generally, a simpler one for *periodic*, symmetric and balanced words has been proven in Chap. 6, see also Brauner et al. [68], which indicates that the balanced words can indeed be very rare generally and as the model delivery sequences in particular.

9.9 Option Delivery Sequences

We now explore the change of the sequence variability measured by its degree of balance down the supply chain.

A supplier s option delivery sequence can be readily obtained from the model delivery sequence S and the supplier content matrix C by *deleting* from S all models i not supplied by s, that is those with $C_{si} = 0$. This deletion increases the degree of balance of the *option* delivery sequences for suppliers in comparison to the model delivery sequence as we show in this section. Let us first introduce some necessary notation.

- $A_s \subseteq \{1, \ldots, n\}$ – the subset of models supplied by s.
- $A_{sj} \subseteq A_s$ – the subset of models requiring option j of supplier s. Different models may require the same option delivered by s.

- $r_{sj} = \dfrac{\sum_{m \in A_{sj}} d_m}{\sum_{m \in A_s} d_m}.$

- $r_{A_{sj}} = \dfrac{\sum_{m \in A_{sj}} d_m}{D} = \sum_{m \in A_{sj}} r_m.$

- $r_{A_s} = \dfrac{\sum_{m \in A_s} d_m}{D} = \sum_{m \in A_s} r_m.$

We notice that

$$r_{sj} = \frac{r_{A_{sj}}}{r_{A_s}}. \tag{9.13}$$

First, we investigate the bottleneck deviation in the *option* delivery sequence of supplier s. Supplier s has total derived demand $\sum_{m \in A_s} d_m$ and the derived demand for its option j equals $\sum_{m \in A_{sj}} d_m$. A model delivery sequence S with x_{mk} copies of model m out of first k copies delivered results in the actual total derived demand $\sum_{m \in A_s} x_{mk}$ for supplier s out of which $\sum_{m \in A_{sj}} x_{mk}$ is the actual demand for option j of s. Therefore, the bottleneck deviation for the option delivery sequence of supplier s equals

$$\max_{j,k} \left| \sum_{m \in A_{sj}} x_{mk} - r_{sj} \sum_{m \in A_s} x_{mk} \right|. \tag{9.14}$$

However, for S with bottleneck deviation B^* we have

$$kr_m - B^* \le x_{mk} \le kr_m + B^* \tag{9.15}$$

for any model m and k, and consequently

$$kr_{A_{sj}} - |A_{sj}|B^* \le \sum_{m \in A_{sj}} x_{mk} \le kr_{A_{sj}} + |A_{sj}|B^*$$

for option j of s and

$$kr_{A_s} - |A_s|B^* \le \sum_{m \in A_s} x_{mk} \le kr_{A_s} + |A_s|B^*.$$

Thus,

$$\sum_{m \in A_{sj}} x_{mk} = kr_{A_{sj}} + \varepsilon_{A_{sj}},$$

where $|\varepsilon_{A_{sj}}| \le |A_{sj}|B^*$ and

$$\sum_{m \in A_s} x_{mk} = kr_{A_s} + \varepsilon_{A_s},$$

where $|\varepsilon_{A_s}| \le |A_s|B^*$.

Therefore, (9.14) becomes

$$\max_{j,k} \left| kr_{A_{sj}} - \frac{r_{A_{sj}}}{r_{A_s}} (kr_{A_s} + \varepsilon_{A_s}) + \varepsilon_{A_{sj}} \right| \tag{9.16}$$

or

$$\max_{j,k} \left| r_{sj} \varepsilon_{A_s} - \varepsilon_{A_{sj}} \right|. \tag{9.17}$$

However,

$$\max_{j,k} |r_{sj}\varepsilon_{A_s} - \varepsilon_{A_{sj}}| \le |A_s|B^*.$$

Therefore,

$$\max_{j,k} \left| \sum_{k \in A_{sj}} x_{mk} - r_{sj} \sum_{m \in A_s} x_{mk} \right| \le |A_s|B^*. \tag{9.18}$$

We have just proved the following theorem.

Theorem 9.10. *The bottleneck deviation of the option delivery sequence for supplier s who supplies $|A_s|$ different models out on n produced increases $|A_s|$ times in comparison with the bottleneck deviation of the model delivery sequence.*

Theorem 9.9 shows that the model delivery sequences minimizing bottleneck deviation are 3-balanced. However, Theorem 9.10 proves that the bottleneck deviation of the option delivery sequence of supplier s grows proportionally to the number of models s supplies. Therefore, the option delivery sequence becomes less balanced. We have the following result.

Theorem 9.11. *The option delivery sequence for supplier s is $\lfloor 4|A_s|B^* \rfloor$-balanced.*

Proof. For supplier s consider k and k_Δ, $\Delta \ge 1$ such that between k and k_Δ there are exactly Δ copies of models requiring some option from s. That is

$$\sum_{m \in A_s} x_{mk_\Delta} - \sum_{m \in A_s} x_{mk} = \Delta.$$

We then have by (9.18)

$$-|A_s|B^* \le \sum_{k \in A_{sj}} x_{mk} - r_{sj} \sum_{m \in A_s} x_{mk} \le |A_s|B^*,$$

and

$$-|A_s|B^* \le \sum_{k \in A_{sj}} x_{mk_\Delta} - r_{sj} \sum_{m \in A_s} x_{mk_\Delta} \le |A_s|B^*,$$

which results in

$$-2|A_s|B^* \le \sum_{k \in A_{sj}} x_{mk_\Delta} - \sum_{k \in A_{sj}} x_{mk} - r_{sj}\Delta \le 2|A_s|B^*$$

for each k. Therefore, the numbers of option j occurrences in any two supplier s delivery subsequences of length Δ differ by at most $\lfloor 4|A_s|B^* \rfloor$. □

9.10 Periodic Synchronized Delivery

This section considers supply chains where supplies are regularly delivered to level L1 a given number of times a day by using a constant order cycle, variable order size framework. The pull material flow mechanism of just-in-time systems requires direct suppliers to supply only what is needed by level L1 and only when it is needed. This requirement is determined by the final model delivery sequence of models produced at level L1. Monden [22] discusses how Toyota Motor Corporation uses this framework with its suppliers. Aigbedo [123] points out that for example Japan Glass Sheet Company delivers parts 10 times and 16 times daily to Toyota's Motomachi plant and Tsutsumi plant, respectively, and he goes on to point out that similar arrangements are common in North America, where part delivery frequency is usually agreed on between the Original Equipment Manufacturer (OEM) and the suppliers, by considering factors such as distance and transportation costs between the supplier's plant and OEM plant.

9.10.1 The Model

We now present an approach to building a model delivery sequence that minimizes order size variability for a constant order cycle and thus minimizing safety stocks. Suppose that a supplier s is to deliver all its supplies periodically at the instants kcT_s, $k = 0, 1, \ldots$, where c is the assembly cycle time and T_s is the total number of copies produced in one period, for instance an intended bi-hourly or half-day production. Let us begin with an example based on the example of supply chain from Sect. 9.7.

Example 9.12. The deliver periods are $T_1 = T_2 = 2$ for both suppliers at $L2$ and $T_3 = T_4 = 5$ for both suppliers at $L3$. Thus, supplier 1 at $L2$ delivers according to the sequence (9.10)

Supplier 2 at $L2$ delivers according to the sequence (9.9)
Supplier 4 at $L3$ delivers according to the sequence (9.11)
Supplier 3 at $L3$ delivers according to the sequence (9.12). □

Now let us develop a general model of this problem. Consider supplier s. Given that the demand for model i equals $d_i, i = 1, \ldots, n$, the total demand for the supply j of s equals

$$D_{(s,j)} = \sum_{i=1}^{n} b_{i,(s,j)} d_i \qquad (9.19)$$

and the average demand for supply j of s, or simply supply (s, j), per copy of the model delivery sequence equals

$$R_{(s,j)} = \frac{D_{(s,j)}}{D},$$

where D is the total demand for all models. Therefore, on average, supplier s expects to deliver

$$T_s R_{(s,j)} = T_s \frac{D_{(s,j)}}{D} = T_s \frac{\sum_{i=1}^{n} b_{i,(s,j)} d_i}{D} = T_s \sum_{i=1}^{n} r_i b_{i,(s,j)}$$

copies of its supply j in each delivery with period T_s. Since the $T_s R_{(s,j)}$ may not be an integer, we assume that on average the order size should be either

$$\lfloor T_s R_{(s,j)} \rfloor \quad \text{or} \quad \lceil T_s R_{(s,j)} \rceil$$

Table 9.4 Supplier 1 at L2 order sizes in each of the five periods, $T_2 = 2$

Period/delivery	$(1,1)$	$(1,2)$
1	2	1
2	1	1
3	2	1
4	1	1
5	2	1

Table 9.5 Supplier 2 at L2 order sizes in each of the five periods, $T_2 = 2$

Period/delivery	$(2,1)$	$(2,2)$	$(2,3)$
1	2	1	1
2	1	1	1
3	2	1	1
4	1	1	1
5	2	1	1

Table 9.6 Supplier 4 at L3 order sizes in each of the two periods, $T_4 = 5$

Period/delivery	$(4,1)$	$(4,2)$	$(4,3)$
1	7	6	4
2	6	7	6

Table 9.7 Supplier 3 at L3 order sizes in each of the two periods, $T_3 = 5$

Period/delivery	$(3,1)$
1	3
2	2

copies of supply j in each delivery period for supplier s. Thus, to ensure that each delivery of supplier s matches the actual average demand for its supplies as closely as possible we request that the model delivery sequence requires no less than $\lfloor T_s R_{(s,j)} \rfloor$ and no more than $\lceil T_s R_{(s,j)} \rceil$ units of supply j in each *delivery period*

$$[(l-1)T_s, lT_s]$$

of length T_s for supplier s. The deliveries of supplier s take place at moments

$$0, T_s, 2T_s, \ldots, (m_s - 1)T_s,$$

where

$$m_s = \left\lceil \frac{D}{T_s} \right\rceil,$$

is the number of deliveries for supplier s required to meet the total demand D with delivery period T_s. Let

$$\delta_0 = 0 < \delta_1 < \cdots < \delta_p = D, \quad p \le \sum_{s=0}^{S} m_s + 1$$

be all different values among kT_s and D, where $k = 0, 1, \ldots, (m_s - 1)$ and $s \in \{0, 1, \ldots, S\}$. We shall refer to these values as the *possible delivery dates*. Define the k−th *inter-delivery interval* $[\delta_{k-1}, \delta_k]$, and the k−th *inter-delivery duration* $\Delta_k = \delta_k - \delta_{k-1}$ for $k = 1, \ldots, p$. Define variable

$z_{ik}=$ the number of model i copies produced in $[\delta_{k-1}, \delta_k]$.

Moreover, let

$$I_{l,s} = \{k : (l-1)T_s \le \delta_{k-1} < \delta_k \le lT_s\}$$

for $1 \le l < m_s$ be the set of all inter-delivery intervals included in delivery period $[(l-1)T_s, lT_s]$ of supplier s. The existence of a model delivery sequence of models that ensures the least possible variability of interval order sizes for each supplier s and its supply j, that is the variability of at most 1 for the interval order sizes are either $\lfloor T_s R_j \rfloor$ or $\lceil T_s R_j \rceil$ for supply j, can be checked by solving the following set of integer linear inequalities:

$$\sum_{k=1}^{p} z_{ik} = d_i \qquad i = 1, \ldots, n$$

$$\lfloor T_s R_{(s,j)} \rfloor \le \sum_{i=1}^{n} b_{i,(s,j)} Z_{il} \le \lceil T_s R_{(s,j)} \rceil \qquad (9.20)$$

$$\text{for} \quad l = 0, \ldots, m_s - 1, s \in \{0, 1, \ldots, S\}, \ j = 1, \ldots, n_s$$

$$\sum_{i=1}^{n} z_{ik} = \Delta_k \qquad k = 1, \ldots, p$$

where

$$Z_{il} = \sum_{k \in I_{l,s}} z_{ik}$$

is the total production, or the number of copies, of model i in the delivery interval $[(l-1)T_s, lT_s]$. Let us return to Example 9.12 to illustrate the concepts just introduced.

Example 9.13. Notice that the sequence (9.8) produces solution that meets constraint (9.20) for all suppliers and supplies with the exception of supply $(4,3)$ of supplier 4 where it results in the deliver of size 4 in the first period and 6 in the second whereas ideally it should deliver the average of 5 in each delivery. $\square$

Table 9.8 Average order sizes

(s,j)	$(1,1)$	$(1,2)$	$(2,1)$	$(2,2)$	$(2,3)$	$(3,1)$	$(4,1)$	$(4,2)$	$(4,3)$
$D_{(s,j)}$	8	5	8	5	5	5	13	13	10
$T_s \times R_{(s,j)}$	1.6	1	1.6	1	1	2.5	6.5	6.5	5

We have the following lemma to bound the variability of the order size in model delivery sequences with minimum bottleneck deviation.

Lemma 9.14. *Any sequence with minimum bottleneck gives a solution z_{ik} with no more than three copies of any model difference between delivery periods l and l' of supplier s. That is*

$$|Z_{il} - Z_{il'}| \leq 3$$

for $i = 1, \ldots, n$ and any two delivery periods l and l' of supplier s.

Proof. By Theorem 5.8 the minimum bottleneck $B^* < 1$, and by Theorem 9.9 any solution with bottleneck $B < 1$ is 3−balanced. Therefore, any factor of length T_s has no more than three occurrences of any model i than any other factor of the same length, which proves the lemma.   □

Thus, we get the following bound on the order size variability, see (9.20), for the supply j of supplier s

$$\left| \sum_{i=1}^{n} b_{i,(s,j)} Z_{il} - \sum_{i=1}^{n} b_{i,(s,j)} Z_{il'} \right| \leq B_{(s,j)} |Z_{il} - Z_{il'}| \leq 3B_{(s,j)},$$

where

$$B_{(s,j)} = \sum_{i=1}^{n} b_{i,(s,j)}.$$

Notice that since

$$\sum_{i=1}^{n} Z_{il} - \sum_{i=1}^{n} Z_{il'} = T_s - T_s = 0,$$

then $B_{(j,s)}$ can in fact be replaced by the sum of the $\left\lceil \frac{n}{2} \right\rceil$ largest among $b_{1,(s,j)}, \ldots, b_{n,(s,j)}$ and thus strengthen the bound.

Finally, since the problem (9.20) might be infeasible, then we consider the following always feasible modification

$$\sum_{k=1}^{p} z_{ik} = d_i \qquad i = 1, \ldots, n \tag{9.21}$$

$$\lfloor T_s R_j \rfloor - q_{js} \leq \sum_{i=1}^{n} b_{i,(j,s)} Z_{il} \leq \lceil T_s R_j \rceil + Q_{js}$$

$$\text{for} \quad l = 0, \ldots, m_s - 1, s \in \{0, 1, \ldots, \mathcal{S}\}, j = 1, \ldots, n_s$$

$$\sum_{i} z_{ik} = \Delta_k \qquad k = 1, \ldots, p$$

where the $Q_{js} \geq 0$ and $\lfloor T_s R_j \rfloor \geq q_{js} \geq 0$ are the bounds on maximum shortage and oversupply of option j in a single delivery period of supplier s. The bounds may of course be minimized giving rise to a family of optimization problems involving Q_{js} and q_{js}.

9.10.2 The complexity

We have the following complexity result that states that minimizing order size variability in the periodic synchronized supply chains is NP-hard in the strong sense, in fact even the feasibility test for the system of inequalities (9.20) is NP-hard in the strong sense.

Theorem 9.15. *The minimization of order size variability in the periodic synchronized supply chain problem is NP-hard in the strong sense.*

Proof. Let the positive integers $a_1, \ldots, a_{3m}$ and A make up an instance of the 3-partition problem. We define a two level supply chain with $n = 3m$ models at L1 and a single supplier at L2. The supplier content **b** of each model is given in Table 9.9 where model i simply requires a_i units of supply $(1,1)$. The demand for each model equals 1, and we assume period $T_1 = 3$ for the supplier. Then, $D_{(1,1)} = mA$, $D = 3m$, and $T_1 R_{(1,1)} = A$. For a 3-partition $P_1, \ldots, P_m$ of the set $\{1, \ldots, 3m\}$ the sequence

$$P_1 \rightarrow \cdots \rightarrow P_m$$

where the permutation in each set P_i is arbitrary solves the periodic synchronized supply chain problem since the order size in each delivery cycle equals A for $\sum_{j \in P_i} a_j = A$. Now, let the sequence of models

$$j_1 \rightarrow j_2 \rightarrow j_3 \rightarrow \cdots \rightarrow j_{3m}$$

meet the constraint (9.20). Then, the sets $P_i = \{j_{3(i-1)+1}, j_{3(i-1)+2}, j_{3i}\}$ make up the required 3-partition of the set $\{1, \ldots, 3m\}$. This transformation is pseudo-polynomial, thus the periodic synchronized supply chain problem is NP-hard in the strong sense. $\square$

9.10.3 Model-Supplier One-to-One Case

We close this section by showing a special case of the periodic synchronized delivery problem (9.21) solvable in polynomial time. The case is called model-supplier

Table 9.9 The **b** matrix corresponding to the instance of 3-partition problem

$m \backslash (s,j)$	$(1,1)$
1	a_1
$\vdots$	$\vdots$
i	a_i
$\vdots$	$\vdots$
$3m$	a_{3m}

one-to-one case to emphasize that the matrix $\mathbf{b}$ in this case has exactly one non-zero entry in each column and each row.

If each supply (s, j) is model unique, that is there exists exactly one model i such that $b_{i,(s,j)} \neq 0$, then the problem can be simplified as follows. Let $i^*(s, j)$ the unique model that requires the supply (s, j). Then, (9.19) can be recast as follows

$$D_{(s,j)} = \sum_{i=1}^{n} b_{i,(s,j)} d_i = b_{i^*(s,j),(s,j)} d_{i^*(s,j)},$$

and

$$T_s R_{(s,j)} = T_s \frac{D_{(s,j)}}{D} = T_s \frac{b_{i^*(s,j),(s,j)} d_{i^*(s,j)}}{D} = T_s r_{i^*(s,j)} b_{i^*(s,j),(s,j)}.$$

We thus have

$$\sum_{i=1}^{n} b_{i,(s,j)} Z_{il} = b_{i^*(s,j),(s,j)} Z_{i^*(s,j)l},$$

for any $l = 0, \ldots, m_s - 1$, $s \in \{0, 1, \ldots, \mathcal{S}\}$, and $j = 1, \ldots, n_s$. Consequently, we would require to ideally have

$$b_{i^*(s,j),(s,j)} Z_{i^*(s,j)l} = T_s r_{i^*(s,j)} b_{i^*(s,j),(s,j)}$$

or

$$Z_{i^*(s,j)l} = T_s r_{i^*(s,j)}. \tag{9.22}$$

However, since the right hand side of (9.22) is integer and the left might not be then the constraint (9.20) is replaced by

$$\lfloor T_s r_{i^*(s,j)} \rfloor \leq Z_{i^*(s,j)l} \leq \lceil T_s r_{i^*(s,j)} \rceil \tag{9.23}$$
$$\text{for} \quad l = 0, \ldots, m_s - 1, s \in \{0, 1, \ldots, \mathcal{S}\}, j = 1, \ldots, n_s$$

Additionally, we observe that if the models are supplier disjoint as well, that is for each model i there is exactly one supply $i(s, j)$, that is there is exactly one nonzero entry in each row of matrix $\mathbf{b}$, then constraint (9.23)

$$\lfloor T_{s^*} r_{i^*} \rfloor \leq Z_{i^* l} \leq \lceil T_{s^*} r_{i^*} \rceil$$
$$\text{for} \quad l = 0, \ldots, m_{s^*} - 1, i^* = 1, \ldots, n$$

where s^* is a unique supplier for i^* and the other way round, that is the i^* is the unique model supplied by s^*. Then, the system (9.21) becomes

$$\sum_{k=1}^{p} z_{ik} = d_i \quad i = 1, \ldots, n \tag{9.24}$$
$$\lfloor T_{s^*} r_{i^*} \rfloor \leq Z_{i^* l} \leq \lceil T_{s^*} r_{i^*} \rceil$$
$$\text{for} \quad l = 0, \ldots, m_{s^*} - 1, i^* = 1, \ldots, n$$
$$\sum_i z_{ik} = \Delta_k \quad k = 1, \ldots, p$$

and it can be solved by the network flow algorithms.

9.11 Synchronized Delivery

An alternative approach to the leveling off demand problem follows the bottleneck problem for the model level L1. However, now smoothing at each supplier is complicated by the fact that the demand for its supplies is not independent as it is the case for L1 but rather depends on the model delivery sequence. This makes the leveling off problem much more challenging. We now give details of this approach.

Let x_{ik} be the cumulative production of i for the k-prefix of the model delivery sequence. Then, the delivery sequence-dependent total demand for (s, j) originating from this k-prefix is

$$Y_{(s,j),k} = \sum_{i=1}^{n} b_{i,(s,j)} x_{ik}.$$

Consequently, the total demand for supplier s originating from this k-prefix equals

$$Y_{s,k} = \sum_{(s,j) \in S_s} Y_{(s,j)} = \sum_{(s,j) \in S_s} \sum_{i=1}^{n} b_{i,(s,j)} x_{ik}.$$

On the other hand, the total demand for supplier s as a result of demands $d_1, \ldots, d_n$ for models $1, \ldots, n$ respectively, can be derived as follows

$$D_s = \sum_{(s,j) \in S_s} D_{(s,j)} = \sum_{(s,j) \in S_s} \sum_{i=1}^{n} b_{i,(s,j)} d_i \qquad (9.25)$$

out of which

$$D_{(s,j)} = \sum_{i=1}^{n} b_{i,(s,j)} d_i \qquad (9.26)$$

is demand for (s, j) only. Thus, the ratio

$$r_{(s,j)} = \frac{D_{(s,j)}}{D_s}, \qquad (9.27)$$

for the supply (s, j). Let us illustrate these definitions with an example based on the example of chain from Sect. 9.7.

Example 9.16. For the matrix **b** given in Table 9.3 the ratio calculations are shown in Tables 9.10 and 9.11. □

Table 9.10 Matrix $\sum_{(s,j) \in S_s} b_{i,(s,j)}$

$i/(s,j)$	(1,1)	(1,2)	(2,1)	(2,2)	(2,3)	(3,1)	(4,1)	(4,2)	(4,3)
1	2		1			1	3		
2	1		3			0	5		
3	0		2			0	3		

Thus, ideally the fraction $r_{(s,j)}$ of $Y_{s,k}$ should be equal to $Y_{(s,j)}$. In other words, we minimize

$$\min_{(s,j),k} \left| Y_{(s,j),k} - r_{(s,j)} Y_{s,k} \right|. \tag{9.28}$$

We can rewrite the deviation in (9.28) as follows

$$Y_{(s,j),k} - r_{(s,j)} Y_{s,k} = \sum_{i=1}^{n} b_{i,(s,j)} x_{ik} - r_{(s,j)} \sum_{i=1}^{n} \sum_{(s,j) \in S_s} b_{i,(s,j)} x_{ik}$$

$$= \sum_{i=1}^{n} \left(b_{i,(s,j)} - r_{(s,j)} \sum_{(s,j) \in S_s} b_{i,(s,j)} \right) x_{ik}$$

$$= \sum_{i=1}^{n} \rho_{i,(s,j)} x_{ik}$$

where

$$\rho_{i,(s,j)} = b_{i,(s,j)} - r_{(s,j)} \sum_{(s,j) \in S_s} b_{i,(s,j)}. \tag{9.29}$$

The following example illustrates the calculations defined by (9.29).

Example 9.17. For the matrix **b** given in Table 9.3 and the ratios in Table 9.11 we have the matrix ρ in Table 9.12.

Thus, the problem is to minimize

$$\min_{(s,j),k} \left| \sum_{i=1}^{n} \rho_{i,(s,j)} x_{ik} \right| \tag{9.30}$$

Subject to

$$\sum_{i=1}^{n} x_{ik} = k \quad k = 1, \dots, D \tag{9.31}$$

$$0 \leq x_{ik+1} - x_{ik} \leq 1 \quad i = 1, \dots, n; k = 1, \dots, D-1 \tag{9.32}$$

$$x_{iD} = d_i \quad i = 1, \dots, n. \tag{9.33}$$

Kubiak et al. [124] prove that the synchronized delivery problem, that is the minimization of (9.30) subject to (9.31)–(9.33), is NP-hard in the strong sense and they give a dynamic programming algorithm to solve it. Their algorithm runs in time $O(n^2 \prod_{i=1}^{n} (d_i + 1))$. Its experimental performance evaluation can be found in [124].

We now show a bound on the objective (9.30). We begin with the following observation.

Table 9.11 Ratios $r_{(s,j)}$

(s,j)	$(1,1)$	$(1,2)$	$(2,1)$	$(2,2)$	$(2,3)$	$(3,1)$	$(4,1)$	$(4,2)$	$(4,3)$
$D_{(s,j)}$	8	5	8	5	5	5	13	13	10
D_s	13		18			5	36		
$r_{(s,j)}$	$\frac{8}{13}$	$\frac{5}{13}$	$\frac{8}{18}$	$\frac{5}{18}$	$\frac{5}{18}$	1	$\frac{13}{36}$	$\frac{13}{36}$	$\frac{10}{36}$

Table 9.12 The matrix ρ

$i/(s,j)$	$(1,1)$	$(1,2)$	$(2,1)$	$(2,2)$	$(2,3)$	$(3,1)$	$(4,1)$	$(4,2)$	$(4,3)$
1	$-\frac{3}{13}$	$\frac{3}{13}$	$\frac{10}{18}$	$-\frac{5}{18}$	$-\frac{5}{18}$	0	$\frac{33}{36}$	$-\frac{3}{36}$	$-\frac{30}{36}$
2	$\frac{5}{13}$	$-\frac{5}{13}$	$-\frac{6}{18}$	$\frac{3}{18}$	$\frac{3}{18}$	0	$-\frac{29}{36}$	$\frac{7}{36}$	$\frac{22}{36}$
3	0	0	$-\frac{16}{18}$	$\frac{8}{18}$	$\frac{8}{18}$	0	$-\frac{39}{36}$	$-\frac{3}{36}$	$\frac{42}{36}$

Lemma 9.18. *We have*

$$\sum_{i=1}^{n} \rho_{i,(s,j)} r_i k = 0.$$

Proof. We get

$$\sum_{i=1}^{n} \rho_{i,(s,j)} r_i = \frac{1}{D}\left(\sum_{i=1}^{n} b_{i,(s,j)} d_i - r_{(s,j)} \sum_{i=1}^{n} \sum_{(s,j)\in \mathcal{S}_s} b_{i,(s,j)} d_i\right)$$

by (9.29). Thus, by definitions (9.26), (9.25) and (9.27) we have

$$\sum_{i=1}^{n} \rho_{i,(s,j)} r_i = \frac{1}{D}(D_{(s,j)} - r_{(s,j)} D_s) = 0.$$

$\square$

For optimal solution x_{ik} to the bottleneck problem we have

$$x_{ik} = k r_i + \varepsilon_{ik},$$

where $-1 < -B^* \le \varepsilon_{ik} \le B^* < 1$. Thus, we have

$$\left|\sum_{i=1}^{n} \rho_{i,(s,j)} x_{ik}\right| = \left|\sum_{i=1}^{n} \rho_{i,(s,j)} k r_i + \sum_{i=1}^{n} \rho_{i,(s,j)} \varepsilon_{ik}\right|,$$

by Lemma 9.18 the bound is as follows

$$\left|\sum_{i=1}^{n} \rho_{i,(s,j)} x_{ik}\right| = \left|\sum_{i=1}^{n} \rho_{i,(s,j)} \varepsilon_{ik}\right| \le B^* \sum_{i=1}^{n} \left|\rho_{i,(s,j)}\right| \le B^* B_s,$$

where

$$B_s = \sum_{(s,j)\in \mathcal{S}_s}^{n} B_{(s,j)} = \sum_{(s,j)\in \mathcal{S}_s}^{n} \sum_{i=1}^{n} b_{(s,j),i}.$$

9.12 Exercises

Exercise 9.19. Prove that there is no feasible solution for the following instance given in Table 9.13. Hint: See Gent [111].

Table 9.13 An infeasible instance of the temporal supplier capacity problem

Demand d_i	$p_s : q_s$ 1:2	2:3	1:3	2:5	1:5	Model i
2	0	0	0	1	1	1
2	0	0	1	0	1	2
5	0	1	1	1	0	3
4	0	0	0	1	0	4
4	0	1	0	1	0	5
1	1	1	0	0	1	6
3	1	1	1	0	1	7
4	0	0	1	0	0	8
19	0	1	0	0	0	9
7	1	1	0	1	0	10
10	1	0	0	0	0	11
1	0	0	1	1	0	12
5	1	1	1	1	0	13
2	1	0	1	1	0	14
6	1	1	0	0	0	15
4	1	1	1	0	0	16
8	1	0	0	1	0	17
1	1	0	0	0	1	18
4	0	1	1	0	0	19
2	0	0	0	0	1	20
4	0	1	0	0	1	21
1	1	1	0	1	1	22
1	0	1	1	0	1	23

Exercise 9.20. Prove that the car sequencing problem is NP-hard in the strong sense even each capacity constraint is 1:2 and demand for each model is unit, that is $d_i = 1$ for $i = 1, \ldots, n$

Exercise 9.21. Based of the formulation in (9.4), (9.5) and (9.6) for the car sequencing problem give a formulation for the generalized pinwheel scheduling.

Exercise 9.22. Consider an instance of the car sequencing problem and a just-in-time sequence S, possibly violating some capacity constraints, for this instance. Find an efficient way of determining the minimum number of positions in S than need to be turned in blanks – a blank is a fictitious models with no options – to turn S into a sequence that meets all option capacity constraints but not necessarily just-in-time. What is the complexity to this problem?

9.13 Comments and References

The car sequencing problem was shown NP-hard in the strong sense by an elegant transformation from the Hamiltonian path problem by Gent [111], though his transformation requires different capacity constraints for different car options. The

problem later was shown NP-hard in the strong sense even if each capacity requirement is either 1:4 or 2:3, see Kis [125], however yet again this transformation requires that the demands for different models differ. The IP formulation of the car
sequencing problem is based on Gravel et al. [114], and Drexl and Kimms [115].
The former presents a quite comprehensive reference list of paper that solve the car
sequencing problem by the ant colony optimization technique. The proof of Theorem 9.2 is from Kubiak [126], so is the mixed-model, pull supply chain framework
from Sect. 9.7 and the discussion of model deliver sequences as balanced words in
Sect. 9.8. Theorem 9.9 was shown in Jost [121]. More on word combinatorics can
be found in Vuillon [46], Altman et al. [7], Tijdeman [60], and on balanced words
in Berthe and Tijdeman [122].

Section 9.11 is based on approach to the leveling off demand from Miltenburg
and Sinnamon [127], Miltenburg and Goldstein [128], Kubiak et al. [124], and
Kubiak [23]. This last refers to the problem as the Output Rate Variation (ORV)
problem.

Chapter 10
Fair Queueing and Stride Scheduling

10.1 Introduction

Fairness related objectives appear to have gained their prominence through fair queueing algorithms and stride scheduling – both essential building blocks of today's information technology. This chapter focusses on the fundamental issue of defining and quantifying fairness for these two applications rather than on the technical details of their implementation and performance which can be readily found in the literature, see for instance Keshav [129], Bertsekas and Gallager [130], and Waldspurger and Weihl [131].

We first observe that though both basic fair queueing and stride scheduling use essentially the same algorithms based on the Jefferson's method of apportionment their approach to defining and quantifying fairness have been rather different. While the former is based on the max–min criterion, and the relative as well as the absolute fairness bounds, the latter is based on the bottleneck and variance minimization. Despite these differences we propose here a common ground for both approaches to fairness in fair queueing and stride scheduling. This approach is based on the apportionment theory and the just-in-time optimization. It opens possibilities for alternative fair queueing and stride scheduling algorithms. One such clear alternative is the Webster's method of apportionment not previously used in either fair queueing or stride scheduling contexts. This method is the only one that results in a peer-to-peer fairness consistent with a standard two-client solution – a new and promising in the context of technical systems concept of fairness.

Section 10.2 presents an illustrative example that serves as a general framework for our discussion of fair queueing algorithms and stride scheduling. Section 10.3 provides an introduction to fair queueing and its connection to the apportionment parametric methods. Section 10.4 presents different measures of fairness used historically in fair queueing. In particular the max–min fairness criterion, the relative fairness bound, and the absolute fairness bound. We show that the uniform apportionment methods naturally satisfy the max–min fairness criterion though with different objective functions. We show that the relative fairness bound is minimized by

W. Kubiak, *Proportional Optimization and Fairness,* International Series in Operations Research & Management Science 127, DOI 10.1007/978-0-387-87719-8_10,

an algorithm well-known in the apportionment theory since the sixties. The algorithm, however, is not house monotone, thus there is no on-line queueing algorithm to minimize the relative fairness bound. We also show a good approximation of the Generalized Processor Sharing policy which is based on the weighted bottleneck deviation minimization algorithm developed in Chap. 3. The Generalized Processor Sharing policy is considered as an ideal fair scheduling policy in the networking literature. This policy is a benchmark against which to compare any other queueing algorithm in the absolute fairness measure.

Section 10.5 introduces stride scheduling. We show there that the basic stride scheduling is the Jefferson's divisor method of apportionment problem. We argue that the quota divisor methods would greatly reduce the throughput error if used instead of the Jefferson's method in stride scheduling, however, at the cost of loosing population monotonicity of the Jefferson's method. We also show that the bottleneck algorithm would minimize the throughput error. We finally discuss the minimization of the response time variability in stride scheduling.

Section 10.6 proposes another approach to fairness in fair queueing and stride scheduling. This one is called peer-to-peer fairness and it is essentially based on the Webster's apportionment method. In a nutshell, this approach produces a solution which is consistent with the standard two-client solution being arguably ideal for any two flows or clients competing for resources – either the bandwidth or the central processor unit time.

10.2 The Story of Tiles: The Start, The Finish, or The In-Between

We now introduce a common framework for our discussion of fairness in fair queueing and stride scheduling.

The framework is based on a simple model with n types of tiles. Each tile of type i is ℓ_i units long, $i = 1, \ldots, n$. The ℓ_i can be the packet size (in bits) of flow i in fair queueing or the stride length, the reciprocal of the number of tickets, of client i in stride scheduling. The tiles of each type are arranged into an infinite strip of type i where tiles of type i are placed one next to another without gaps, see Fig. 10.1. This

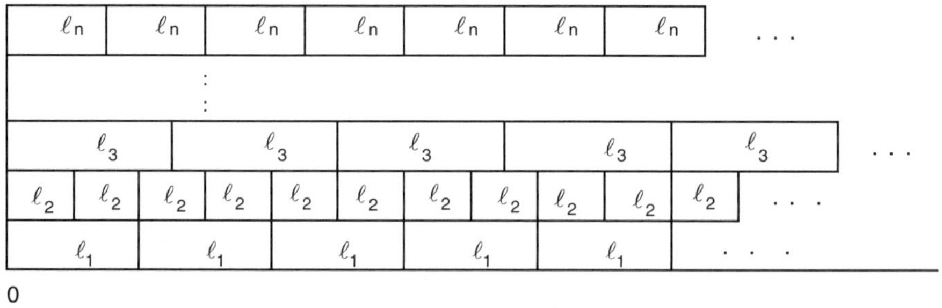

Fig. 10.1 The framework for fair queueing and stride scheduling

strip models backlogged packets from flow i waiting for bandwidth allocation in the fair queueing context or backlogged strides of client i waiting for quanta allocation in the stride scheduling context. Thus, the ends of tiles in the strip of type i are at points $k\ell_i$, $k = 1, 2, 3, \ldots$ The set of these points is then

$$E_i = \{k\ell_i : k = 1, 2, 3, \ldots\}.$$

Let

$$f_1 \leq f_2 \leq \cdots$$

be the ascending order of the points of the multiset $\bigcup_{i=1}^{n} E_i$. Let us consider the sequence

$$S = s_1 s_2 s_3 \cdots$$

where

$$s_k = i \text{ if and only if } f_k \in E_i.$$

The sequence S is the Jefferson's sequence. To see this let us consider the point in position k. Let $x_{i,k}$ be the number of the ends from E_i on the list

$$s_1 s_2 \cdots s_k$$

and consider

$$x_{i,k}\ell_i.$$

Then, position $k + 1$ of S is occupied by the end from the list i^* such that

$$l_j + x_{j,k}l_j \geq l_{i^*} + x_{i^*,k}l_{i^*} \text{ for all } j = 1, \ldots, n$$

or

$$\frac{\frac{1}{\ell_{i^*}}}{x_{i^*,k} + 1} \geq \frac{\frac{1}{\ell_j}}{x_{j,k} + 1}$$

which is exactly the Jefferson's parametric method for the population vector

$$\left(\frac{1}{\ell_1}, \ldots, \frac{1}{\ell_n} \right),$$

see Theorem 2.2 and Table 2.1. Notice that without loss of generality the populations can be made integer by taking $p_i = \frac{\text{lcm}(\ell_1, \ldots, \ell_n)}{\ell_i}$ instead of $\frac{1}{\ell_i}$.

Similarly, we observe that by considering the starts $k\ell_i$, $i = 1, \ldots, n$, $k = 0, 1, 2, 3, \ldots$,

$$S_i = \{k\ell_i : k = 0, 1, 2, 3, \ldots\}$$

we can build the Adams's sequence, and by considering the middle points $k\ell_i + \frac{1}{2}\ell_i$, $i = 1, \ldots, n$, $k = 0, 1, 2, 3, \ldots$,

$$M_i = \{k\ell_i + \frac{1}{2}\ell_i : k = 0, 1, 2, 3, \ldots\}$$

we can obtain the Webster's sequence. Finally, by taking $0 \leq \delta \leq 1$ and points $k\ell_i + \delta\ell_i$, $i = 1,\ldots,n$, $k = 0,1,2,3,\ldots$, we obtain the sequence for the parametric method ϕ^δ, see Sect. 2.4.2. We illustrate these sequences, that is the Jefferson's, the Adams's, and the Webster's, by the following example.

Example 10.1. For $\ell_a = 2, \ell_b = 3$, and $\ell_c = 5$ we have

$$E_a = \{2,4,6,8,10,12,14,16,18,20,22,24,26,28,30,32,\ldots\}$$
$$E_b = \{3,6,9,12,15,18,21,24,27,30,33,\ldots\}$$
$$E_c = \{5,10,15,20,25,30,35,\ldots\}$$

and

$$M_a = \{1,3,5,7,9,11,13,15,17,19,21,23,25,27,29,31,\ldots\}$$
$$M_b = \{1.5,4.5,7.5,10.5,13.5,16.5,19.5,22.5,25.5,28.5,31.5,\ldots\}$$
$$M_c = \{2.5,7.5,12.5,17.5,22.5,27.5,32.5,\ldots\}$$

and

$$S_a = \{0,2,4,6,8,10,12,14,16,18,20,22,24,26,28,30,\ldots\}$$
$$S_b = \{0,3,6,9,12,15,18,21,24,27,30,\ldots\}$$
$$S_c = \{0,5,10,15,20,25,30,\ldots\}$$

Thus, the sequence for the finish times, the Jefferson's sequence, is

$$(abacababacababcaabacbaabcabaabc)^\infty$$

the sequence for the midpoints, the Webster's sequence, is

$$(abcabaabcabacababacababcaabacba)^\infty$$

the sequence for the start times, the Adams's sequence, is

$$(abcabacababacababcaabacbaabcaba)^\infty.$$

Each sequence has the cycle of length 31 since $\mathrm{lcm}(2,3,5) = 30$ and the three populations are $\frac{30}{2} = 15$, $\frac{30}{3} = 10$, and $\frac{30}{5} = 6$ respectively, summing up to 31.   □

10.3 Fair Queueing

In a packet-switching or datagram network, the Internet being one, fair queueing algorithms set the rules of bandwidth allocation at the gateways of the network so that a good user behavior is encouraged and enforced throughout the network. Demers et al. [132] argue, following the game-theoretic view of the bandwidth allocation first proposed by Nagle [12], that a formal game-theoretic analysis of simple gateway models suggests that fair queueing algorithms may be "the only reasonable queueing algorithms to ... make self-optimizing source behavior result in fair, protective, non-manipulable, and stable networks." We show here that the fair queueing algorithms, proposed in the literature and implemented in practice, to determine the rules of the gateway's bandwidth allocation are essentially two basic parametric methods of apportionment: the Adams's and the Jefferson's methods. Following this new insight, we argue that the Webster's method of apportionment, not known to the author to be used as a queueing algorithm thus far, might be even a better choice for a fair queueing algorithm than the other two. Let us now look closer at the issue of fair queueing at a gateway of a packet-switching network.

A gateway has a number of incoming and outgoing links through which packets from various sources are routed to various destinations. Each outgoing link maintains separate queues associated with it, one queue for each individual source–destination pair. The nonempty queues are cycled through so that in each cycle the packet at the head of one queue is selected for transmission on the outgoing link. The packets of a single source–destination queue are all of the same size. However, the sizes of packets for different source–destination pairs, often referred to as users, flows or queues, may be different. For instance, short packets may result from users engaging in remote login sessions using the Telnet protocol whereas long packets may result from users transferring files with the FTP protocol. Furthermore, the packets are assembled at the source and disassembled at the destination, thus preempting a packet in any intermediate node would be in violation of the communications protocol and it is not permitted. Maintaining separate queues for different source–destination pairs protects well-behaved pairs from ill-behaved ones. The latter by sending their packets too quickly simply increase the length of their own queues without actually increasing their share of the bandwidth of the outgoing link or delaying other pairs. This ensures that the ill-behaved pairs can get no more than their fair share of the bandwidth. Though the model with the separate queues provides certain protection against antisocial behavior, it leaves open the question of a fair share and the way to ensure it so that the whole network remains stable. Nagel describes the problem elegantly in his seminal paper [12] as follows:

This game-theory view of datagram networks leads us to a digression on the stability of multiplayer games. Systems in which the optimal strategy for each player is suboptimal for all players are known to tend towards the suboptimal state. The well-known prisoner's dilemma problem in game theory is an example of a system with this property. But a closer analog is the tragedy of the commons problem in economics. Where each individual can improve his own position by using more of a free resource, but the total amount of the resource degrades as the number of users increases, self-interest leads to overload of the resource and collapse. Historically, this analysis was applied to the use of common grazing lands; it also applies to such diverse resources as air quality and timesharing systems. In general, experience indicates that many player systems with this type of instability tend to get into serious trouble. Solutions to the tragedy of the commons problem fall into three classes: cooperative, authoritarian, and market. Cooperative solutions, where everyone agrees to be well behaved, are adequate for small numbers of players, but tend to break down as the number of players increases. Authoritarian solutions are effective when behavior can be easily monitored, but tend to fail if the definition of good behavior is subtle. A market solution is possible only if the rules of the game can be changed so that the optimal strategy for players results in a situation that is optimal for all. Where this is possible, market solutions can be quite effective. The above analysis is generally valid for human players. In the network case, we have the interesting situation that the player is a computer executing a preprogrammed strategy. But this alone does not ensure good behavior; the strategy in the computer may be programmed to optimize performance for that computer, regardless of network considerations. A similar situation exists with automatic redialing devices in telephony where the user's equipment attempts to improve performance over an overloaded network by rapidly redialing failed calls. Since call-setup facilities are scarce resources on telephone systems, this can seriously impact the network; there are countries that have been forced to prohibit such devices (Brazil, for one). This solution by administrative fiat is sometimes effective and sometimes not, depending on the relative power of the administrative authority and the users. As transport protocols become more commercialized and competing systems are available, we should expect to see attempts to tune the protocols in ways that may be optimal from the point of view of a single host, but suboptimal from the point of view of the entire network. We already see signs of this in the transport protocol implementation of one popular workstation manufacturer. So, to return to our analysis of a pure datagram internetwork, an authoritarian solution, would order all hosts to be "well behaved" by fiat; this might be difficult since the definition of a well-behaved host in terms of its externally observed behavior is subtle. A cooperative solution faces the same problem, along with the difficult additional problem of applying the requisite social pressures in a distributed system. A market solution requires that we make it pay to be well behaved. To do this, we will have to change the rules of the game. see Nagle [12], pp. 436–437.

The authoritarian solution refereed to in the Nagel's comments has perhaps unknowingly been found in the apportionment solution to the fair representation problem of meeting the ideal of one man, one vote. Here is why.

The following two queueing algorithms have been formulated in the literature in response to the Nagel work:

1. Fair queueing based on starting times by Greenberg and Madras, [133], stipulates that whenever a packet finishes transmission on outgoing link the next one transmitted on that link is the one with the earliest start time.
2. Fair queueing based on finishing times by Demers et al. [132], stipulates that whenever a packet finishes transmission on outgoing link the next one transmitted on that link is the one with the earliest finish time.

These two work essentially as follows. Let ℓ_i be the packet size of queue i, and let a_i be the number of packets from that queue that got transmitted on the outgoing link thus far. When a packet from some queue finishes transmission, then the former algorithm selects a next packet from the queue (flow) i^* such that

$$\ell_j a_j \geq \ell_{i^*} a_{i^*} \text{ for all } j \tag{10.1}$$

whereas the latter selects a next packet from the queue i^* such that

$$\ell_j(a_j + 1) \geq \ell_{i^*}(a_{i^*} + 1) \text{ for all } j. \tag{10.2}$$

Thus, the fair queueing based on starting times is clearly the Adams's method of apportionment and the fair queueing based on finishing times is the Jefferson's method, see Chap. 2 for details of these methods, where the population of state i is $\frac{1}{\ell_i}$.

Consequently, the fair queueing based on starting times maximizes the minimum share of the bandwidth, that is it optimizes

$$\max_i \min \frac{a_i}{p_i} = \max \min_i \ell_i a_i, \tag{10.3}$$

whereas the fair queueing based on finishing times minimizes the maximum share of the bandwidth, that is it optimizes

$$\min_i \max \frac{a_i}{p_i} = \min \max_i \ell_i a_i, \tag{10.4}$$

see Balinski and Young [2].

However, the fair queueing based on the midpoints rather than either starting or finishing times may prove even better choice for a fair queueing algorithm since it results in the Webster's method of apportionment, which as we argued throughout this book, see Chaps. 2 and 5, provides a number of advantages over the other two, the Adams's and the Jefferson's methods. The fair queueing based on the midpoint selects a next packet from the queue i^* such that

$$\ell_j \left(a_j + \frac{1}{2} \right) \geq \ell_{i^*} \left(a_{i^*} + \frac{1}{2} \right) \text{ for all } j. \tag{10.5}$$

The fair queueing based on the midpoint, or the Webster's method, minimizes the following, different from both (10.3) and (10.4), objective. The queue i share of the bandwidth is $\frac{a_i}{p_i} = \ell_i a_i$ whereas the ideal share for all the queues at the gateway is

$$\frac{\sum a_i}{\sum p_i} = \frac{a}{\sum \frac{1}{\ell_i}} = \frac{a}{P}.$$

The Webster's method minimizes the weighted squared difference between each queue share and the ideal share defined as follows

$$\sum_i p_i \left(\frac{a_i}{p_i} - \frac{a}{P} \right)^2. \tag{10.6}$$

Notice that the weight for the queue i is $\frac{1}{\ell_i}$ where ℓ_i is the queue i packet size in bits, thus the Webster's method minimizes the weighted squared difference between each flow's share and the ideal share summed over all queues, see Balinski and Young [2], and Young [3].

All three methods, the Adams's and the Jefferson's known from the existing queueing literature and the Webster's suggested here for the fair queueing algorithm, are the parametric methods of apportionment and thus they are all uniform. Therefore, they ensure that a fair bandwidth allocation remains a fair bandwidth allocation for every subset of queues. For the Adams's method, this ensures that minimum share is maximized and that holds even if we remove any queue and reduce the total bandwidth accordingly by taking away all this queues's bandwidth allocations. The Jefferson's method ensures that the maximum share is minimized and this condition remains recursively true as we remove any queue and again reduce the total bandwidth accordingly by taking away all this queues's bandwidth allocations. Finally, the Webster's method minimizes the weighted squared difference between each queue's share and the ideal share summed over all queues and again this holds recursively true as we remove any queue.

10.4 Which Queueing Fairness?

10.4.1 Max–Min Fairness Criterion

The selection of the most appropriate objective for fair queueing algorithms has not received its due attention in the literature except for the max–min fairness criterion, see Gafni and Bertsekas [134], which has been historically yet somewhat arbitrarily tied up with fair queueing. The criterion can be defined as follows.

Consider the allocation of a single resource among n users. There are λ units of this resource and each of the users requests ρ_i and receives λ_i units. The max–min fairness criterion stipulates that an allocation is fair if

1. No user receives more than its request, that is $\rho_i \geq \lambda_i$
2. No other allocation scheme that satisfies (1) has a higher minimum allocation
3. Condition (2) remains recursively true as we remove the minimal user and reduce the total resource accordingly, that is

$$\lambda := \lambda - \lambda_{\min}$$

As we have shown, the Adams's, the Jefferson's and the Webster's methods satisfy the conditions (2) and (3) of this definition albeit with different objectives (10.3), (10.4), and (10.6) respectively. Actually, the Adams's method maximizes the minimum allocation, see (10.3), thus literally meeting condition (2). The condition (1) simply provides an upper bounds for user allocations and it is irrelevant for backlogged flows. Thus, we conclude that the well known and widely used for more than 20 years max–min fairness criterion results in the Adams's method of apportionment. Though this method's quality has been widely recognized, the method may prove not the best queueing algorithm available since it may be bettered by the qualities of the Webster's method as we argued in Chaps. 2 and 5.

The choice of an objective to optimize in the condition (2) of the max–min criterion remains an open question. Though, the fair queueing aims may not be exactly the same as those of the apportionment problem, this still remains to be seen, the choice of the objective function for a fair queueing algorithm may actually be a wrong question to ask in a much the same way as it has been argued to be a wrong question to ask for the apportionment problem, see Balinski and Young [2]. The latter asks instead what is the apportionment method, if any, that meets all desirable properties without actually explicitly asking what is the objective it optimizes. The approach is thus axiomatic, and the focus is on the desirable properties instead.

10.4.2 Relative Fairness Bound

Relative fairness will be determined for flows that are all backlogged in time interval $(0,t)$, where t is time in seconds, see also Zhou and Seth [135]. For a queueing algorithm Q the relative fairness bound, RFB, is defined as follows.

$$RFB = \max_{(0,t),i,j} \left| \frac{S_i^Q(t)}{w_i} - \frac{S_j^Q(t)}{w_j} \right|, \tag{10.7}$$

where w_i is an integer weight of queue i, and $S_i^Q(t)$ is the number of bits (in complete packets) send by Q in time interval $(0,t)$ from flow (queue) i. Since bits from each queue are sent in complete packets only, the $S_i^Q(t)$ is as follows

$$S_i^Q(t) = x_{i,(0,t)} \ell_i, \tag{10.8}$$

where $x_{i,(0,t)}$ is the number of packets send by Q in $(0,t)$ from i. Thus,

$$\frac{S_i^Q(t)}{w_i} = \frac{x_{i,(0,t)}}{\frac{w_i}{\ell_i}} = \frac{x_{i,(0,t)}}{p_i}, \tag{10.9}$$

where the population $p_i = \frac{w_i}{\ell_i}$ is simply weight per packet bit for the queue i. The $x_{i,(0,t)}$ is a counting function that increases its value by 1 at points, referred to as step points,

$$0 < t_{1,i} < \cdots < t_{k,i} < \cdots$$

in time, and it remains constant in between any two consecutive step points. That is

$$x_{i,(0,t_{k,i})} - x_{i,(0,t_{k,i}-\varepsilon)} = 1$$

and

$$x_{i,(0,t_{1,i}-\varepsilon)} = 0$$

for any sufficiently small $\varepsilon > 0$. Observe that

$$t_{k+1,i} - t_{k,i} \geq \frac{\ell_i}{C}$$

where $\frac{\ell_i}{C}$ is the minimum amount of time to send one packet from queue i, and C is the transmission rate in bits per second. Let

$$0 < T_1 < T_2 < \cdots < T_m < \cdots$$

be all step points of these counting functions. Let us denote this set of counting functions by **C**.

We can rewrite (10.7) as follows

$$RFB = \max_{(0,t),i,j} \left| \frac{x_{i,(0,t)}}{p_i} - \frac{x_{j,(0,t)}}{p_j} \right| = \max_{m,i,j} \left| \frac{x_{i,m}}{p_i} - \frac{x_{j,m}}{p_j} \right| \tag{10.10}$$

where

$$x_{i,(0,m)} = x_{i,(0,T_m)}.$$

Our goal is to find the $x_{i,(0,m)} = x_{i,m}$ that minimizes (10.10). We have the following theorem.

Theorem 10.2. *There exists no queueing algorithm that is house monotone and minimizes the RFB measure.*

Proof. Balinski and Shahidi [16], and Ibaraki and Katoh [19] show that there exists no house monotone apportionment method minimizing objective function in (10.10). □

Theorem 10.2 proves that it is impossible to build an on-line algorithm that minimizes the RFB measure. That is it may always happen that the sequence that minimizes the RFB measure for the packets transmitted so far becomes suboptimal by extending it by one more packet. However, if the total set of packets is known a priory, an optimal sequence can be found by an off-line algorithm based on a dynamic programming algorithm proposed by Burt and Harris [136].

10.4.3 Absolute Fairness Bound

The absolute fairness bound deals with the absolute deviation between the flow send from a queue by a given queueing algorithm Q and the ideal amount of flow send from the queue by the Generalized Processor Sharing (GPS) policy. The GPS works in a weighted round robin fashion, see Parekh and Gallager [137] for the details and analysis of the GPS, sending from each queue an amount of flow proportional to the weight associated with the queue. The flow send by the GPS is divisible which is not the case for Q. The GPS policy meets the following condition, for any two queues i and j in interval $(0,t)$

$$\frac{F_i^G(t)}{F_j^G(t)} = \frac{w_i}{w_j}, \tag{10.11}$$

where $F_i^G(t)$ is the flow send from queue i in interval $(0,t)$. Summing up (10.11) over all n queues, we get the following

$$F_i^G(t) = tC\frac{w_i}{W} \tag{10.12}$$

where C is the transmission rate in flow units per second and $W = \sum_{i=1}^{n} w_i$. In practice, the GPS policy can not be implemented since the data in packet-switching networks is send in packets rather than in divisible flows. However, the GPS serves as a benchmark in the absolute fairness bound. The absolute fairness bound, AFB, for a given queueing algorithm Q is defined as follows.

$$AFB = \max_{(0,t),i} \left| \frac{S_i^Q(t)}{w_i} - \frac{F_i^Q(t)}{w_i} \right|.$$

Since the data is send by Q in packets rather than in divisible flows the following condition holds for Q

$$S_i^Q(t) = x_{i(0,t)} \ell_i, \tag{10.13}$$

where $x_{i(0,t)}$ denotes the number of complete packets sent by Q from queue i in the interval $(0,t)$. Combining (10.12) and (10.13) with (10.13), we can write the following formula for the AFB measure.

$$AFB = \max_{(0,t),i} \left\{ \frac{1}{w_i} \left| x_{i(0,t)} \ell_i - tC \frac{w_i}{W} \right| \right\}, \tag{10.14}$$

which is equivalent to

$$AFB = \max_{(0,t),i} \left\{ \frac{\ell_i}{w_i} \left| x_{i(0,t)} - tC \frac{w_i}{W \ell_i} \right| \right\}. \tag{10.15}$$

Now, let us define k as follows

$$k = \sum_{i=1}^{n} tC \frac{w_i}{W \ell_i} = \frac{tCD}{WL}. \tag{10.16}$$

and the rate r_i

$$r_i = \frac{d_i}{D} = \frac{\frac{Lw_i}{\ell_i}}{D}, \tag{10.17}$$

where the demand $d_i = \frac{\mathrm{lcm}(\ell_1,...,\ell_n) w_i}{\ell_i} = \frac{Lw_i}{\ell_i}$ for queue i, and

$$D = \sum_{i=1}^{n} \frac{Lw_i}{\ell_i}.$$

Thus, the GPS could potentially send

$$\lfloor kr_i \rfloor = \left\lfloor tC \frac{w_i}{W \ell_i} \right\rfloor \tag{10.18}$$

packets from queue i in $(0,t)$. We can thus rewrite (10.15) as follows

$$AFB = \max_{(0,t),i} \left\{ \frac{1}{d_i} \left| x_{i(0,t)} - kr_i \right| \right\}. \tag{10.19}$$

Let $AFB(\mathbf{C})$ be the minimum absolute fairness bound over all counting functions in $\mathbf{C}$ defined in Sect. 10.4.2. Let $\mathbf{C}^M \subset \mathbf{C}$ be a subset of the counting functions where each $t_{m,i}$ is a linear integer combination of $\frac{\ell_j}{C}, j = 1,\ldots,n$, that is there are non-negative integers $\alpha_j, j = 1,\ldots,n$, dependent on m and i such that

$$t_{m,i} = \alpha_1 \frac{\ell_1}{C} + \cdots + \alpha_n \frac{\ell_n}{C}.$$

Let $AFB(\mathbf{C}^M)$ be the minimum absolute fairness bound over all counting functions in $\mathbf{C}^M$. Obviously, $AFB(\mathbf{C}) \le AFB(\mathbf{C}^M)$. Now, assume that all counting functions $x_{i(0,t)}$ are in $\mathbf{C}^M$ for $i = 1,\ldots,n$. Let

$$0 < T_1 < T_2 < \cdots < T_m < \cdots$$

be all step points of these counting functions. All of them are different, and the total

$$\sum_{i=1}^{n} x_{i(0,t)} = m$$

for any $T_m \le t < T_{m+1}$. Moreover,

$$T_m = \alpha_1 \frac{\ell_1}{C} + \cdots + \alpha_n \frac{\ell_n}{C}$$

thus there are

$$m = \alpha_1 + \cdots + \alpha_n$$

complete packets send in $(0, T_m)$ by Q. Hence, we have

$$\left|x_{i(0,t)} - k r_i\right| \le \max\left\{\left|x_{i(0,T_m)} - k' r_i\right|, \left|x_{i(0,T_m)} - k'' r_i\right|\right\} \qquad (10.20)$$

for $T_m \le t < T_{m+1}$, where

$$k' = \frac{CD}{WL} T_m \text{ and } k'' = \frac{CD}{WL} T_{m+1}.$$

Consider the counting function defined as follows $y_{i(0,m)} = x_{i(0,T_m)}$ with the step points at $1, 2, \ldots, m, \ldots$ Clearly, the $y_{i(0,m)}$ is in $\mathbf{C}^M$. For simplicity, we use the notation $y_{i,m}$ for $y_{i(0,m)}$. We have

$$\max_{m,i}\left\{\frac{1}{d_i} \left|y_{i,m} - k r_i\right|\right\} \qquad (10.21)$$

$$\le \max_{m,i}\left\{\frac{1}{d_i} \max\{\left|y_{i,m} - m r_i + (m - k') r_i\right|, \left|y_{i,m} - m r_i + (m - k'') r_i\right|\}\right\} \qquad (10.22)$$

Let $y'_{i,m}$ minimize the bottleneck on the right hand side of the inequality (10.21). We then have

$$\max_{m,i}\left\{\frac{1}{d_i}\max\{|y'_{i,m}-mr_i+(m-k')r_i|,|y'_{i,m}-mr_i+(m-k'')r_i|\}\right\}$$

$$\leq \max_{m,i}\left\{\frac{1}{d_i}\max\{|y^*_{i,m}-mr_i+(m-k')r_i|,|y^*_{i,m}-mr_i+(m-k'')r_i|\}\right\}, \quad (10.23)$$

for $y^*_{i,m}$ being a solution to the bottleneck problem (3.37)–(3.41) in Chap. 3 and d_i as in (10.17). However, for $y^*_{i,m}$ we have by (10.18)

$$\sum_{i=1}^{n}\lceil k'r_i\rceil \geq m \geq \sum_{i=1}^{n}\lfloor k'r_i\rfloor,$$

thus,

$$|m-k'| < n.$$

The $|m-k'|$ is the difference between the throughputs of Q and the GPS in the time interval $(0,T_m)$ measured in the number of complete packets send by the two. Similarly,

$$\sum_{i=1}^{n}\lceil k''r_i\rceil \geq m+1 \geq \sum_{i=1}^{n}\lfloor k''r_i\rfloor,$$

thus

$$|m-k''| \leq n.$$

Thus, (10.23) does not exceed

$$\max_{m,i}\left\{\frac{1}{d_i}|y^*_{i,m}-mr_i|+\frac{n}{D}\right\}.$$

The bottleneck assignment problem (3.42) leads to a counting function $y^*_{i,m}$ that minimizes

$$\max_{m,i}\left\{\frac{1}{d_i}|y^*_{i,m}-mr_i|\right\},$$

and this optimal solution has value $\alpha^* \leq \frac{B}{d_{i^*}}$ for some i^* and $B < 1$, by Theorem 5.8. Thus, we have the following upper bound of the optimal absolute fairness bound

$$AFB = \max_{(0,t),i}\left\{\frac{1}{d_i}|x^*_{i(0,t)}-kr_i|\right\} \leq \frac{B}{d_{i^*}}+\frac{n}{D},$$

where $x^*_{i(0,t)}$ is optimal in $\mathbf{C}$. Finding, an optimal $x^*_{i(0,t)}$ is an open question as is proving that $x^*_{i(0,t)}$ is simply $y^*_{i,m}$.

10.5 Stride Scheduling

The market solution first alluded to by Nagle [12] as an alternative to the cooperative and authoritarian solutions to stabilize the multiplayer games played played to obtain resources in computer systems has been proposed in the form of stride scheduling. The stride scheduling was introduced by Waldspurger and Weihl in their 1995 reports [13, 131], and independently by Maheshwari [138], as a universal, deterministic scheduling paradigm.

The stride scheduling allows clients to buy, sell and trade tickets according to the rules of a computational market, for instance the Spawn distributed computational economy proposed by Waldspurger et al. [11]. The number d_i of tickets acquired by a client i then determines the rate at which the client will receive the resources needed for its job, which competes for resources with jobs of other clients. The total number of tickets issued equals $D = \sum_{1 \leq i \leq n} d_i$.

The number of tickets issued to a client quantifies the client's resource rights. The tickets can be issued in various currencies and distributed in a modular fashion. The clients can be issued tickets in the base currency and then each of them can allocate tickets in its own currency to its various tasks. For instance, a client who obtained 10 tickets in the base currency may allocate 40 tickets in its own currency to its task A, 50 to its task B, and 10 to its task B. Thus, allocating 4, 5, and 1 tickets, respectively in the base currency.

The resources are allocated to clients in discrete time slices $1, 2, \ldots$, called quanta. The main idea of stride scheduling is to calculate a *stride* $\ell_i = \frac{1}{d_i}$, inversely proportional to number of client's tickets, that client must wait between successive resource allocations. A client with a shorter stride will get resource allocations more frequently than a client with a longer stride. For instance a client with double the stride of another client will execute twice as slowly getting just half of the resource allocations of the client with the shorter stride. We assume for the simplicity of exposition that the stride ℓ_i of client i equals $\frac{L}{d_i}$, where the constant L is only used to obtain high-precision fixed-point integer representation of the strides ℓ_i for individual clients, see Waldspurger and Weihl [13] for a discussion of this technical problem. We assume L to be equal $1\mathrm{cm}(d_1, \ldots, d_2)$. The client i^* to be allocated resources in quantum $k + 1$, $k = 0, 1, 2, \ldots$, is calculated as follows

$$\frac{L(x_{j,k} + 1)}{d_j} \geq \frac{L(x_{i^*,k} + 1)}{d_{i^*}} \qquad (10.24)$$

where $x_{i,k}$ is the client's i number of quantum allocations received during the first k quantum allocations. We assume that initially $x_{i,0} = 0$ for $i = 1, \ldots, n$, and the ties are broken using the ascending order of the client's index $i = 1, \ldots, n$. It is obvious from (10.24) that the sequence of allocations produced by the stride scheduling is independent of L. Therefore, from now on, we shall assume without loss of generality that i^* is calculated, equivalently, as follows.

$$i^* = \arg\max_i \left\{ \frac{d_i}{x_{i,k} + 1} \right\}, \qquad (10.25)$$

Thus, the basic stride scheduling is just the Jefferson's method of apportionment. Because of the parameter $\delta = 1$ for the Jefferson's parametric method, see Table 2.1, we call it the 1-stride scheduling.

We now discuss main metrics of the stride schedules: the throughput error and the response time variability.

10.5.1 Throughput Error

The stride scheduling keeps track of the number of resource allocations each client i has received out of the first k allocations, $x_{i,k}$, and uses the criterion in (10.24) to decide which client gets the next quantum. Therefore there is a negligible overhead for this rather efficient scheduling algorithm. Waldspurger and Weihl [13] view the stride scheduling as a mechanism to keep the progress of each client i job as close to the ideal progress defined by the straight line

$$k\frac{d_i}{D}$$

as possible. It is worth noticing that it is essentially the same goal as in the solution to the Liu–Layland problem given in Chap. 8.

The deterministic stride scheduling has been originally conceived as an alternative to a random lottery scheduling which results in a more erratic progress of a client job. To see this let as have a closer look at the lottery scheduling first. The lottery scheduling determines each allocation by holding a lottery where client i with d_i tickets has probability $\pi_i = \frac{d_i}{D}$ of winning it. We use the notation π_i instead of the traditional p_i for the probability since the latter is set aside for the population in the apportionment problem. Thus, the lottery is a Bernoulli process with a probability of success π_i for client i, and consequently the number of lotteries w_k won by i in k identical, independent lotteries has a binomial distribution. Therefore, the expected number of wins $E(w_k)$ is

$$E(w_k) = k\pi_i = k\frac{d_i}{D},$$

and the variance $V(w_k)$ is

$$V(w_k) = \sigma^2(w_k) = k\pi_i(1 - \pi_i) = k\frac{d_i(D - d_i)}{D^2} \leq \frac{k}{4}.$$

Given these, let us estimate the expected throughput error for the lottery scheduling defined as the expected value of the following absolute deviation

$$E\left(\left|w_k - k\frac{d_i}{D}\right|\right)$$

by Waldspurger and Weihl [13]. By the Markov's inequality, see Ross [66], we have

$$P\left(\left|w_k - k\frac{d_i}{D}\right| \geq \sigma(w_k)\right) \leq \frac{E\left(\left|w_k - k\frac{d_i}{D}\right|\right)}{\sigma(w_k)}.$$

By the Central Limit Theorem

$$P\left(\frac{\left|w_k - k\frac{d_i}{D}\right|}{\sigma(w_k)} \geq 1\right)$$

tends to a constant 0.3174 as k tends to ∞. Moreover,

$$E\left(\left|w_k - k\frac{d_i}{D}\right|\right) \leq \frac{\sqrt{k}}{2}, \tag{10.26}$$

see Exercise 10.5. Thus, we have

$$\frac{E\left(\left|w_k - k\frac{d_i}{D}\right|\right)}{\sigma(w_k)} \leq 1.$$

Therefore,

$$E\left(\left|w_k - k\frac{d_i}{D}\right|\right) = \Theta(\sqrt{k}),$$

that is $\sqrt{k}$ is both an upper and lower bound on the expected throughput error $E\left(\left|w_k - k\frac{d_i}{D}\right|\right)$. Thus, the throughput error of the lottery scheduling can grow without bound as k increases. This is certainly the main disadvantage of the lottery scheduling. On the other hand, the stride scheduling reduces the throughput error considerably. In fact it may keep the throughput error below 1 as long as we replace the Jefferson's method in it by a bottleneck optimal algorithm or quota-divisor methods. We show this now.

Waldspurger and Weihl [13] define the throughput error of stride scheduling, called also accuracy, as the maximum absolute or bottleneck deviation

$$\min_{} \max_{i,k} |x_{i,k} - k\frac{d_i}{D}|$$

introduced also in Chap. 5 in (5.2). Waldspurger and Weihl realize that their basic 1-stride scheduling may produce sequences with a large absolute throughput error. We now know that this is because a parametric method with $\delta = 1$, that is the Jefferson's method, always advances the allocations of the high-throughput clients, that is clients with large number of tickets. Waldspurger and Weihl also point out that a similar behavior, resulting in large throughput error, has been exhibited by similar to the stride scheduling fair queueing algorithms. Again, we know now that these algorithms are essentially the Jefferson's method of apportionment as well.

To illustrate this problem with the parametric methods consider an instance with 101 clients and the following ticket allocations

$$d_1 = 100, d_2 = \ldots = d_{101} = 1.$$

The 1-stride algorithm would result in a cycle where client 1 receives the first 100 quanta followed by the remaining 100 clients receiving 1 quantum each. The throughput error of this cycle is 50. Notice that since $x_{1,k} = 100$ for $k = 100$ and $r_1 = \frac{1}{2}$, then $|x_{1,k} - kr_1| = 50$ for $k = 100$. The $\frac{1}{2}$-stride scheduling, the Webster's method, would reduce this error twice. It would result in a cycle where client 1 receives the first 50 quanta followed by the remaining 100 clients receiving 1 quantum each, and followed by 50 quanta allocated to client 1. Clearly, the throughput error of this cycle is 25. Notice that since $x_{1,k} = 50$ for $k = 50$ and $r_1 = \frac{1}{2}$, then $|x_{1,k} - kr_1| = 25$ for $k = 50$. However, it follows from Theorem 5.8 that the optimal throughput error does not exceed $1 - \frac{1}{200} = \frac{199}{200}$, a cycle that attains this throughput error allocates all odd quanta between 1 and 200 to client 1, and all even quanta to the remaining clients in an arbitrary way. In fact, an optimal cycle results in a throughput error of $\frac{198}{200}$. This optimal cycle allocates all odd quanta between 1 and 100, and all even quanta between 101 and 200 to client 1 and all the remaining quanta to the remaining clients in an arbitrary way. Notice that a more sophisticated hierarchical stride scheduling, we refer the reader to Waldspurger and Weihl [13] for details, is able to reduce the throughput error for the instance discussed above to 4.5 only. However, further reduction is possible and follows the ideas developed in Chap. 5. This reduction can be summarized as follows.

Theorem 10.3. *The algorithm defined in Sect. 5.2 minimizes the throughput error. The optimal throughput error is always less than 1. Moreover, all quota-divisor methods of apportionment keep the throughput error below 1.*

By the Impossibility theorem, Theorem 2.4, respecting priorities, which is a key axiom in the axiomatic approach to the apportionment problem, can only be achieved at the cost of increased throughput error. Therefore, by attempting to minimize the throughput error the stride scheduling may compromise population monotonicity.

10.5.2 Response Time Variability

Waldspurger and Weihl [13] define the response time variability as the elapsed time from a client's completion of one quantum up to and including its completion of the next. In the lottery scheduling the number of lotteries needed until next success of client i, α_i, is a geometric random variable with the expected value

$$E(\alpha_i) = \overline{\alpha}_i = \frac{D}{d_i}$$

and the variance

$$Var(\alpha_i) = \overline{\alpha}_i(\overline{\alpha}_i - 1) = \frac{D(D - d_i)}{d_i^2}.$$

For the stride scheduling, a natural measure of the response time variability for a client is the variance of its response time. For cyclic sequences this variance for client i can be defined as the variance of response time for the first $d_i + 1$ allocations. More formally, let α_j^i be the number of quanta between the completion of the jth allocation and the completion of the $j + 1$-st, $j = 1, \ldots, d_i$. Then the response time variability for client i is

$$RTV_i = \sum_{1 \leq j \leq d_i} (\alpha_j^i - \overline{\alpha}_i)^2,$$

and

$$Var_i = \frac{1}{d_i} \sum_{1 \leq j \leq d_i} (\alpha_j^i - \overline{\alpha}_i)^2 = \frac{1}{d_i} RTV_i.$$

Notice that the average response time $\overline{\alpha}_i$ is constant and equal to $\frac{D}{d_i}$ for client i. For any solution with throughput error less than 1 we have, see Exercise 10.6,

$$1 \leq \alpha_j^i \leq 2\overline{\alpha}_i, \tag{10.27}$$

thus

$$Var_i \leq \overline{\alpha}_i^2 \tag{10.28}$$

Though the upper bound in (10.28) does not imply the response time variability reduction by the stride scheduling in comparison to the lottery scheduling, Waldspurger and Weihl [13] observe in their computational experiments that the stride scheduling produces much less response time variability than the lottery scheduling. As well, Corominas et al. [82] observe a substantial reduction in the response variability obtained by the exchanges on a bottleneck optimal sequence. The improvement however usually comes at a slight increase in the throughput error, see Fig. 7.2.

Unfortunately, the problem of minimizing the total response time variability defined as follows

$$RTV = d_1 Var_1 + \cdots + d_n Var(n). \tag{10.29}$$

is computationally more difficult than the problem of minimizing throughput error. We have Theorem 7.15 proven in Chap. 7 to show that the problem of minimizing the response time variability is NP-hard. However, it is open whether the problem of minimizing the response time variability is NP-hard in the strong sense. See Chap. 7 for more results on the response time variability minimization.

10.6 Peer-To-Peer Fairness

We now present the peer-to-peer fairness model. First, we give some motivation behind the model using a simple example of the stride scheduling problem, though the model itself is general and equally well applies to the fair queueing.

Consider two clients a and b. Suppose client a obtains 3 tickets and client b obtains 6, then b expects advancing its task at a rate which is twice the rate of a.

Thus, if clients a and b were the only two clients competing for shared resources, then the infinite cyclic sequence

$$(abb)^\infty$$

with the cycle abb, would be *fair* for both and neither would have a *casus* for complaining since out of any three consecutive quanta two are allocated to b, with 6 tickets, and one to a, with 3 tickets. Notice that $\frac{3}{6} = \frac{1}{2}$ and consequently $1 + 2 = 3$ is the smallest number of quanta to consider for clients a and b in any mathematically *sound* discussion of what constitute a fair allocation of quanta for the couple. Any smaller number of quanta would obviously be *biased* towards one of the clients and the $\frac{1}{2}$ ratio could not then be achieved.

However, if another client, say c, joins the couple with 4 tickets in their competition for the shared resources, then the sequence

$$(cbabcbabcbabc)^\infty \qquad (10.30)$$

with the bottleneck deviation $B = \frac{9}{13}$, for the three becomes

$$(bab)^\infty$$

for a and b, which ensures that out of any three consecutive quantum allocations two are made to b and one to a. This should certainly be *fair* for the two. Also, for clients a and c it becomes

$$(cacacac)^\infty$$

which ensures that out of any seven consecutive quanta 3 are allocated to a and 4 to c. Again, it is fair to both a and c as their ticket ratio is $\frac{3}{4}$, and $3 + 4 = 7$ is the length of the shortest cycle where this ratio is achievable. Finally, for clients b and c the sequence becomes

$$(cbbcbbcbbc)^\infty$$

which means that out of any five consecutive quanta client b is allocated at least 3, but sometimes 4. This could make client c feel that it does not receive its fair share of allocations with respect to client c as it has as many as $\frac{2}{3}$ of the number of tickets client b has, though it sometimes gets only $\frac{1}{5}$ of consecutive 5 quanta. This potential perception of *unfairness* can be avoided by the sequence

$$(bcbacbbcabcba)^\infty \qquad (10.31)$$

with the bottleneck deviation $B = \frac{11}{13}$, as it becomes

$$(bcbcb)^\infty$$

for clients b and c. Thus out of any consecutive 5 quanta, 3 are allocated to b and 2 to c. Furthermore, the sequence becomes

$$(bba)^\infty$$

for clients a and b, which ensures that out of any three consecutive quantum allocations two are made to b and one to a, and

$$(caccaca)^\infty$$

which ensures that out of any seven consecutive quanta 3 are allocated to a and 4 to c. The sequence (10.31), has an obvious advantage of being *peer-to-peer* fair though it is not optimal from the bottleneck deviation stand point. The sequence (10.30) we begun with is not peer-to-peer fair but has lower bottleneck deviation of $\frac{9}{13}$. The question then is can we have a peer-to-peer fair sequence for a, b, c which at the same time minimizes maximum deviation. In our example of clients a, b, and c the answer is positive for the sequence

$$(bcabcbabcbacb)^\infty \tag{10.32}$$

minimizes bottleneck deviation, its optimal value is $\frac{7}{13}$, and is peer-to-peer fair: we have

$$(bcbcb)^\infty$$

for b and c,

$$(bab)^\infty$$

for a and b, and finally

$$(cacacac)^\infty$$

for a and c. Observe that the sequence (10.32) is the Webster's sequence. Notice also that this sequence is not a balanced word since it includes subsequences bb with two bs and ca with no b. In fact since the Fraenkel's conjecture holds for $n \leq 6$, see Tijdeman [5], no balanced word for clients a, b, c with tickets 3, 6, and 4 respectively is possible.

We have the following definition.

Definition 10.4 (Peer-To-Peer Fair Sequencing Problem). Given the clients $1, \ldots, n$ with tickets $d_1 \leq \cdots \leq d_n$ respectively. Define the peer-to-peer ratio for clients i and j, $i < j$, as $f_{ij} = \frac{d_i}{d_j} = \frac{\alpha_{ij}}{\beta_{ij}}$, where α_{ij} and β_{ij} are relatively prime. Find an infinite periodic sequence S^∞ with the cycle S which has as small a bottleneck deviation as possible, and which is peer-to-peer fair. That is client i occurs exactly d_i times in S, and for each couple i, j, $i < j$, of clients any subsequence of S_{ij} obtained from S by deleting all clients except i and j, whose relative positions remain as in the original sequence S, with the length $\alpha_{ij} + \beta_{ij}$ has exactly α_{ij} client i allocations and exactly β_{ij} client j allocations. □

The peer-to-peer fair sequencing problem remains open. We make however a number of observations. First, it is tempting to conjecture that for any instance of the peer-to-peer fair sequencing problem there always exists solution that minimizes bottleneck deviation. An instance with three clients a, b and c and their tickets 3, 5, and 15, respectively, makes a simple counterexample to this conjecture. Observe, however, that the following sequence for these three clients

$$(cbcaccbcccabcccbccacbcc)^{\infty}$$

is peer-to-peer fair and it has bottleneck deviation $B = \frac{19}{23}$ less than 1, which follows from Theorem 2.6 as the sequence is Webster's. The minimum bottleneck deviation for this instance is $B^* = \frac{14}{23}$. Thus, the peer-to-peer fair sequence with bottleneck not exceeding 1 is possible after all for this instance though its bottleneck is not optimal.

Though this last observation may hold true for many instances, it does not hold generally if we insists on the sequences which are not only peer-to-peer fair but also remain consistent with the standard two-client (state) solution. Then, by Theorem 2.15, the Webster's method is the unique apportionment method that is pairwise consistent with the standard two-client (state) solution. This standard solution ensures possibly the fairest treatment of any two clients, see Chaps. 2 and 5. Therefore, the Webster's method ensures not only that for each couple i, j, $i < j$, of clients any subsequence of S_{ij} obtained from S by deleting all clients except i and j, whose relative positions remain as in the original sequence S, with the length $\alpha_{ij} + \beta_{ij}$ has exactly α_{ij} client i allocations and exactly β_{ij} client j allocations but also that these allocation will be made as fair as possible in the S_{ij}. The uniformity of other parametric methods would ensure peer-to-peer fairness as well however these methods would not ensure the standard two-client solution. However, by the Impossibility Theorem 2.4, the Webster's method does not satisfy quota and thus may result in the bottleneck deviation higher that 1. Recall from Theorem 5.8 that there always is a sequence with bottleneck deviation less than 1.

Finally, Balinski and Young [2] point out that there is no uniform and anonymous apportionment method that satisfies quota. Consequently, the solutions to the peer-to-peer fair sequencing will most likely have their bottleneck deviations higher than 1 for some instances. This is the case for the anonymous methods which ignore the names or other than the number of tickets characteristics of clients.

10.7 Exercises

Exercise 10.5. Show that the expected absolute error under lottery scheduling is $O(\sqrt{k})$.

Exercise 10.6. Prove that (10.27) holds.

Exercise 10.7. Develop the quota-Jefferson's stride scheduling and fair queueing algorithms.

Exercise 10.8. Develop the quota-Adams's stride scheduling and fair queueing algorithms.

Exercise 10.9. Develop the quota-Webster's stride scheduling and fair queueing algorithms.

Exercise 10.10. Develop an algorithm that produces a peer-to-peer fair sequence with minimum bottleneck.

Exercise 10.11. What are the RFB and AFB for the Jefferson's method?

Exercise 10.12. What are the RFB and AFB for the Adam's method?

Exercise 10.13. What is the RFB and AFB for the Webster's method?

10.8 Comments and References

The observation that the stride scheduling is the Jefferson's was made in Kubiak [107]. The stride scheduling has been used for scheduling resource allocations in the Linux kernel, Waldspurger and Weihl [13], in network routers by the Click modular router, Kohler et al. [139], and in storage appliances by the NeST software-only storage appliance, Bent et al. [140].

For the tragedy of the commons see the classic paper by Hardin [141].

The fairness measures used in fair queueing are discussed in Zhou and Seth [135] who provide more references on the topic.

The idea of fair, meaning proportional, resource allocation is widely used in many areas of information technology. It can be found for instance in network of workstations (NOW) which is defined as a loose collection of workstations physically scattered across users' desk by Theimer et al. [142]. The current advances in local networks, especially in low-latency, high-bandwidth switches enabled networks of workstations to resemble more closely massively parallel processors. By connecting many workstations one can build incrementally scalable, cost-effective and highly available cluster that can act as a shared server. An application can use resources of the whole cluster such as: processors, memory and disks that may not otherwise be utilized. It is desirable that each user of NOW receives a fair share of the available resources, see Arpaci-Dusseau and Culler [143]. Another example where the idea of proportional resource allocation is used is metacomputing or grid environment of independent geographically dispersed sites connected by wide-area networks, in which parallel applications can be run, see Nabrzyski et al. [144] on the grid resource management. Metacomputing enables the user to effectively benefit from all resources it has access to. The basic motivation for metacomputing is resource trading, see Cirne and Marzullo [145]. Cirne and Marzullo [145] propose proportional resource scheduling algorithm for scheduling applications running on geographically dispersed clusters that form a metacomputer called Computational Co-op.

Proportional resource allocation is required also in operating systems. Especially, the multithread systems allot a resource to threads according to their relative importance. Many systems solve this problem using priority-based schedulers. Such schedulers ignore the lower priority threads as long as there exists a higher priority thread which can be run. This situation can lead to a starvation of the lower priority thread. Some priority-based schedulers solve this problem by the reduction of priorities to allow all threads to be executed. However, in this case the problem of reducing the priorities of threads in a predictable way needs to be addressed.

A stride scheduling algorithm of Waldspurger and Weihl [13] was proposed to solve this problem. This algorithm provides a way to control the relative execution rates of threads. This control over relative computation rates is required to achieve the service rate objectives for the user and the application.

Józefowska et al. [146] observe that algorithms for a proportional resource allocation proposed in Arpaci-Dusseau and Culler [143] for NOW environment as well as in Cirne and Marzullo [145] for grid computing use in fact the idea of stride scheduling of Waldspurger and Weihl [13].

Kumar and Kleinberg [147] study the so-called max–min fair vectors of allocations. The vectors borrow the main idea of recursively maximizing the minimum allocation from the max–min criterion, considered already in the seventies by Megiddo [148], and are used to various combinatorial problems: bandwidth allocation, scheduling, and facility location. However, finding the max–min fair vectors is computationally intractable thus the focus is on the approximation algorithms.

Chapter 11
Smoothing and Batching

11.1 Introduction

Mixed-model production systems are widely adopted by manufacturing companies in a broad range of industries where customers demand a variety of models of the same product, Monden [22], Sawik [149], and Boysen et al. [150]. Automotive and electronics industries are two well-known examples with well-established presence of different options that create a tremendous variety of models of a single product. At the same time the demand for a particular model is often insufficient to warrant dedicated production resources for the model, instead the resources have to be shared by a whole collection of models. Thus, operating such mixed-model manufacturing systems efficiently is a challenging problem that has been widely studied in the operations management literature particularly with the goal to reduce their inherent variability.

The last few decades have witnessed a fast growing adaptation of the just-in-time manufacturing philosophy by numerous companies, including those that operate mixed-model manufacturing systems. These companies sooner or later come to realize that the variability reduction in its various forms is what makes their just-in-time manufacturing systems designed for mixed-model production tick, see a recent study of the automotive industry by Harbour [151]. The Toyota Motor Corporation is credited with being the pioneer in variability reduction, in particular *production smoothing* through level scheduling, see Monden [22]. The underlying idea, see for instance Jones [152], is the *three M* concept where large variability results in *mura*, that is undesirable variability in productivity and quality, and *muri*, that is overburden of resources, managers, and workers. The two together create *muda*, that is waste. The answer to the production smoothing problem is known as heijunka, see McBride [153] and Jones [152], that is leveling the load and in particular leveling production by properly selecting and sequencing the product mix. To explain the concept let us start with an example. Suppose a demand for five models A, B, C, D and E are $d_A = 8, d_B = 16, d_C = 4, d_D = 6$ and $d_E = 2$ respectively. Since the $\gcd(2, 4, 6, 8, 16) = 2$, then the solution for $\frac{d_A}{2} = 4, \frac{d_B}{2} = 8, \frac{d_C}{2} = 4, \frac{d_D}{2} = 3, \frac{d_E}{2} = 1$ will be repeated twice. This solution according to the Webster's method is

W. Kubiak, *Proportional Optimization and Fairness,* International Series in Operations
Research & Management Science 127, DOI 10.1007/978-0-387-87719-8_11,
© Springer Science+Business Media LLC 2009

BADBCBABDEBABCBDAB.

Suppose moreover that the system capacity allows to release at most five orders (Kanbans in a typical just-in-time system) every 20 min, then the sequence can be split as follows:

BADBC | BABD | EBABC | BDAB

and the heijunka box, see Jones [152], filled in as follows

Model\time	7:00	7:20	7:40	8:00	8:20	8:40	9:00	9:20
Model *A*	*A*	*A*	*A*	*A*	*A*	*A*	*A*	*A*
Model *B*	*BB*	*BB*	*BB*	*BB*	*BB*	*BB*	*BB*	*BB*
Model *C*	*C*		*C*		*C*		*C*	
Model *D*	*D*	*D*		*D*	*D*	*D*		*D*
Model *E*			*E*				*E*	

Thus starting at 7:00 the orders are withdrawn from the Heijunka box every 20 min following the time line in the first row and released for production.

Hence, in the first 20 min a copy of model A, two copies of model B, one copy of model C, and one copy of model D will be produced by the system. It is worth observing that by Theorem 9.9 there always exists a sequence such that the numbers of copies of each model in each 20 min time column interval differ by at most 4, actually they differ by at most 3 if the number of orders released in each interval are equal, see Exercise 11.9.

The production smoothing problem has been studied mainly for synchronized assembly lines where each model takes exactly one unit of processing time and setup times are being assumed negligible, see Chaps. 3 and 9. Under these assumptions any permutation of models gives a feasible sequence as long as the sequence meets demands for models, and the problem reduces to uniformly dispersing the models over the sequence, see Monden [22] and Miltenburg [21]. However, many manufacturing systems with processing and setup times significantly varying among the models are far from being synchronized assembly lines. Lummus [154] investigates the operation of manufacturing systems with high model variability under the just-in-time manufacturing principles. Her limited simulation study of a nine-station production system where different models have different setup and processing time requirements shows that the smooth production schedule obtained by the methods primarily designed for synchronized assembly lines is as good as a random schedule. This result has underpinned and motivated a number of subsequent studies of mixed-model manufacturing systems with different models having different setup and processing time requirements.

Nevertheless the concept of Heijunka can still be successfully used with variable setup times as follows. Suppose that the weekly demand calls for producing $d_A = 1{,}000$ units of model A, $d_B = 600$ units of model B, and $d_C = 400$ units of model C. Suppose moreover that each unit of each model takes 1 min to produce, however, it takes 3 min to setup production for A, 1 min for B, and 4 min for C. The Heijunka would find the $\gcd(1,000, 600, 200) = 200$ and then a sequence for 5 units of A, 3 units of B, and 2 units of C. The sequence could be for instance

$$ABCAABACBA \tag{11.1}$$

which minimizes the bottleneck deviation, see Example 5.3 in Chap. 5. The sequence requires eight setups, three for A, three for B, and two for C. Observe that the sequence starts and ends with the A. Thus, the total setup per sequence is $3 \times 3 + 3 \times 1 + 2 \times 4 = 20\,\text{min}$. Therefore, at most 20 setups are allowed weekly consuming the total of $400\,\text{min}$ of the $2,400\,\text{min}$ available. Then the smallest feasible cycle based on the sequence (11.1) is as follows

$$A)^{20}B^{20}C^{20}(AA)^{20}B^{20}A^{20}C^{20}B^{20}(A.$$

This chapter studies the production smoothing problem arising in mixed-model just-in-time manufacturing systems where processing and setup times significantly vary among the models. Though this chapter's real-life motivation comes from just-in-time operations at a leading US automotive pressure hose manufacturer it can apply its results to any mixed-model just-in-time manufacturing systems where the final production stage or the bottleneck stage of the system can be reasonably modeled as a single machine where different models are allowed to have different setups and processing times. The solution found for this machine can then be spread through a pull just-in-time control mechanism to the entire system or even further to the whole supply chain.

The chapter presents two methods to solve the production smoothing problem in mixed-model just-in-time systems. One method finds all Pareto-optimal solutions that minimize total production rate variation (we use this term instead of total deviation in this chapter) of models and Work-In-Process, and maximize the system utilization and responsiveness. These Pareto-optimal solutions are found efficiently in $O(D^4)$ time, where D is the total demand for all models. The other relies on the Webster's method of apportionment for production smoothing which produces periodic, uniform and reflective production sequences that can improve operations management of the just-in-time systems. The method can also be used as a heuristic for the Pareto-optimization in $O(nD^2)$ time.

The performance of the algorithms is tested through an extensive computational study. The study shows that the proposed algorithms are computationally efficient and they produce solutions with provable characteristics desirable for operations management. The results also show that both methods are effective in minimizing the total variation in production rates of different models and the WIP; and maximizing system utilization and responsiveness. More specifically, they result in solutions with low WIP levels, and high machine utilization and system responsiveness, enabling them to be used as efficient and effective tools of operations management in just-in-time manufacturing systems.

The remainder of this chapter is organized as follows. In Sect. 11.2 the real-life operation will be briefly described. In Sect. 11.3, we formulate the problem. Section 11.4 proposes a procedure to find all Pareto-optimal solutions to the problem. Section 11.5 presents the axiomatic approach to the total production rate variation and proves its main properties. Section 11.6 discusses the axiomatic approach in the context of optimization. Section 11.7 presents the design and the results of our computational experiment.

11.2 A Real-Life System

This chapter's real-life motivation comes from the just-in-time operations at the leading US automotive pressure hose manufacturer first studied by Yavuz et al. [155]. The manufacturer produces various types of pressure hoses for the automotive industry. The hoses pass through a six-stage process demonstrated in Fig. 11.1 where the first three stages are heavier processes that use large batch-sizes, whereas the latter three stages are more model-specific operations that use smaller batch-sizes. The latter stages are separated from the former ones by a buffer with partially processed hoses. This operation is far from being a synchronized assembly line, therefore, under the just-in-time philosophy it requires a new production smoothing method. Yavuz et al. [155] propose focusing on the assembly stage, the last stage of the process. This stage can be modeled as a single-machine batching and scheduling problem. The solution to this problem can then be spread through a pull just-in-time control mechanism to the two preceding stages. The focus of this approach is on the bottleneck of the system, that is the assembly stage, thus its other stages can be scheduled with relative ease.

The chapter proposes a new model and a new solution to the problem. First, the existing line of research adopts a two-phase solution methodology that first determines the batch sizes (the first phase) and then sequences those batches (the second

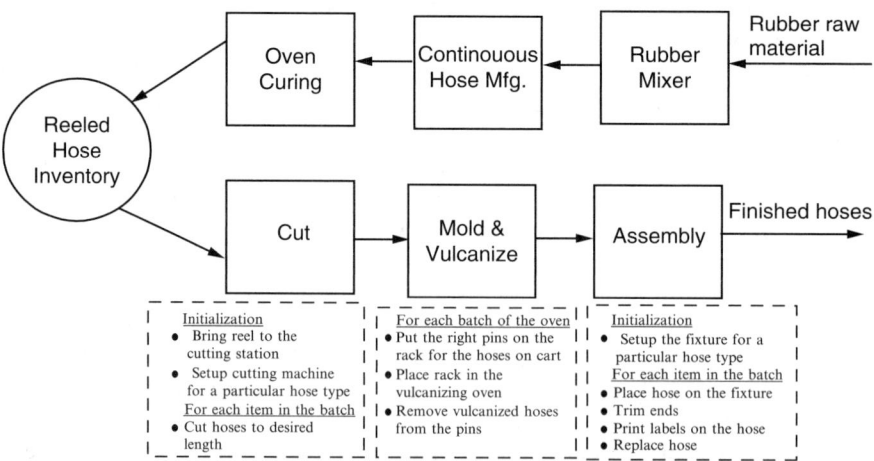

Fig. 11.1 Process flow at the automotive pressure hose manufacturing plant

phase). Since it solves the two phases sequentially and separately, the existing approach possibly yields a sub-optimal solution for the entire system. In contrast, this chapter attacks the problem as a whole. Both the model and the solution approach developed here broadly apply to any mixed-model just-in-time manufacturing system where the final production stage or the bottleneck stage of the system can practically reduce the entire system to a single machine. This reduction approach is commonly used in practice and theory of planning and scheduling of manufacturing and service systems, see Pinedo [156].

Second, it argues that the choice of objective function responsible for smoothing out production of models will always remain somewhat arbitrary since there does not seem to ever exist any criterion to choose one over another production smoothing objective function. More specifically, the model proposed in this chapter is based on four criteria: the minimization of the total production rate variation of the models and Work-In-Process, and the maximization of the system utilization and responsiveness. Among different ways of handling multi-objective optimization problems, see T'kindt and Billaut [157], we adopt Pareto-optimization. This optimization advocates keeping all the objectives at hand without any priorities or weights and seeking a set of optimal solutions instead of only one optimal solution being sought by the other approaches. The set of optimal solutions in this approach is called the set of *non-dominated* or *Pareto-optimal* solutions. A Pareto-optimal solution by definition can not be improved on any of its objective function values without worsening some other of these values. Our optimization approach in this chapter seeks to find Pareto-optimal solutions.

Third, in addition to the optimization approach for the minimization of the total production rate variation, we also propose an approach inspired by an axiomatic method used in solving the apportionment problem, see Chap. 2. The axiomatic approach looks for an algorithm that possesses certain desirable properties (or axioms), for instance periodicity and scalability, or proves that such an algorithm does not exist. Both the optimization method, the workhorse of the operations research approach, and the axiomatic method, novel yet practically absent from the solutions of problems in operations management are equally valid and practically useful for production smoothing. Thus, both will be exploited in this chapter. The axiomatic method used in this chapter is based on the Webster's method of apportionment. We show its desirable properties and its relationship to the optimization problem for which it can *also* be used as a very efficient heuristic.

11.3 Problem Definition

11.3.1 Preliminaries

Let n be the number of models to produce, and $i = 1, 2, \ldots, n$ be the models themselves. The demand for model i equals $d_i > 0$ and the sequence-independent setup and processing times for the model equal $s_i \geq 0$ and $\tau_i > 0$, respectively. The planning horizon is $T > 0$ time units long. All data are integers. Finally, let $D = \sum_{i=1}^{n} d_i$ be the total demand for all models over T.

In order to smooth out the natural variability present in the values of setup and processing times our approach first calls for dividing the time horizon T into Q *time-buckets* of *equal length* $t = \frac{T}{Q}$ and then filling in some of these time-buckets, possibly all, with models so that each non-empty time-bucket begins with a setup and it is filled in with copies of the same model. The non-empty time-buckets will be referred to as *batches*. The empty time-buckets will be referred to as empty batches or *blanks*. The number of blanks is denoted by $q_0 \geq 0$. Clearly, t must be sufficiently long to be able to fit a setup and at least a single copy of a model. Therefore, without loss of any generality, we assume $t \geq s_i + \tau_i$ for any $i = 1, 2, \ldots, n$. Consequently, t is feasible if and only if there exist positive integers, the numbers of batches for each model,

$$q_i = \left\lceil \frac{d_i}{\left\lfloor \frac{t-s_i}{\tau_i} \right\rfloor} \right\rceil, i = 1, \ldots, n, \tag{11.2}$$

such that

$$q_0 = Q - \sum_{i=1}^{n} q_i \geq 0. \tag{11.3}$$

The inequality (11.3) ensures that the total number of batches is sufficiently *small* so that it does not exceed the capacity of the system defined by T. At the same time, the integers (11.2) ensure that the number of batches of each model is sufficiently *large* so that the demand for each model is met exactly. To see this we observe that there always is an r_i such that

$$d_i = \left\lfloor \frac{d_i}{q_i} \right\rfloor q_i + r_i, \ i = 1, \ldots, n,$$

where $0 \leq r_i < q_i$. Therefore, filling in r_i batches with $B_i = \left\lceil \frac{d_i}{q_i} \right\rceil$ copies of model i each, and the remaining $q_i - r_i$ batches with $b_i = \left\lfloor \frac{d_i}{q_i} \right\rfloor$ copies of model i each results in a total of exactly

$$B_i r_i + b_i(q_i - r_i) = d_i$$

copies of model i produced. Notice that the average number of model i copies per batch is thus

$$\bar{\gamma}_i = \frac{d_i}{q_i},$$

for $i = 0, \ldots, n$. We assume $d_0 = 0$ for a fictitious model 0 making up blanks, and set $\bar{\gamma}_0 = 0$. Moreover, we have

$$t \geq s_i + B_i \tau_i, \ i = 1, \ldots, n,$$

which holds since by (11.2)

$$\frac{t - s_i}{\tau_i} \geq \left\lfloor \frac{t - s_i}{\tau_i} \right\rfloor \geq \frac{d_i}{q_i}.$$

Thus, t is sufficiently long to accommodate the B_i (or $b_i \leq B_i$) copies of model i, $i = 1, \ldots, n$. Finally, we observe that $t \geq s_i + \tau_i$ in (11.2) implies $q_i \leq d_i$ for all $i = 1, \ldots, n$. Hence, no excess of i will be ever produced. To summarize, the feasible integer Q (or equivalently $t = \frac{T}{Q}$) are those and only those that satisfy the following two inequalities

$$Q - \sum_{i=1}^{n} \left\lceil \frac{d_i}{\left\lfloor \frac{\frac{T}{Q} - s_i}{\tau_i} \right\rfloor} \right\rceil \geq 0, \tag{11.4}$$

$$\frac{T}{\max_i \{ s_i + \tau_i \}} \geq Q. \tag{11.5}$$

The solution to the set of inequalities will be found by an enumeration method shown in Algorithm 1. To our knowledge no other method for solving the nonlinear inequality (11.4) exists, and finding a more efficient method for its solution remains a challenging open problem.

Example 11.1. Consider an example with $n = 2$ models and with the following data: $d_1 = 50$, $d_2 = 80$, $s_1 = 4$, $s_2 = 2$, $\tau_1 = 1$, $\tau_2 = 3$ and $T = 450$. For $Q = 15$, we obtain $t = 450/15 = 30$, $q_1 = 2$, $q_2 = 9$, $b_1 = B_1 = 25$, $b_2 = 8$, $B_2 = 9$, $r_1 = 0$, and $r_2 = 8$. Thus, the demand for model 1 is met by two batches of 25 units each, and for model 2 by one batch of 8 units and eight batches of 9 units each. There are $q_0 = 4$ blank batches that make 120 units of time available. The $Q = 15$ batches can be sequenced in $\frac{15!}{4!2!9!} = 75,075$ different ways, two of which are demonstrated in Fig. 11.3a, b along with two sequences for $Q = 10$ depicted in Fig. 11.3c, d. □

Any feasible $t = \frac{T}{Q}$ thus defines a vector $q = (q_0, q_1, \ldots, q_n)$ specifying feasible number of batches for each model and the number of blanks. However, the sequence in which the Q batches are produced is crucial for smoothing out model production in just-in-time systems, Monden [22]. The selection of Q (or, equivalently t) and sequencing of the emerging batches will be Pareto-optimized using the following four objective functions.

11.3.2 Selection of the Objectives

- The first objective, Z_1, aims at minimizing the total variation in the production rates of the models. This production rate variation is often expressed by a *total deviation* of the actual cumulative model production levels from their ideal levels over the whole time horizon T. The total deviation has been widely used as the

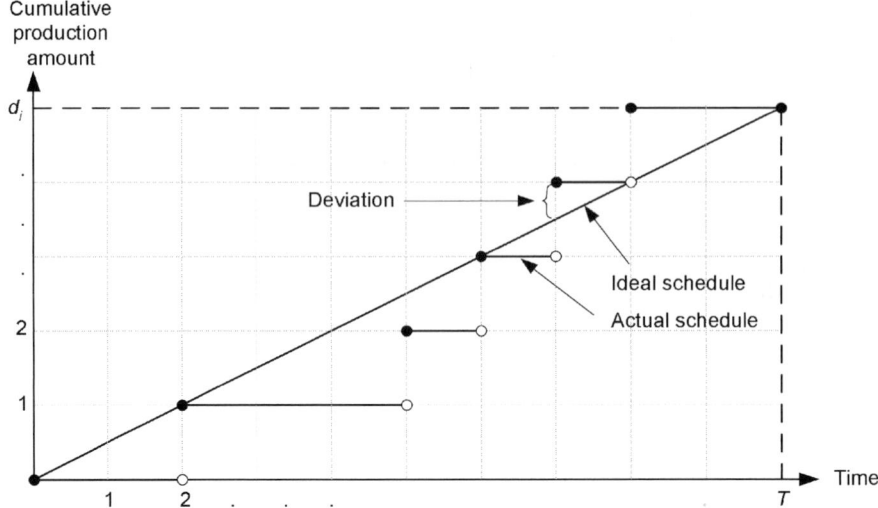

Fig. 11.2 The ideal and actual schedules

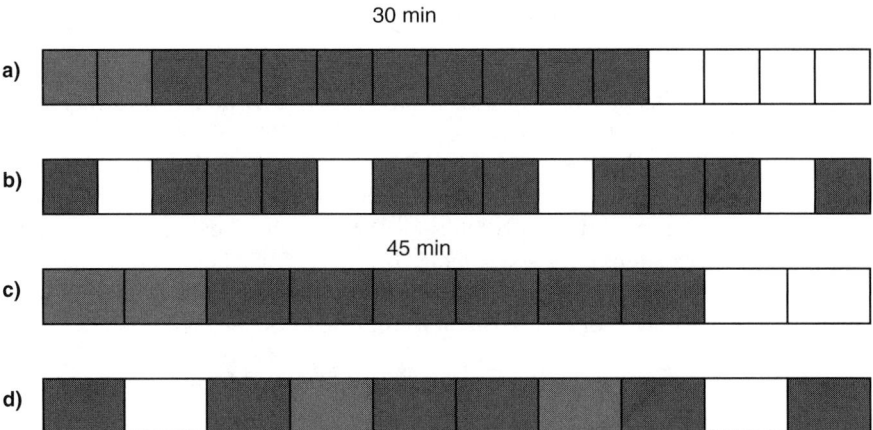

Fig. 11.3 Four alternative solutions in Example 11.1

primary objective in the production smoothing literature following Miltenburg [21] and Monden's [22] work, and it is illustrated in Fig. 11.2. Its application in just-in-time manufacturing systems relies on the assumption that demand for models is uniformly distributed over the planning horizon, and, hence, the goal of meeting the model demand with minimal inventories (or just-in-time) calls for distributing the production of each model as uniformly as possible over the planning horizon as well. Ideally, one would produce a model with a constant rate, thereby making the cumulative production of the model to progress along a straight cumulative demand line with its slope equal to the demand rate of the model, as seen in Fig. 11.2. The actual cumulative production level for the model, on the other hand, would progress along a stair-case like curve that includes the discrete values marked with a "•" but excludes the ones with a "○" in the figure. The actual production levels depend on the sequence in which batches of different models are produced as well as on their sizes. However, the batch sizes, that is b_i and B_i, of model i differ by at most one in our model, therefore it is a reasonable approximation to assume that all batches of the same model are of the same average size, $\overline{\gamma}_i$ for $i = 1,\ldots,n$. This assumption permits us to define the actual cumulative production level of model i as simply $\overline{\gamma}_i x_{i,k}$, where the actual cumulative number of model i batches in the first k time buckets is denoted by $x_{i,k}$ and it is sequence dependent. On the other hand the ideal number, perhaps unattainable, of time buckets allocated to model i out of the first k time buckets is the $\frac{q_i}{Q}$ fraction of k, see Fig. 11.2. Therefore, the ideal production level for model i is defined as $\overline{\gamma}_i \frac{q_i}{Q} k$. Consequently, the total squared deviation between the two production levels for model i over the Q time buckets equals $\sum_{k=1}^{Q} \overline{\gamma}_i^2 (x_{i,k} - k\frac{q_i}{Q})^2$ and the total deviation for all models equals

$$Z_1 = \sum_{k=1}^{Q} \sum_{i=0}^{n} \overline{\gamma}_i^2 \left(x_{i,k} - k\frac{q_i}{Q} \right)^2$$

over the time horizon T. We choose the total squared deviation in this chapter since it is a standard measure of deviation in the well-known least squares method; it has also been suggested as a measure of deviation by Miltenburg [21] and Moden [22].

- The second objective, Z_2, aims at minimizing the *total time lost*. The variability in setup and processing times results in time-buckets being filled in to various degree. This varying degree of the time-bucket fill-up is the cost paid for achieving a smoother production schedule. Therefore, there will be unavoidably some time lost in each batch. Moreover, the setup time itself is in fact time lost. However, meeting all demand does not necessarily mean filling in the whole time horizon T with batches. It would be unreasonable to advocate that being busy all the time is preferable to having time to spare sometime, see Bertrand Russell [158] on general advantages of idleness, which holds especially true for the just-in-time systems, Monden [22]. A just-in-time system does not need to be run at all times, on the contrary, due to its pull control mechanism the system can remain idle awaiting the best moment to produce in order to meet demand. The model i uses $d_i\tau_i$ time units productively out of q_it time in all batches with model i. Therefore, the *total time lost* is

$$\sum_{i=1}^{n} (tq_i - d_i\tau_i) = t\sum_{i=1}^{n} q_i - \sum_{i=1}^{n} d_i\tau_i$$

$$= \frac{T}{Q}(Q - q_0) - \sum_{i=1}^{n} d_i\tau_i = T\left(1 - \frac{q_0}{Q}\right) - \sum_{i=1}^{n} d_i\tau_i$$

which is minimized by maximizing the ratio

$$Z_2 = \frac{q_0}{Q}.$$

The q_0 blanks are not considered the time lost as they can potentially be used productively, if need be, for instance they can be filled in with any model $i = 1, \ldots, n$ should there be an increase in demand for the model.

- The third objective, Z_3, aims at improving the system responsiveness by minimizing the average time between blanks. The average time referred to as the *average response time* equals

$$\frac{Q}{q_0} t = \frac{Q}{q_0} \frac{T}{Q} = \frac{T}{q_0}$$

which is minimized by maximizing

$$Z_3 = q_0.$$

The blanks can be used to perform maintenance or other tasks aimed at improving the system performance, or they can be filled in with some unexpected orders, or they can provide spare capacity should quality problems occur in batches. In few words, the blanks provide a desired flexibility.

- The fourth objective, Z_4, is to minimize the Work-In-Process (WIP) inventory and it is based on the Little's Law (see Hopp and Spearman [159, p. 223]) that ties up the flow time t, the average flow rate $\sum_{i=1}^{n} \overline{\gamma}_i \frac{q_i}{Q}$ per t and the WIP as follows

$$WIP = t \left(\sum_{i=1}^{n} \overline{\gamma}_i \frac{q_i}{Q} \right) / t = \frac{D}{Q}.$$

Thus, the minimization of WIP is equivalent to the maximization of

$$Z_4 = Q.$$

In Fig. 11.3 the sequences (a) and (c) group the batches of the same model together, whereas the sequences (b) and (d) disperse the batches over the planning horizon in a more uniform fashion. The objective Z_1 favors sequences (b) and (d) over (a) and (c). The Z_2 favors (a) and (b) with the value 4/15 over (c) and (d) with the value 2/10. Finally, the Z_3 and Z_4, favor (a) and (b) with the values 4 and 15 of Z_3 and Z_4 respectively over (c) and (d) with the values 2 and 10. Thus, sequence (b) in this example is optimal according to all four objectives.

11.3.3 A Mathematical Programming Formulation

We now formulate the batching and smoothing problem as a multi-objective, non-linear mathematical programming model, as follows.

$$\text{minimize} \quad Z_1 = \sum_{k=1}^{Q} \sum_{i=0}^{n} \left(\frac{d_i}{q_i}\right)^2 \left(x_{i,k} - \frac{kq_i}{Q}\right)^2 \tag{11.6}$$

$$\text{maximize} \quad Z_2 = \frac{q_0}{Q} \tag{11.7}$$

$$\text{maximize} \quad Z_3 = q_0 \tag{11.8}$$

$$\text{maximize} \quad Z_4 = Q \tag{11.9}$$

subject to

$$x_{i,k} - x_{i,k-1} \in \{0,1\}, \qquad i = 0,1,\ldots,n; \quad k = 1,2,\ldots,Q \tag{11.10}$$

$$x_{i,0} = 0, \qquad i = 0,1,\ldots,n \tag{11.11}$$

$$x_{i,Q} = q_i, \qquad i = 0,1,\ldots,n \tag{11.12}$$

$$q_i = \left\lceil \frac{d_i}{\left\lfloor \frac{\frac{T}{Q}-s_i}{\tau_i} \right\rfloor} \right\rceil, \qquad i = 1,2,\ldots,n \tag{11.13}$$

$$Q - \sum_{i=1}^{n} q_i \geq 0, \tag{11.14}$$

$$q_0 = Q - \sum_{i=1}^{n} q_i, \tag{11.15}$$

$$\frac{T}{\max_i \{s_i + \tau_i\}} \geq Q, \tag{11.16}$$

$$Q \in \mathbb{Z}^+. \tag{11.17}$$

Constraints (11.10–11.12) guarantee that each model is assigned as many time-buckets as the number q_i of batches for that model as well as they define a unique sequence of batches that minimizes Z_1 for any feasible vector $q = (q_0, q_1, \ldots, q_n)$. Constraints (11.13–11.17) guarantee feasibility of the vector $q = (q_0, q_1, \ldots, q_n)$.

11.4 Pareto Optimization

Our solution to the model of Sect. 11.3.3 is presented in Algorithm 1. Step 1 finds the largest possible Q value, Q_U, for which a feasible solution may still exist. This value is determined by the constraint (11.16). Then, steps 2–4 simply enumerate all possible Q, $Q = n, \ldots, Q_U$. Step 3 calculates for each such Q the minimum number of batches q_i necessary to meet demand d_i for each model i. Then, Step 4 checks if the current Q is sufficiently large to accommodate the q_is.

Algorithm 1 Pseudo-code for Algorithm E

1: $Q_U = \left\lfloor \frac{T}{\max_i(s_i + \tau_i)} \right\rfloor$;

2: **for all** Q such that $n \le Q \le Q_U$ **do**

3: $\quad t = \frac{T}{Q}$;

4: $\quad q_i = \left\lceil \frac{d_i}{\left\lfloor \frac{t - s_i}{\tau_i} \right\rfloor} \right\rceil$, $\quad i = 1, \ldots, n$;

5: $\quad$ **if** $q_0 = Q - \sum_{i=1}^{n} q_i \ge 0$ **then**

6: $\quad\quad$ sequence batches using the algorithm for total deviation minimization given in Chap. 3

7: $\quad$ **end if**

8: **end for**

The first four steps thus find all feasible Q values and their corresponding feasible vectors $\mathfrak{q} = (q_0, q_1, \ldots, q_n)$ and they do it in $O(nD)$ time. Observe that the first four steps reveal all possible values of Z_2, Z_3, and Z_4, as well. For each feasible $\mathfrak{q} = (q_0, q_1, \ldots, q_n)$, Step 5 finds a sequence that minimizes Z_1, by transforming the minimization of Z_1 for a given $\mathfrak{q} = (q_0, q_1, \ldots, q_n)$ to the assignment problem, see Chap. 3 for the solution to the total deviation problem. The transformation defines an *ideal position*, see (3.11), for each batch of each model and a cost function which increases as a batch deviates from its ideal position. The details of this transformation are as follows. Define P_j^{i*} to be the ideal position of jth batch of model i

$$P_j^{i*} = \left\lceil \frac{(2j-1)Q}{2q_i} \right\rceil. \tag{11.18}$$

Define $C_{j,k}^i$ to be the cost of assigning jth batch of model i to the kth position as follows

$$C_{j,k}^i = \begin{cases} \overline{\gamma}_i^2 \sum\limits_{l=k}^{P_j^{i*}-1} \Psi_{j,l}^i, & \text{if } k < P_j^{i*} \\ 0, & \text{if } k = P_j^{i*} \\ \overline{\gamma}_i^2 \sum\limits_{l=P_j^{i*}}^{k-1} \Psi_{j,l}^i, & \text{if } k > P_j^{i*} \end{cases}$$

$$\text{where} \quad \Psi_{j,l}^i = \left| \left(j - l\frac{q_i}{Q} \right)^2 - \left(j - 1 - l\frac{q_i}{Q} \right)^2 \right|.$$

The cost $\Psi_{j,l}^i$ represents the excess cost, either excess inventory ($l < P_j^{i*}$) or excess shortage ($l \ge P_j^{i*}$) costs of having j batches of model i produced by period (position) l *instead of* having $j-1$ batches of i by l. Consequently, if the jth batch of model i is produced too *early*, that is $k < P_j^{i*}$, then the excess inventory costs Ψ_{jl}^i are incurred in periods $l = k, \ldots, P_j^{i*} - 1$, adding up to C_{jk}^i. On the other hand if the jth batch of i is produced too *late*, that is $k > P_j^{i*}$, then the excess shortage costs Ψ_{jl}^i are incurred in periods $l = P_j^{i*}, \ldots, k-1$, again adding up to C_{jk}^i. Finally, if the jth batch of i is produced in its ideal position P_j^{i*} then no excess costs are incurred and thus $C_{jk}^i = 0$.

Now, let $Y_{j,k}^i \in \{0,1\}$ be the decision variable equal 1 if the jth batch of model i is assigned to the kth position of the sequence. The assignment problem formulation of the Z_1 minimization is then given by, see Chap. 3:

$$(AP) \quad Minimize \quad F = \sum_{i=0}^{n} \sum_{j=1}^{q_i} \sum_{k=1}^{Q} C_{j,k}^i Y_{j,k}^i \tag{11.19}$$

subject to

$$\sum_{k=1}^{Q} Y_{j,k}^i = 1 \quad i = 0, 1, \ldots, n; \ j = 1, 2 \ldots, q_i \tag{11.20}$$

$$\sum_{i=0}^{n} \sum_{j=1}^{q_i} Y_{j,k}^i = 1 \quad k = 1, 2, \ldots, Q \tag{11.21}$$

$$Y_{j,k}^i \in \{0,1\} \tag{11.22}$$

Constraints (11.20) ensure that each batch of each model is assigned to exactly one position. Similarly, constraints (11.21) ensure that each position is occupied by exactly one batch or blank. Note that $\bar{\gamma}_0 = 0$, thus the selection of positions for blanks does not affect the objective function. This assignment problem (AP) with $2Q$ nodes can be solved in $O(Q^3)$ time using the well-known Hungarian method or any other optimization algorithm for the assignment problem, Kuhn et al. [160, 161], see also Burkard et al. [25] for a comprehensive review of assignment algorithms. We shall refer to the Algorithm E with the optimal solution to the assignment problem in Step 5 as the optimization method. Therefore, all Pareto-optimal solutions can be found in $O(D(n + Q^3)) = O(DQ^3)$ time, since $Q^3 >> n$. Moreover, since $Q \leq D$, the complexity is $O(D^4)$.

11.5 Axiomatic Approach

This section presents an axiomatic approach to the product rate variation. We give an algorithm that always produces sequences having certain desirable properties (axioms). Furthermore, we conjecture that there is no other algorithm that ensures meeting these properties all at the same time. Thus, we conjecture that the algorithm is in fact unique. We begin with the definition of the properties followed by the definition of the method and the proofs of its properties.

We propose that the sequence (or sequences, given that the ties can be broken in many different ways) should be

1. *Periodic.* A method is periodic if it produces a sequence S^m whenever applied to a batch vector $mq = (mq_0, mq_1, \ldots, mq_n)$, for any integer $m \geq 1$, where S is the sequence produced by the method for $q = (q_0, q_1, \ldots, q_n)$. Thus, solution to a smaller size instance can be repeated sufficiently many times to obtain a solution for a larger instance – clearly a desirable property in practice and also one that is closely related to the well-known concept of the periodic scheduling of the *minimum part set*, see Pinedo [156], often used in manufacturing.

2. *Uniform or scalable.* Let $M \subseteq \{0, 1, \ldots, n\}$ be a subset of models and $\overline{M}$ be its complement. Also, for a given $q = (q_0, q_1, \ldots, q_n)$ vector, let $q(\overline{M})$ be the vector obtained from q by deleting coordinates corresponding to the models in M, let S be the sequence produced by a method for q, and let $S(\overline{M})$ be the sequence obtained by deleting from S all models in M.

   A method is uniform if it produces the sequence $S(\overline{M})$ for $q(\overline{M})$. Moreover, if $S'(\overline{M})$ is another sequence produced by the method for $q(\overline{M})$, then the sequence S' obtained from S by replacing the sequence of models in $\overline{M}$ by $S'(\overline{M})$ is also produced by the method for vector q. Therefore, a cancellation of all orders for any model in a uniform sequence leaves the remaining sequence uniform, thus there is no need to re-run the method on a reduced vector $q(\overline{M})$ – again a clear practical advantage for operations management.

3. *Reflective.* The method is reflective if it produces *quasi-palindromes*. This has an important practical consequence for production planning since it ensures that the second half of any production plan designed according to the reflective method looks like its first half if read backwards. The reflective methods may create a desirable *déjà vu* effect helping the learning process.

Example 11.2. Consider the following instance $q_1 = 7, q_2 = 5, q_3 = 4$. The reflective method could produce the following sequence

$$1231213123121321$$

which is the same whether one reads it forwards or backwards except for the middle subsequence 1–2. The middle sub-sequences consist of exactly one batch of each model with an odd number of batches to produce, they can be sequenced essentially in any order which depends on a tie breaking rule. □

We now present the method itself. The method that we claim is periodic, uniform and reflective consists in sorting batches of models in non-decreasing order of

$$\frac{(2j-1)Q}{2q_i},\tag{11.23}$$

$j = 1, \ldots, q_i$ and $i = 0, 1, \ldots, n$, for a given vector $q = (q_0, q_1, \ldots, q_n)$ and then allocating batches to the Q positions according to this order. We refer to this algorithm as the Earliest Ideal Position Method or EIP Method (Algorithm 2). We later prove in Sect. 11.6 that this algorithm is equivalent to the Webster's method of apportionment.

Algorithm 2 EIP Method

1: $k = 1$;
2: $j_i = 1$, $i = 0, 1, \ldots, n$;
3: **while** $k \leq Q$ **do**
4:     $i^* = \arg\min_i \{ \frac{(2j_i - 1)Q}{2q_i} \}$;
5:     **if** i^* is not unique **then**
6:         break the tie as follows: If the current position k is (not) earlier than the common ideal vertex, then assign the model with the smallest (largest) assignment weight $\frac{d_i}{q_i}$ first;
7:     **end if**
8: **end while**

Step 3 describes a tie breaking rule used by the EIP method in our computational experiments. However, no particular tie-breaking rule will be assumed in our discussion of the EIP in this section and Sect. 11.6, where the ties are assumed to be broken arbitrarily.

The following lemmas show, see also Sect. 2.8, that the EIP method has all three desired properties listed at the beginning of this section.

Lemma 11.3. *The EIP is periodic.*

Proof. Follows immediately from the fact that

$$\frac{(2j' - 1)Q'}{2q_i'} = kQ + \frac{(2j - 1)Q}{2q_i}$$

for $Q' = mQ$, $q_i' = mq_i$, $j' = kq_i + j$ and $j = 1, \ldots, q_i$. □

Lemma 11.4. *The EIP is uniform.*

Proof. It suffices to notice that the EIP orders the batches in ascending order of

$$\frac{(2j - 1)Q}{2q_i}$$

$j = 1, \ldots, q_i$ and $i = 0, 1, \ldots, n$, where Q is fixed for a given vector q. Thus, the EIP in fact orders the batches in ascending order of

$$\frac{2j - 1}{2q_i}.$$

This proves the lemma since the coordinates of q are the same as $q(\overline{M})$ for all models in $\overline{M}$. □

Lemma 11.5. *The EIP produces sequences of the following form SMS^R, where M is any permutation of batches of the models with odd qs, and S^R is a mirror reflection of S.*

Proof. It suffices to observe that if $\frac{2j-1}{q_i} \leq \frac{2l-1}{q_k}$, then $\frac{2(q_i+1-j)-1}{q_i} \geq \frac{2(q_k+1-l)-1}{q_k}$. Therefore, if copy j of model i precedes copy l of model k, then copy $q_k + 1 - l$ of model k precedes copy $q_i + 1 - j$ of model i. Notice that all models with odd $q's$ have their ideal position for the middle batch equal to $\frac{Q}{2}$. These models will make the middle subsequence M, where a single batch of each of these models will occur.   □

Lemmas 11.3–11.5 prove that the EIP method has the qualities listed at the beginning of this section, thus, proving the EIP method's clear advantages for operations management. By comparison, the optimization method has only some of the three desired properties. Namely, Chap. 4 shows that periodic schedules are optimal for a generalized objective function $\sum_k \sum_i F_i(\cdot)$, where $F_i(\cdot)$ is a convex and symmetric function with $F_i(0) = 0$. The objective function Z_1 falls clearly into this category. Thus, periodic sequences are optimal for Z_1. However, the following lemma shows that the optimization method is not uniform in general, see also Exercise 11.12.

Lemma 11.6. *The optimization method is not uniform.*

Proof. Consider vector $q = (7,6,4,3,1)$ for an instance $d = (7,6,4,3,1)$. The following sequence

$$12312413215231421321 \tag{11.24}$$

is optimal for q and d with the value of total absolute deviation equal to 9.55. On the other hand, the sequence

$$2324325234232$$

obtained by deleting model 1 from (11.24) is not optimal for $q' = (6,4,3,1)$ and $d' = (6,4,3,1)$ since its total absolute deviation equals 4.923, whereas the optimal solution for q' and d' is

$$2342325232432$$

with the total absolute deviation of 4.615 units.   □

Finally, it remains open whether the optimization method is reflective. However, we can prove the following weaker property.

Lemma 11.7. *If solution S minimizes Z_1, then so does its mirror reflection S^R.*

Proof. By definition

$$S_{Q+1-k} = S_k^R \text{ for } k = 1,\dots,Q.$$

Thus, denoting the number of batches sequenced in the first k stages of S and S^R by $x_{i,k}$ and $y_{i,k}$, respectively; we have

$$y_{i,k} = q_i - x_{i,Q-k} \text{ for all } i \text{ and } k = 1,\dots,Q,$$

where $x_{i,0} = y_{i,0} = 0$ for all i. Let $k' = Q + 1 - k$, $k = 1,\dots,Q$ be the one-to-one correspondence between k' and k. Then,

$$\left(\frac{d_i}{q_i}\right)^2 \left(y_{i,k'} - k'\frac{q_i}{Q}\right)^2 = \left(\frac{d_i}{q_i}\right)^2 \left(y_{i,Q+1-k} - (Q+1-k)\frac{q_i}{Q}\right)^2$$

$$= \left(\frac{d_i}{q_i}\right)^2 \left(q_i - x_{i,Q-(Q+1-k)} - q_i + (k-1)\frac{q_i}{Q}\right)^2$$

$$= \left(\frac{d_i}{q_i}\right)^2 \left(x_{i,k-1} - (k-1)\frac{q_i}{Q}\right)^2.$$

Since $\left(\frac{d_i}{q_i}\right)^2 (x_{i,Q} - Q\frac{q_i}{Q})^2 = 0$ for all i, then

$$\sum_{k=1}^{Q}\sum_{i=0}^{n}\left(\frac{d_i}{q_i}\right)^2 \left(y_{i,k} - \frac{kq_i}{Q}\right)^2 = \sum_{k'=1}^{Q}\sum_{i=0}^{n}\left(\frac{d_i}{q_i}\right)^2 \left(y_{i,k'} - \frac{k'q_i}{Q}\right)^2$$

$$= \sum_{k=1}^{Q}\sum_{i=0}^{n}\left(\frac{d_i}{q_i}\right)^2 \left(x_{i,k} - \frac{kq_i}{Q}\right)^2,$$

which proves the lemma. $\square$

11.6 EIP Method and Optimization

This section will have a closer look at the EIP method as an optimization tool. We begin with the following lemma.

Lemma 11.8. *The EIP method minimizes the following objective*

$$E_0(x) = \sum_{k=1}^{Q}\sum_{i=0}^{n}\frac{1}{q_i}\left(x_{i,k} - k\frac{q_i}{Q}\right)^2,$$

where we assume $\frac{1}{q_0} = 0$ for $q_0 = 0$.

Proof. The EIP method orders the batches in ascending order of

$$\frac{(2j-1)Q}{2q_i}$$

$j = 1,\ldots,q_i$ and $i = 0,1,\ldots,n$, and then assigns the positions in the sequence according to this order. Equivalently, the model ℓ gets position $k+1, k = 0,\ldots,Q-1$ if

$$\frac{q_\ell}{x_{\ell,k} + \frac{1}{2}} \geq \frac{q_i}{x_{i,k} + \frac{1}{2}}$$

for all i, where $x_{i,k}$ is the number of batches of model i in the first k positions. This shows that the EIP method is the Webster's method of apportionment, Balinski and Young [2]. Let us consider a fixed position $k = 1,\ldots,Q$, then the vector $(x_{1,k},\ldots,x_{n,k})$ minimizes

$$\sum_{i=0}^{n} \frac{1}{q_i} \left(x_{i,k} - k\frac{q_i}{Q} \right)^2,$$

see Balinski and Shahidi [16]. Moreover, the EIP method ensures that $x_{i,k+1} \geq x_{i,k}$ for all $i = 0, 1, \ldots, n$ and $k = 0, \ldots, Q-1$. Therefore, the EIP method minimizes

$$\sum_{k=1}^{Q} \sum_{i=0}^{n} \frac{1}{q_i} \left(x_{i,k} - k\frac{q_i}{Q} \right)^2.$$

$\square$

The original objective Z_1 and E_0 are clearly different, the former uses the squared average model i batch size $(\frac{d_i}{q_i})^2$ as the weight for model i deviations, the latter uses just $\frac{1}{q_i}$ instead. (There will be no difference from the optimization stand point if it used $\frac{(\bar{d})^2}{q_i\bar{q}}$, where $\bar{d} = \frac{D}{n}$ and $\bar{q} = \frac{Q}{n+1}$, instead of $\frac{1}{q_i}$ since $\bar{d}$ is constant for any instance and $\bar{q}$ is constant for any vector q. Consequently, if the standard deviations of demands $d_1, \ldots, d_n$ and batch numbers $q_1, \ldots, q_n$ are negligible, then Z_1 and E_0 could become practically equal.)

The EIP method, being the Webster's method of apportionment, as shown in Lemma 11.8 is *near quota*, Balinski and Young [2], that is at any position k no model i can be brought closer to its ideal level $k\frac{q_i}{Q}$ without bringing another model j further from its ideal level $k\frac{q_j}{Q}$. On the other hand, the assignment method used in Step 5 of Algorithm E shows how to do the exchanges between the models optimally so as to minimize the total deviation. This explains why the EIP method often produces production sequences with the total deviation higher than the optimum in our experiments. The optimization method, however, causes the production sequences to lose some of their desirable characteristics as we have shown in Lemma 11.6.

Although the EIP is near the quota it does not actually *meet the quota*, that is it does not satisfy $\lfloor k\frac{q_i}{Q} \rfloor \leq x_{i,k} \leq \lceil k\frac{q_i}{Q} \rceil$ for all k and i. Interestingly, the optimization method does not satisfy these inequalities either. However, the quota violations by either the optimization or EIP methods are rare. Thus both methods will meet the quota in practice quite often, see the experimental evidence in Balinski and Young [2], Kubiak et al. [39], Corominas and Moreno [38], and Lebacque et al. [40]

11.7 Computational Experiment

11.7.1 Experimental Design

We consider small, medium and large-size instances with $n = 5$, 10, and 15 models, respectively, in our computational experiment. The small-size instances are weekly ($T = 2{,}250$ min) with total demand of $D = 1{,}000$ units. The medium-size instances are monthly ($T = 9{,}900$ min) with total demand of $D = 3{,}000$ units; and the large-size instances are quarterly ($T = 39{,}150$ min) with total demand of $D = 5{,}000$ units.

For each problem size, we randomly create test instances based on two parameters: α and β. The α is the expected setup to processing time ratio ($0.9 \times \alpha \le \frac{s_i}{\tau_i} \le 1.1 \times \alpha$, $i = 1, 2, \ldots, n$) and β is the ratio of total available time T to minimum required time for setups and processing of the models to meet the demands ($\beta = T / (\sum_{i=1}^{n} s_i + d_i \tau_i)$). We let $\alpha \in \{50, 10, 5\}$ and $\beta \in \{2, 1.75, 1.5\}$, which results in nine problem sets for each problem size. We then create ten test instances for each set. Therefore, we have 90 test instances for each problem size. The instance generator ensures that a feasible solution exists for every test instance. Our computational experiment relies on synthetic data generated over a wide spectrum. This allows us for broader representation of possible scenarios observed in practice and more thorough performance tests of the algorithms.

11.7.2 Methods

We experiment with the Algorithm E that produces the list of Pareto-optimal solutions. The Hungarian method and the EIP method are used in Step 5 of Algorithm E to solve the resulting assignment problem. All the methods and the random instance generator are coded in Microsoft Visual C#. NET language and run on a PC with an Intel Pentium 4 3.4 GHz CPU and 2 GB of memory.

11.7.3 Results

We summarize the average and maximum run times taken by both methods in Table 11.1. The table shows that the optimization method takes approximately 23 min on average and it can take up to 3 h and 40 min to solve large-sized instances. This time, however, is just below 0.6% of the duration of the time horizon T for which it finds an optimum batching and scheduling solution. The EIP method takes approximately 21 s in the worst case. Thus the EIP method runs much faster than the optimization method. The table also shows the average cardinality of the set of Pareto-optimal (P-opt) solutions for each of the two methods.

Table 11.1 Solution times and the numbers of Pareto-optimal solutions

n	Method	Solution time (s)		Avg. # of	Average relative (%)		
		Avg.	Max.	P-opt sol.	Overall	Void	Extra
5	Opt	2.280	22.422	21.0			
	EIP	0.019	0.094	18.7	92.8	11.9	5.5
10	Op	147.224	1,380.560	65.8			
	EIP	0.378	1.797	53.6	86.7	15.0	2.0
15	Opt	1,360.348	13,158.300	112.2			
	EIP	2.240	20.750	98.9	92.0	10.1	2.3

The summary of differences between the sets of Pareto-optimal solutions found by the optimization method and the EIP method are given in the last three columns of Table 11.1. The first column is the average percentage of the number of Pareto-optimal solutions found by the EIP method to that of the optimization method. This percentage is around 90 for all problem sizes. The second column is the average percentage of Pareto-optimal solutions found by the optimization method but missed by the EIP method. We call these solutions *void* solutions and report the percentage of the void solutions in the Pareto-optimal set of solutions found by the optimization method. The third column is the average percentage of *mistaken* Pareto-optimal solutions. That is the solutions identified as Pareto-optimal by the EIP method but not by the optimization method. We call these solutions *extra* and report the percentage of the extra solutions in the Pareto-optimal set of solutions found by the EIP method. The results show that the extra solutions are under 6%, and the void solutions are under 15%.

The optimization and EIP methods differ in the way they solve the batch sequencing problem for a given vector $q = (q_0, q_1, \ldots, q_n)$. However, the objectives Z_2, Z_3, and Z_4 depend solely on Q and q_0 but not of the sequence of batches. Thus, they are not affected by either the actual sizes of batches $q_1, \ldots, q_n$ or the method employed in batch sequencing. Since Algorithm E enumerates all possible Q and q_0 values neither method will miss the best possible values of Z_2, Z_3, and Z_4. Furthermore, our analysis of the Pareto-optimal sets obtained by the two methods in the experiment reveals that the Q value that yields the smallest possible value of Z_1 is always included in the Pareto-optimal set obtained the EIP method. These observations indicate that the extreme solutions of the Pareto-optimal set are never missed by the EIP approach. Furthermore, our experimental observations indicate that the Pareto-optimal solutions missed by the EIP method are some intermediate solutions whose absence do not affect the diversity of the Pareto-optimal set. Consequently, we claim that the EIP is an excellent heuristic for Z_1 in addition to having very desirable properties for operations management.

Finally, the results for the objectives Z_2, Z_3, and Z_4 are reported in Table 11.2. The objective Z_2 represents the percentage of time lost, which is under 6% for both methods in our experiment. The time lost is a major issue in operating mixed-model manufacturing systems, a recent study of mixed-model assembly lines by Mendes

Table 11.2 Summary of average lost and response time as well as WIP

n	Method	Average percentage of		
		Lost time	Response time	WIP
5	Optimization	5.99	16.49	3.96
	EIP	5.93	16.77	3.96
10	Optimization	5.98	15.45	1.54
	EIP	5.91	15.99	1.56
15	Optimization	5.76	14.98	0.93
	EIP	5.72	15.20	0.92

et al. [162] reports the average station utilization around 70%. This comparison demonstrates that our approach is rather effective in machine utilization as well.

The Z_3 objective shows the responsiveness of the system. We observe that the average response time is approximately 15% of the planning horizon for both methods in our experiments. This result shows that an average schedule contains approximately six blanks dispersed over the horizon. From a practical perspective, these blanks introduce flexibility into operations which can be used to respond to possible demand increases or quality problems while maintaining a smooth production.

The minimization of WIP level, i.e., Z_4, is also used in our model. The results in Table 11.2 show that both methods yield approximately the same average WIP levels. The relative WIP is under 4%, which is rather low, furthermore it decreases with the size of the problem.

Note that, the percentages of Table 11.2 are found using all Pareto-optimal solutions, a careful selection of the Pareto-optimal solution that will be implemented may lead to even further improvement of the performance measures in practice.

11.8 Exercises

Exercise 11.9. Show that there is Heijunka sequence with the difference of no more that 4 for each model and time interval.

Exercise 11.10. A subsequence of at least two batches of the same model i, $i = 1,\ldots,n$, is referred to as a *cluster*. A sequence of batches with clusters is refereed to as a *clustered* sequence. Obviously, each cluster needs only a single setup for all batches in the cluster, that is all other setups in the cluster's batches can be removed. Without loss of generality, let us assume that $q_1 \geq q_2 \geq \cdots \geq q_n$. Show that the EIP can produce clustered sequences even if $q_1 \leq \frac{Q}{2}$, this happens for the following instance $q_1 = 9, q_2 = 5, q_3 = 4$.

Exercise 11.11. Prove that if $q_1 - q_2 \leq 1$, then no cluster is produced by the EIP.

Exercise 11.12. Lemma 11.6 gives a counterexample to prove that the optimization method is not uniform. The counterexample is based on the total absolute deviation. Find a counterexample based on Z_1.

Exercise 11.13. Find an example showing that the optimization method is not reflective.

Exercise 11.14. Consider a production smoothing problem with the objective to minimize a weighted sum of the number of setups and Z_1. What is the computational complexity of the problem. Hint: This is an open problem.

Exercise 11.15. Show the following upper bound on Z_1: $Z_1(x^*) \leq \frac{q^2}{Q} \sum_{i=1}^{n} (\frac{d_i}{q_i})^2$ for some $q < Q$.

11.9 Comments and References

This chapter is based on Kubiak and Yavuz [163].

Kurashige et al. [164] address the problem of scheduling mixed-model assembly lines under differing assembly time requirements, and propose a modification of Toyota's goal chasing method Monden [22], namely the time-based goal-chasing method (TBGCM), for its solution.

Drexl and Kimms [115], and Drexl et al. [116] combine the production smoothing problem with a car sequencing problem where processing times of the models are embedded into hard constraints that prevent models with lengthy operations from being sequenced consecutively.

Aigbedo and Monden [165] allow different models to have different processing time requirements both at the final assembly line and supplying processes, and propose a heuristic procedure to solve the emerging multi-criteria problem. McMullen et al.'s [166] approach aims at minimizing the variability in the production sequence and the number of setups simultaneously, see McMullen [167], and McMullen et al. [166] for a weighted sum of the two, and McMullen [168] for a bi-criteria model. Although the incorporation of setups in the optimization model captures an important variability dimension of the production smoothing problem, the number of setups minimization does not reduce variability by itself. Moreover, these approaches ignore setup times and their variability, and thus they do not consider batching as a smoothing tool at all.

Yavuz et al. [155] study operations at a leading U.S. automotive pressure hose manufacturer Yavuz et al. [155], and Yavuz and Tufekci [169].

References

1. Gardels, N.: Lunch with the FT: He has seen the future. Financial Times **August 19/August 20** (2006) Weekend W3
2. Balinski, M., Young, H.: Fair Representation: Meeting the Ideal of One Man, One Vote. Yale University Press, New Haven and London (1982)
3. Young, H.P.: Equity. In Theory and Practice. Princeton University Press, Princeton, NJ (1994)
4. Balinski, M.: Le scrutin. Pour La Science **294** (2002) 46–51
5. Tijdeman, R.: Fraenkel's conjecture for six sequences. Discrete Mathematics **222** (2000) 223–234
6. Hajek, B.: Extremal splittings of point processes. Mathematics of Operations Research **10** (1985) 543–556
7. Altman, E., Gaujal, B., Hordijk, A.: Balanced sequences and optimal routing. Journal of the ACM **47** (2000) 752–775
8. Monden, Y.: Toyota Production System. Industrial Engineering and Management Press, Norcross, GA (1983)
9. Hopcroft, J., Ullman, J.: Introduction to Automata Theory, Languages and Computation. Addison-Wesley, Reading, MA (1979)
10. Graham, R., Knuth, D., Potashnik, O.: Concrete Mathematics, second edn. Addison-Wesley, Reading, MA (1994)
11. Waldspurger, C., Hogg, T., Huberman, B., Kephart, J., Stornetta, W.: Spawn: A distributed computational economy. IEEE Transactions on Software Engineering **18** (1992) 103–117
12. Nagle, J.: On packet switches with infinite storage. IEEE Transactions on Communications **Com-35** (1987) 435–438
13. Waldspurger, C., Weihl, W.: Stride scheduling: Deterministic proportional-share resource management. Technical report mit/lcs/tm-528, MIT Laboratory for Computer Science, Massachusetts Institute of Technology, Cambridge, MA (1995)
14. Balinski, M., Rachev, S.: Rounding proportions: Methods of rounding. Mathematical Scientist **22** (1997) 263–279
15. Balinski, M., Ramirez, V.: Parametric methods of apportionment, rounding and production. Mathematical Social Sciences **37** (1999) 107–122
16. Balinski, M., Shahidi, N.: A simple approach to the product rate variation problem via axiomatics. Operations Research Letters **22** (1998) 129–135
17. Leyvraz, J.P.: Le Probleme de larepartition proportionnelle. Ph.D. thesis, Ecole Polytechnique Federale de Lausanne, Lausanne, Switzerland (1977)
18. di Cortona, P.G., Manzi, C., Pennisi, A., Ricca, F., Simeone, B.: Evaluation and Optimization of Electoral Systems. SIAM Monographs on Discrete Mathematics and Its Applications, Philadelphia (1999)

19. Ibaraki, T., Katoh, N.: Resource Allocation Problems: Algorithmic Approaches. MIT, Cambridge, MA (1988)
20. Bautista, J., Companys, R., Corominas, A.: A note on the relation between the product rate variation (PRV) problem and the apportionment problem. Journal of the Operational Research Society **47** (1996) 1410–1414
21. Miltenburg, J.: Level schedules for mixed-model assembly lines in just-in-time production systems. Management Science **35** (1989) 192–207
22. Monden, Y.: Toyota Production System: An Integrated Approach to Just-In-Time, third edn. Engineering & Management Press, Norcross, GA (1998)
23. Kubiak, W.: Minimizing variation of production rates in just-in-time systems: A survey. European Journal of Operational Research **66** (1993) 259–271
24. Burkard, R., Hahn, W., Zimmermann, U.: An algebraic approach to assignment problems. Mathematical Programming **12** (1977) 318–327
25. Burkard, R., Dell'Amico, M., Martello, S.: Assignment Problems. SIAM Monographs on Discrete Mathematics and Its Applications, Philadelphia (2008)
26. Jozefowska, J., Jozefowski, L., Kubiak, W.: Characterization of Just in Time Sequencing via Apportionment. In: Yan, H., Yin, G., Zhang, Q., eds.: Stochastic Processes, Optimization, and Control Theory. Springer, Berlin (2006) 175–200
27. Kubiak, W., Sethi, S.: A note on "level schedules for mixed-model assembly lines in just-in-time production systems". Management Science **37** (1991) 121–122
28. Kubiak, W., Sethi, S.P.: Optimal just-in-time schedules for flexible transfer lines. The International Journal of Flexible Manufacturing Systems **6** (1994) 137–154
29. Bautista, J., Companys, R., Corominas, A.: Modelling and solving the product rate variation problem. TOP **5** (1997) 221–239
30. Steiner, G., Yeomans, J.: Optimal level schedules in mixed-model, multi-level jit assembly systems with pegging. European Journal of Operational Research **95** (1996) 38–52
31. Burkard, R., Klinz, B., Rudolf, R.: Perspectives of Monge properties in optimization. Discrete Applied Mathematics **70** (1996) 95–161
32. Inman, R., Bulfin, R.: Sequencing JIT mixed-model assembly lines. Management Science **37** (1991) 901–904
33. Grigoriev, A.: High Multiplicity Scheduling Problems. Ph.D. thesis, Maastricht University, Maastricht, the Netherlands (2003)
34. Bondy, J., Murty, U.: Graph Theory with Applications. Elsevier, Amsterdam (1976)
35. Rockafellar, R.T.: Convex Analysis. Princeton University Press, Princeton, NJ (1997)
36. Bautista, J., Companys, R., Corominas, A.: Note on cyclic sequences in the product variation problem. European Journal of Operational Research **124** (2000) 468–477
37. Kubiak, W.: Cyclic just-in-time sequences are optimal. Journal of Global Optimization **27** (2003) 333–347
38. Corominas, A., Moreno, N.: On the relations between optimal solutions for different types of min-sum balanced JIT optimisation algorithms. Information Processing and Operational Research **41** (2003) 333–339
39. Kovalyov, M., Kubiak, W., Yeomans, J.: A computational analysis of balanced JIT optimization algorithms. Information Processing and Operational Research **39** (2001) 299–316
40. Lebacque, V., Jost, V., Brauner, N.: Simulataneous optimization of classical objectives in jit scheduling. European Journal of Operational Research **182** (2007) 29–39
41. Steiner, G., Yeomans, S.: Level schedules for mixed-model, just-in-time processes. Management Science **39** (1993) 728–735
42. Brauner, N., Crama, Y.: The maximum deviation just-in-time scheduling problem. Discrete Applied Mathematics **134** (2004) 25–50
43. Tijdeman, R.: The chairman assignment problem. Discrete Mathematics **32** (1980) 323–330
44. Kubiak, W.: On small deviation conjecture. Bulletin of the Polish Academy of Sciences **51** (2003) 189–203
45. Gaujal, B.: Optimal allocation sequences of two processes sharing a resource. Discrete Event Dynamic Systems: Theory and Applications **7** (1997) 327–354

46. Vuillon, L.: Balanced words. Technical report 3, LIAFA CNRS, Universite Paris 7 (2003)
47. Gaujal, B., Jafari, M., Baykal-Gurso, M., Gulgun, A.: Allocation sequences of two processes sharing a resource. IEEE Trnsactions on Robotics and Automation **11** (1995) 748–753
48. Yu, W.: The two-machine flow shop problem with delays and the one-machine total tardiness prpblem. Ph.D. thesis, Eindhoven University of Technology, Eindhoven, the Netherlands (1996)
49. Murata, T.: Petri nets: Properties, analysis and applications. Proceedings of The IEEE **77** (1989) 541–580
50. Ramamoorthy, C., Ho, G.: Performance evaluation of asynchronous concurrent systems using petri nets. IEEE Trnsactions on Software Engineering **SE-6** (1980) 440–449
51. Glover, F.: Maximum matching in a convex bipartite graph. Naval Research Logistics Quarterly **4** (1967) 313–316
52. Lipski, W. Jr., Preparata, F.: Efficient algorithms for finding maximum matching in convex bipartite graphs and related problems. Acta Informatica **15** (1981) 329–346
53. Frederickson, G.: Scheduling unit-time tasks with integer release times and deadlines. Information Processing Letters **16** (1983) 171–173
54. Gallo, G.: An O(NlogN) algorithm for the convex bipartite matching problem. Operations Research Letters **3** (1984) 31–34
55. Gabow, H., Tarjan, R.: A linear-time algorithm for a special case of disjoint set union. Journal of Computer and System Science **30** (1985) 209–221
56. Meijer, H.: On a distribution problem in finite sets. Nederlands Akademie Wetenschappen Indagationes Mathematicae **35** (1973) 9–17
57. Beatty, S.: Problem 3173. American Mathematical Monthly **33** (1926) 159, Solutions, ibid. **34** (1927) 159.
58. Morikawa, R.: On eventually covering families generated by the bracket fuction. Bulletin of the Faculty of Liberal Arts (Nagasaki University), Natural Science **23** (1982) 17–22
59. Morikawa, R.: On eventually covering families generated by the bracket fuction. Bulletin of the Faculty of Liberal Arts (Nagasaki University), Natural Science **25** (1985)
60. Tijdeman, R.: Exact covers of balanced sequences and frankel's conjecture. In: Halter-Koch, F., Tichy, R., eds.: Algebraic Number Theory and Diophantine Analysis. Walter de Gruyter, Berlin (2000) 467–483
61. Simpson, R.: Disjoint covering systems of rational beatty sequences. Discrete Mathematics **92** (1991) 361–369
62. Newman, N.: Roots of unity and covering sets. Mathematics Annals **191** (1971) 279–282
63. Uspensky, J.: On a problem arising out of the theory of a certain game. American Mathematical Monthly **34** (1927) 516–521
64. Graham, R.L.: Covering the positive integers by disjoint sets of the form $[n\alpha + \beta]$. Journal of Combinatorial Theory, Series A **15** (1973) 354–358
65. Altman, E., Gaujal, B., Hordijk, A.: Multimodularity, convexity and optimization properties. Technical report 3181, RUL-TW-97-07, INRIA and Leiden University, the Netherlands (June 1997)
66. Ross, S.: Introduction to Probability Models. Academic, New York (2007)
67. Altman, E., Gaujal, B., Hordijk, A.: Discrete-Event Control of Stochastic Networks: Multimodularity and Regularity. Springer, Berlin (2003)
68. Brauner, N., Jost, V., Kubiak, W.: On Symmetric Fraenkel's and Small Deviations Conjectures. Les cahiers du Laboratoire Leibniz-IMAG, no 54, Grenoble, France (2002)
69. Wilf, H.: Generatingfunctionology, second edn. Academic, New York (1994)
70. Stolarsky, K.: Beatty sequences, continued fractions and certain shift operartors. Canadian Mathematical Bulletin **19** (1976) 473–482
71. Weisstein, E.W.: Beatty sequence. From MathWorld–A Wolfram Web Resource. http://mathworld.wolfram.com/BeattySequence.html (2007)
72. Lothaire, M.: Algebraic Combinatorics on Words. Cambridge University Press, Cambridge (2002)
73. Altman, E., Gaujal, B., Hordijk, A.: Admission control in stochastic event graphs. Technical report 3179, RUL-TW-97-06, INRIA and Leiden University, the Netherlands (June 1997)

74. Han, C., Lin, K., Hou, C.: Distance-constrained scheduling and its applications in real-time systems. IEEE Transactions on Computers **45** (1996) 814–826

75. Herrmann, J.W.: Generating cyclic fair sequences using aggregation and stride scheduling. Technical report, University of Maryland, College Park, MD (2007)

76. Wei, W., Liu, C.: On a periodic maintenance problem. Operations Research Letters **2** (1983) 90–93

77. Anily, S., Glass, C., Hassin, R.: The scheduling of maintenance service. Discrete Applied Mathematics **82** (1998) 27–42

78. Brauner, N., Crama, Y., Grigoriev, A., de Klundert, J.V.: On the complexity of high-multiplicity scheduling problems. Journal of Combinatorial Optimization **9** (2005) 313–323

79. Bar-Noy, A., Bhatia, R., Naor, J., Scheiber, B.: Minimizing service and operation costs of periodic scheduling. Mathematics of Operations Research **27** (2002) 518–544

80. LeVeque, W.J.: Topics in number theory. Dover, New York (2002)

81. Garey, M., Johnson, D.: Computers and Intractability: A Guide to the Theory of NP-Completeness. W. H. Freeman and Company, New York (1979)

82. Corominas, A., Kubiak, W., Moreno, N.: Response time variability. Journal of Scheduling **10** (2007) 97–110

83. Moreno, N.: Solving the product rate variation problem (PRVP) of large dimensions as an assignment problem. Ph.D. thesis, Department D'Organizacio D'Empreses, UPC, Barcelona (2002)

84. Wagner, H.: Principles of operations research-with applications to managerial decisions. Prentice-Hall, Englewood Cliffs, NJ (1969)

85. Corominas, A., Kubiak, W., Pastor, R.: Mathematical programming modelling of the response time variability problem. Working paper ioc-dt-p-2006-17, UPC (2006)

86. Giaro, K.: Private communication. (2005)

87. Liu, C., Layland, J.: Scheduling algorithm for multiprogramming in hard-real-time environment. Journal of ACM **20** (1973) 46–61

88. Cheng, A.: Real-time systems: scheduling, analysis, and verification. Wiley, New York (2005)

89. Buttazzo, G.: Predictable Scheduling Algorithms and Applications, Hard Real-Time Computing Systems. Kluwer, Dordrecht (1997)

90. Holte, R., Mok, A., Rosier, L., Tulchinsky, I., Varvel, D.: The pinwheel: A real-time scheduling problem. In: Proceedings of the 22nd Hawaii International Conference on System Science. (1989) 693–702

91. Baruah, S., Bestavros, A.: Timely and fault-tolerant data access from broadcast disks: A pinwheel-based approach. In: DART'96. (1996)

92. Dertouzos, M.: Control robotics: The procedural control of physical processes. In: Proceedings of IFIP Congress. (1974)

93. Blazewicz, J., Ecker, K., Pesch, E., Schmidt, G., Weglarz, J.: Scheduling Computer and Manufacturing Processes, second edn. Springer, Berlin (1996)

94. Bratley, P., Florian, M., Robillard, P.: Scheduling with earliest start and due date constraints. Naval Research Logistics Quarterly **18** (1971) 511–517

95. Lawler, E.: Combinatorial Optimization. Networks and Matroids. Holt, Rinehart and Winston, New York (1976)

96. Still, J.: A class of new methods for congressional apportionment. SIAM Journal on Applied Mathematics **37** (1979) 401–418

97. Fishburn, P., Lagarias, J.: Pinwheel scheduling: Achievable densities. Algorithmica **34** (2002) 14–38

98. Chan, M., Chin, F.: Schedulers for larger classes of pinwheel instances. Algorithmica **9** (1993) 425–462

99. Lin, S.S.L., Lin, K.J.: A pinwheel scheduler for three dinstinct numbers with a tight schedulability bound. Algorithmica **19** (1997) 411–426

100. Leung, J.Y.T., ed.: Hanbook of Scheduling: Algorithms, Models, and Performance Analysis. Chapman & Hall/CRC, London/Boca Raton (2004)

101. Rabin, M.O.: Efficient dispersal of information for security, load balancing, and fault tolerance. Journal of ACM **36** (1989) 335–348
102. Baruah, S., Lin, S.S.: Pfair scheduling of generalized pinwheel task systems. IEEE Transactions on Computers **47** (1998) 812–816
103. Holte, R., Rosier, L., Tulchinsky, I., Varvel, D.: Pinwheel scheduling with two distinct numbers. Theoretical Computer Science **100** (1992) 105–135
104. Devillers, R., Goossens, J.: Liu and layland's schedulability test revisited. Information Processing Letters **73** (2000) 157–161
105. Kubiak, W.: Solution to the Liu-Layland problem via bottleneck just-in-time sequencing. Journal of Scheduling **8** (2005) 295–302
106. Jozefowska, J., Jozefowski, L., Kubiak, W.: Apportionment methods and the liu-layland problem. European Journal of Operational Research **193** (2009) 857–864, **doi: 10.1016/j.ejor.2007.11.007** (2007)
107. Kubiak, W.: Fair Sequences. In: Handbook of Scheduling. Chapman & Hall/CRC, London/Boca Raton (2004) 19–1 –19 – 21
108. ILOG: ILOG Concert Technology 1.1. (2001)
109. Daganzo, C.: A Theory of Supply Chains. Springer, Berlin (2003)
110. Parello, B., Kabat, W., Wos, L.: Job-shop scheduling using automated reasoning. Journal of Automated Reasoning **2** (1986) 1–42
111. Gent, I.: Two results on car-sequencing problem. Technical report APES-02-1998, Department of Computer Science, University of Strathclyde, Glasgow, UK (1998)
112. Lockledge, J., Mihailidis, D., Sidelko, J., Chelst, K.: Prototype fleet optimization model. Journal of Operational Research Society **53** (2002) 833–841
113. Dincbas, M., Simonis, H., Hentenryck, P.V.: Solving the car-sequencing problem in constraint logic programming. In: ECAI-88. (1988) 290–295
114. Gravel, M., Gagne, C., Price, W.: Review and comparison of three methods for the solution of the car sequencing problem. Journal of Operational Research Society **56** (2005) 1287–1295
115. Drexl, A., Kimms, A.: Sequencing jit mixed-model assembly lines under station-load and part-usage constraints. Management Science **47** (2001) 480–491
116. Drexl, A., Kimms, A., Matthiessen, L.: Algorithms for the car sequencing and the level scheduling problem. Journal of Scheduling **9** (2006) 153–176
117. Desrochers, M., Soumis, F.: A column generation approach to the urban transit crew scheduling problem. Transportation Science **23** (1989) 1–13
118. Solnon, C.: Solving car sequencing problems with artificial ants. In: Werner, H., ed.: ECAI-2000, IOS, Amsterdam, the Netherlands (2000) 118–122
119. Shapiro, J.: Modeling the Supply Chain. Duxbury, North Scituate, MA (2001)
120. Bowersox, D., Closs, D., Cooper, M.: Supply Chain Logistics Management. McGraw-Hill Irwin, New York (2002)
121. Jost, V.: Deux problemes d'approximation diophantine:le partage proportionnel en nombres entires et les pavages equilibres de z. Dea roco, Laboratoire Leibniz-IMAG, Grenoble, France (2003)
122. Berthe, V., Tijdeman, R.: Balance properties of multidimensional words. Theoretical Computer Science **273** (2002) 197–224
123. Aigbedo, H.: Analysis of parts requirements variance for a JIT supply chain. International Journal of Production Research **42** (2004) 417–430
124. Kubiak, W., Steiner, G., Yeomans, J.S.: Optimal level schedules for mixed-model, multi-level just-in-time assembly systems. Annals of Operations Research **69** (1997) 241–259
125. Kis, T.: On the complexity of the car sequencing problem. Operations Research Letters **32** (2004) 331–335
126. Kubiak, W.: Balancing Mixed-Model Supply Chains. In: Avis, D., Hertz, A., Marcotte, O., eds.: Graph Theory and Combinatorial Optimization. Springer, Berlin (2005) 159–189
127. Miltenburg, J., Sinnamon, G.: Algorithms for scheduling multi-level just-in-time production systems. IIE Transactions **24** (1992) 121–130

128. Miltenburg, J., Goldstein, T.: Developing production schedules which balance part usage and smooth production loads for just-in-time production systems. Naval Research Logistics **38** (1991) 893–910

129. Keshav, S.: An Engineering Approach to Computer Networking. Addison-Wesley, Reading, MA (1997)

130. Bertsekas, D., Gallager, R.: Data Networks. Prentice-Hall, Englewood Cliffs, NJ (1991)

131. Waldspurger, C.: Lottery and Stride Scheduling: Flexible Proportional-Share Resource Management. Ph.D. thesis, Massachusetts Institute of Technology, Cambridge, MA (1995)

132. Demers, A., Keshav, S., Shenkar, S.: Analysis and simulation of a fair queueing algorithm. Internetworking: Research and Experience **1** (1990) 3–26

133. Greenberg, A., Madras, N.: How fair is fair queuing. Journal of the ACM **39** (1992) 568–598

134. Gafni, E., Bertsekas, D.: Dynamic control of session input rates in communication networks. IEEE Transactions on Automatic Control **29** (1984) 1009–1016

135. Zhou, Y., Harish, S.: On the relationship between absolute and relative fairness bounds. IEEE Communications Letters **6** (2002) 37–39

136. Burt, O., Harris, C.: Apportionment of the u.s. house of representatives: a minimum range, integer solution, allocation problem. Operations Research **11** (1963) 648–652

137. Parekh, A., Gallager, R.: A generalized processor sharing approach to flow control in integrated services networks: The single-node case. IEEE/ACM Transactions on Networking **1** (1993) 344–357

138. Maheshwari, U.: Charge-based proportional scheduling. Technical memo mit/lcs/tm-529, MIT Laboratory for Computer Science, Massachusetts Institute of Technology, Cambridge, MA (1995)

139. Kohler, E., Morris, R., Chen, B., Jannotti, J., Kaashoek, M.: The click modular router. ACM Transactions on Computer Systems **18** (2000) 263–279

140. Bent, J., Venkateshwaran, V., LeRoy, N., Roy, A., Stanley, J., Arpaci-Dusseau, A., Arpaci-Dusseau, R.H., Livny, M.: Flexibility, managebility and performance in a grid storage appliance. In: Proceedings of the Eleventh IEEE Symposium on High Performance Distributed Computing, Edinburgh, Scotland (2002)

141. Hardin, G.: The tragedy of the commons. Science **162** (1968) 1243–1248

142. Theimer, M., Landtz, K., Cheriton, D.: Preemptable remote execution facilities for the v system. In: Proceedings of the 10th ACM Symposium on Operating Systems Principles. (December 1985)

143. Arpaci-Dusseau, A., Culler, D.: Extending proportional-share scheduling to a network of workstations. In: Proceedings of International Conference on Parallel and Distributed Processing Techniques and Applications. (June 1997)

144. Nabrzyski, J., Schopf, J., Weglarz, J., eds.: Grid Resource Management: State-of-the Art. and Future Trends. Kluwer, Dordrecht (2003)

145. Cirne, W., Marzullo, K.: The computational co-op: Gathering clusters into a metacomputer. In: Proceedings of IEEE International Parallel and Distributed Processing Symposium. (April 1999)

146. Jozefowska, J., Jozefowski, L., Kubiak, W.: Proprtional allocation of discrete resources using divisor methods of apportionment. Foundation of Computing and Decission Sciences **32** (2007) 227–237

147. Kumar, A., Kleinberg, J.: Fairness measures for resource allocation. SIAM Journal on Copmuting **36** (2006) 657–680

148. Megiddo, N.: Optimal flows in networks with multiple sources and sinks. Mathematical Programming **7** (1974) 97–107

149. Sawik, T.: Production Planning and Scheduling in Flexible Assembly Systems. Springer, Berlin (1999)

150. Boysen, N., Fliedner, M., Scholl, A.: Sequencing mixed-model assembly lines: Survey, classification and model critique. Technical report, Fredrich-Schiller-Universitat Jena (2007)

151. Harbour, J.: Automotive competitive challange: Going beyond lean. Harbour-Felax Group Study (2006)

152. Jones, D.: Heijunka: Leveling production. Manufacturing Engineering **137** (2006) 29–36
153. McBride, D.: Heijunka: Leveling the load. Retrieved from http://www.emsstrategies.com/ dm090804article.html on August 6, 2007 (2004)
154. Lummus, R.: A simulation analysis of sequencing alternatives for JIT lines using kanbans. Journal of Operations Management **13** (1995) 183–191
155. Yavuz, M., Akcali, E., Tufekci, S.: Optimizing production smoothing decisions via batch selection for mixed-model just-in-time manufacturing systems with arbitrary setup and processing times. International Journal of Production Research **44** (2006) 3061–3081
156. Pinedo, M.L.: Planning and Scheduling in Manufacturing and Services. Springer, Berlin (2005)
157. T'kindt, V., Billaut, J.C.: Multicriteria Scheduling. Springer, Berlin (2002)
158. Russell, B.: In Praise of Idleness: And other essays. Routledge, London (2004)
159. Hopp, W., Spearman, M.: Factory Physics: Foundations of Manufacturing Management. McGraw-Hill Irwin, New York (2000)
160. Kuhn, H.: The Hungarian method for the assignment problem. Naval Research Logistics Quarterly **2** (1955) 83–97
161. Dell'Amico, M., Toth, P.: Algorithms and codes for dense assignment problems: the state of the art. Discrete Applied Mathematics **100** (2000) 17–48
162. Mendes, A., Ramos, A., Simaria, A., Vilarinho, P.: Combining heuristic procedures and simulation models for balancing a PC camera assembly line. Computers and Industrial Engineering **49** (2005) 413–431
163. Kubiak, W., Yavuz, M.: Just-in-time smoothing through batching. Manufacturing & Service Operations Management **10** (2008) 506–518
164. Kurashige, K., Yanagawa, Y., Miyazaki, S., Kameyama, Y.: Time-based goal chasing method for mixed-model assembly line problem with multiple work stations. Production Planning & Control **13** (2002) 735–745
165. Aigbedo, H., Monden, Y.: A parametric procedure for multicriterion sequence scheduling for just-in-time mixed-model assembly lines. International Journal of Production Research **35** (1997) 2543–2564
166. McMullen, P., Tarasewich, P., Frazier, G.: Using genetic algorithms to solve the multi-product JIT sequencing problem with set-ups. International Journal of Production Research **38** (2000) 2653–2670
167. McMullen, P.: JIT sequencing for mixed-model assembly lines with setups using tabu search. Production Planning & Control **9** (1998) 504–510
168. McMullen, P.: An efficient frontier approach to addressing JIT sequencing problems with setups via search heuristics. Computers & Industrial Engineering **41** (2001) 335–353
169. Yavuz, M., Tufekci, S.: A bounded dynamic programming solution to the batching problem in mixed-model just-in-time manufacturing systems. International Journal of Production Economics **103** (2006) 841–862

Index

Early Titles in the
INTERNATIONAL SERIES IN
OPERATIONS RESEARCH & MANAGEMENT SCIENCE
Frederick S. Hillier, Series Editor, *Stanford University*

Bouyssou et al./ *AIDING DECISIONS WITH MULTIPLE CRITERIA: Essays in Honor of Bernard Roy*

Cox, Louis Anthony, Jr./ *RISK ANALYSIS: Foundations, Models and Methods*

Dror, M., L'Ecuyer, P. & Szidarovszky, F./ *MODELING UNCERTAINTY: An Examination of Stochastic Theory, Methods, and Applications*

Dokuchaev, N./ *DYNAMIC PORTFOLIO STRATEGIES: Quantitative Methods and Empirical Rules for Incomplete Information*

Sarker, R., Mohammadian, M. & Yao, X./ *EVOLUTIONARY OPTIMIZATION*

Demeulemeester, R. & Herroelen, W./ *PROJECT SCHEDULING: A Research Handbook*

Gazis, D.C./ *TRAFFIC THEORY*

Zhu/ *QUANTITATIVE MODELS FOR PERFORMANCE EVALUATION AND BENCHMARKING*

Ehrgott & Gandibleux/ *MULTIPLE CRITERIA OPTIMIZATION: State of the Art Annotated Bibliographical Surveys*

Bienstock/ *Potential Function Methods for Approx. Solving Linear Programming Problems*

Matsatsinis & Siskos/ *INTELLIGENT SUPPORT SYSTEMS FOR MARKETING DECISIONS*

Alpern & Gal/ *THE THEORY OF SEARCH GAMES AND RENDEZVOUS*

Hall/ *HANDBOOK OF TRANSPORTATION SCIENCE - 2nd Ed.*

Glover & Kochenberger/ *HANDBOOK OF METAHEURISTICS*

Graves & Ringuest/ *MODELS AND METHODS FOR PROJECT SELECTION: Concepts from Management Science, Finance and Information Technology*

Hassin & Haviv/ *TO QUEUE OR NOT TO QUEUE: Equilibrium Behavior in Queueing Systems*

Gershwin et al/ *ANALYSIS & MODELING OF MANUFACTURING SYSTEMS*

Maros/ *COMPUTATIONAL TECHNIQUES OF THE SIMPLEX METHOD*

Harrison, Lee & Neale/ *THE PRACTICE OF SUPPLY CHAIN MANAGEMENT: Where Theory and Application Converge*

Shanthikumar, Yao & Zijm/ *STOCHASTIC MODELING AND OPTIMIZATION OF MANUFACTURING SYSTEMS AND SUPPLY CHAINS*

Nabrzyski, Schopf & Węglarz/ *GRID RESOURCE MANAGEMENT: State of the Art and Future Trends*

Thissen & Herder/ *CRITICAL INFRASTRUCTURES: State of the Art in Research and Application*

Carlsson, Fedrizzi, & Fullér/ *FUZZY LOGIC IN MANAGEMENT*

Soyer, Mazzuchi & Singpurwalla/ *MATHEMATICAL RELIABILITY: An Expository Perspective*

Chakravarty & Eliashberg/ *MANAGING BUSINESS INTERFACES: Marketing, Engineering, and Manufacturing Perspectives*

Talluri & van Ryzin/ *THE THEORY AND PRACTICE OF REVENUE MANAGEMENT*

Kavadias & Loch/ *PROJECT SELECTION UNDER UNCERTAINTY: Dynamically Allocating Resources to Maximize Value*

Brandeau, Sainfort & Pierskalla/ *OPERATIONS RESEARCH AND HEALTH CARE: A Handbook of Methods and Applications*

Cooper, Seiford & Zhu/ *HANDBOOK OF DATA ENVELOPMENT ANALYSIS: Models and Methods*

Luenberger/ *LINEAR AND NONLINEAR PROGRAMMING, 2nd Ed.*

Sherbrooke/ *OPTIMAL INVENTORY MODELING OF SYSTEMS: Multi-Echelon Techniques, Second Edition*

Chu, Leung, Hui & Cheung/ *4th PARTY CYBER LOGISTICS FOR AIR CARGO*

Simchi-Levi, Wu & Shen/ *HANDBOOK OF QUANTITATIVE SUPPLY CHAIN ANALYSIS: Modeling in the E-Business Era*

Gass & Assad/ *AN ANNOTATED TIMELINE OF OPERATIONS RESEARCH: An Informal History*

Greenberg/ *TUTORIALS ON EMERGING METHODOLOGIES AND APPLICATIONS IN OPERATIONS RESEARCH*

Weber/ *UNCERTAINTY IN THE ELECTRIC POWER INDUSTRY: Methods and Models for Decision Support*

Figueira, Greco & Ehrgott/ *MULTIPLE CRITERIA DECISION ANALYSIS: State of the Art Surveys*

Printed in the United States of America